The Story of Light Sciencc

Dennis F. Vanderwerf

The Story of Light Science

From Early Theories to Today's Extraordinary Applications

Springer

Dennis F. Vanderwerf
Austin, TX
USA

ISBN 978-3-319-87769-3 ISBN 978-3-319-64316-8 (eBook)
DOI 10.1007/978-3-319-64316-8

Softcover reprint of the hardcover 1st edition 2017

Printed on acid-free paper

This Springer imprint is published by Springer Nature
The registered company is Springer International Publishing AG
The registered company address is: Gewerbestrasse 11, 6330 Cham, Switzerland

All I want to do my whole life is to study light.

—Albert Einstein

Preface

The evolution of light science has followed an irregular but progressive path through the efforts of many investigators—from about the fifth century BC to the current era. Early speculations on the nature of light were usually based on visual observations, leading to a philosophical conclusion, an imagined experiment, or a proposal and implementation of an actual experiment. Over the years, progress in light science has modified earlier widely held views, and future research will inevitably revise these views. Thoughts on the nature of light have led to the formation of theories, and new theories have also arisen from unexpected results of experiments. Often, original theories or proposals to describe the physics of light are rapidly followed by confirming or refuting arguments and experiments. Moreover, successful experiments have often led to practical applications that have generated major new industries. From the earliest recorded theories and experiments to the latest applications in photonic communication and computation, the ways in which light has been put to use are numerous and astounding. Some of the latest endeavors in light science are approaching the realm of science fiction.

Light consists of a broad spectrum of radiation, with only a small spectral region visible to the human eye. Although visible light has been a major factor in the development of photonic technology, much of the understanding of the nature of light was made possible by discoveries occurring outside the visible spectrum. Over the centuries, scientists have attempted to measure, generate, control, and utilize light over specific wavelength ranges. The spatial realm of this research encompasses an almost incomprehensible range—from the size of a photon to the infinite reaches of space. The continuous development of new engineering capabilities has provided a platform for many of these ideas to be realized.

Aided by the extensive communication networks available today, modern innovative experiments using light often involve cooperation between scientists and laboratories in different countries. The defining experiments described in this book have sufficient detail to illustrate the goals, procedures, and conclusions of the investigators and, when possible, are derived from original research papers and reports. The described concepts and experiments are supplemented with many annotated drawings.

Topics in this book include the foundational investigations on the nature of light and ongoing methods to measure its speed. Early experiments on electrodynamics, Maxwell's equations of electromagnetism, and Maxwell's wave equations identified light as an electromagnetic wave. Planck's radiation formula and Einstein's energy equations described light as a fundamental quantized light particle in the form of a packet of electromagnetic energy—the photon. The strong connections between light and relativity were established, and the quantum mechanical properties of photon particles have been experimentally validated.

A progression of electrically driven light sources has developed, including early sustainable arc lamps, modern solid-state lamps, and coherent laser light. The intensity range of light sources available today is remarkable—from intense vortex arc discharge lamps, synchrotron radiators, and petawatt lasers to weak single-photon light generators. The origins and operating principles of the main types of gas and solid-state lasers are described, along with current achievable levels in super-high power and ultra-short pulse duration lasers. In addition, several novel types of lasers have emerged, including dark pulse lasers, time-reversed lasers, and anti-lasers.

Engineering advances in nanotechnology have resulted in the fabrication of novel photonic metamaterials for controlling the propagation of light. Extreme reduction of light speed has been achieved in atomic gases and solid photonic crystal waveguides, along with the stoppage and time reversal of light. Photonic crystal metamaterials have been produced that exhibit negative and zero refractive indices. Structured metamaterial cloaks have enveloped opaque objects and rendered them invisible. Creative experiments and devices have demonstrated both subluminal and superluminal light speeds in free space.

The discovery, production, and utilization of photon quantum entanglement have enabled many new applications, including secure long-distance quantum communication and quantum teleportation. Delayed-choice experiments with entangled photons have been designed to observe the transition between particle and wave behavior of a photon. Quantum memory devices have been developed to receive, stop, store, and re-emit a photon, while preserving the quantum state of the light. Squeezed light at the quantum level was applied in the recent detection of gravitational waves. The photon has emerged as the preferred carrier of information in the rapidly developing field of quantum computing. Recent studies on hypothetical light-based time-like curves have opened new dialogues on time travel possibilities. Lastly, scientific evidence and speculation have produced several scenarios for the cosmological beginning of the photon and its expected fate in a fading universe.

Austin, USA Dennis F. Vanderwerf

Contents

1 Emerging Theories of Light and Measurements of Light Speed 1
1.1 Theories of Light in Classical Antiquity 1
1.2 From Kepler Through Bartholinus 2
1.3 Newton's Corpuscular Theory of Light 2
1.4 Other Investigations on the Nature of Light 3
1.5 Emergence of the Wave Theory of Light 3
1.6 Experimental Speed of Light Measurements 4
1.6.1 Galileo and the Speed of Light 4
1.6.2 The Measurements of Rømer 6
1.6.3 Speed of Light Measurement by Bradley 6
1.6.4 Speed of Light Measurement by Fizeau 8
1.6.5 Speed of Light Measurements by Foucault 9
1.6.6 Speed of Light Measurements by Michelson 10
1.6.7 A Highly-Accurate Speed of Light Measurement 11
References 12

2 Light as an Electromagnetic Wave 13
2.1 The Development of Classical Electrodynamics 13
2.2 The Experiment of Weber and Kohlrausch 15
2.3 Maxwell and Electromagnetism 15
2.3.1 Maxwell's Electromagnetic Equations 15
2.3.2 The Electromagnetic Wave Equations 17
2.4 The Experiments of Hertz 18
2.5 The Discovery of X-Rays 18
2.6 The Electromagnetic Spectrum 19
2.7 Polarization of Light 20
References 21

3 Light and Its Application to Relativity 23
3.1 Light Speed in the Fizeau Moving Medium Experiment 23
3.2 The 1887 Experiment of Michelson and Morley 24
3.3 Lorentz-FitzGerald Contraction 28
3.4 Lorentz Transformations 29
3.5 Einstein Special Theory of Relativity 31
3.6 Minkowski Spacetime 33
3.6.1 The Light Cone 33
3.6.2 The Invariance of Spacetime 35
3.7 The Speed of Light and Einstein's $E = mc^2$ 36
3.7.1 Derivations of $E = mc^2$ 36
3.7.2 Relativistic Energy, Momentum, and Mass of a Photon Particle 37
3.8 Einstein General Theory of Relativity 37
3.9 Experimental Affirmations of Relativity 38
3.9.1 The Kennedy-Thorndike Experiment 38
3.9.2 Transverse Doppler Shift 39
3.9.3 Time Dilation and Muon Lifetime 41
3.9.4 Time Dilation and Moving Atomic Clocks 42
3.9.5 Time Dilation in GPS Satellite Clocks 43
3.9.6 Detection of Lorentz Length Contraction 43
3.9.7 Bending of Light by the Sun 44
3.10 Source Motion and the Constancy of the Speed of Light 45
References 47

4 The Quantum Nature of Light 49
4.1 The Photoelectric Effect 49
4.2 Quantum Blackbody Radiation 50
4.2.1 Wien's Displacement Law 50
4.2.2 Development of the Radiation Distribution Laws 51
4.3 Light as a Quantized Particle 53
4.3.1 The Photon as a Quantized Particle 53
4.3.2 Explanation of the Photoelectric Effect 53
4.4 Stimulated Emission of Einstein 55
4.5 The Experiment of Compton and the Compton Effect 55
4.6 The de Broglie Wavelength 57
4.7 Wave-Particle Duality 57
References 58

5 Natural and Artificial Sources of Light 59
5.1 The Sun 59
5.2 Electrically-Driven Light Sources 61
5.2.1 Arc-Discharge Lamps 61
5.2.2 Incandescent Lamps 63

5.2.3 Light-Emitting Diodes . . . 65
5.2.4 Organic Light-Emitting Diodes . . . 67
5.2.5 Quantum Dot Light-Emitting Diodes . . . 68
5.2.6 Light Emission Using Photonic Crystals . . . 70
5.3 Synchrotron Radiation Light Source . . . 70
References . . . 72

6 Laser Light . . . 75
6.1 Theoretical Foundations . . . 75
6.1.1 Stimulated Emission and Population Inversion . . . 75
6.1.2 Construction of a Laser . . . 77
6.1.3 Temporal and Spatial Coherence . . . 80
6.2 Fundamental Laser Types . . . 81
6.2.1 The Pulsed Ruby Laser . . . 81
6.2.2 The Nd:YAG Laser . . . 83
6.2.3 The Helium-Neon Laser . . . 84
6.2.4 The CO_2 Laser . . . 86
6.2.5 The Argon Ion Laser . . . 86
6.2.6 The Excimer Laser . . . 87
6.2.7 The Dye Laser . . . 89
6.3 Semiconductor Laser Diodes . . . 91
6.3.1 Edge-Emitting and Quantum Well Laser Diodes . . . 91
6.3.2 Vertical Plane Emitting Laser Diodes . . . 93
6.3.3 Laser Diode Pumped Solid-State Lasers . . . 94
6.4 Some Lasers with Special Properties . . . 95
6.4.1 Fiber Lasers . . . 95
6.4.2 Doped Fiber Amplifiers . . . 97
6.4.3 Thin Disc Lasers . . . 98
6.4.4 Quantum Cascade Lasers . . . 100
6.4.5 Tunable Lasers and Linewidth . . . 102
6.4.6 Fast and Ultrafast Pulsed Lasers . . . 104
6.4.7 Surface Plasmons and Nanolasers . . . 108
6.4.8 Ultra-High Power and Energy-Pulsed Lasers . . . 114
6.5 Some Additional Laser Types . . . 118
6.5.1 Free-Electron Lasers . . . 118
6.5.2 Pulsed X-Ray Lasers . . . 119
6.5.3 Visible Color Laser . . . 121
6.5.4 White Light, Supercontinuum, and Multi-frequency Lasers . . . 121
6.5.5 Random Illumination Laser . . . 124
6.5.6 Dark Pulse Lasers . . . 125
6.5.7 Coupled Lasers . . . 126
6.5.8 Biological Laser . . . 127

6.5.9 Time-Reversed Lasing and Anti-lasers 128
6.5.10 Superradiant Laser 128
References 129

7 Variation and Control of Light Propagation Properties 133
7.1 Slowing the Speed of Light 133
7.1.1 Refractive Index 133
7.1.2 Phase Velocity and Group Velocity 133
7.1.3 Some Slow Light Methodologies 135
7.1.4 Thermal deBroglie Wavelength and Bose-Einstein Condensates 137
7.1.5 Extreme Reduction of Light Speed in a Super-Cooled Atomic Gas 140
7.1.6 Ultraslow Light Propagation in Higher Temperature Atomic Gases 142
7.1.7 Slow Light Using Photonic Crystals 144
7.2 Stopped Light and Storage of Light Pulses 146
7.3 Time-Reversed Light Using a Photonic Crystal 147
7.4 Exceeding the Cosmic Speed of Light in a Medium 148
7.4.1 The Speed of Light and Causality 148
7.4.2 Superluminal Light in Dispersive Media 149
7.5 Light Path Manipulation 151
7.5.1 Light Propagation and Negative Refraction 151
7.5.2 Infinite Light Speed Using a Zero Refractive Index Medium 156
7.5.3 Invisibility Cloaking and Transformation Optics 158
7.5.4 Spacetime Hidden Event Cloaking 167
References 170

8 Quantum Mechanics of the Photon 173
8.1 Double-Slit Experiments Using Photon Particles 173
8.2 Delayed-Choice Experiments 174
8.2.1 The Delayed-Choice Experiment of Wheeler 174
8.2.2 Experimental Realization of Delayed-Choice 176
8.2.3 EPR Paradox, Particle Entanglement, and Bell's Inequalities 179
8.2.4 Experimental Verification of Photon Entanglement 179
8.2.5 A Quantum Mechanical Beamsplitter 180
8.2.6 Delayed-Choice Quantum Eraser 182
8.2.7 Quantum Interference Between Indistinguishable Photons 186
8.2.8 Observation of Photon Wave-Particle Transitions 187
8.3 Quantum Electrodynamics and the Photon 189
8.4 The Casimir Effect and Virtual Photons 192

8.5 Quantum Correlations of Single and Multiple Photons 197
8.5.1 Single-Photon Particle Anticorrelation 197
8.5.2 Single-Photon Entanglement 199
8.5.3 Validation of Single-Photon Nonlocality 201
8.5.4 Higher-Order Photon Entanglements 206
8.5.5 Four-Photon Entanglement 209
8.5.6 Six-Photon and Eight-Photon Entanglements 210
8.5.7 Entanglement of Non-coexistent Photons 212
8.6 Photon Tunneling and Superluminality 213
8.7 Two-Photon Interactions 217
8.8 Squeezed Light, Squeezed Vacuum, and Gravitational Wave Detection 220
8.9 Non-destructive Observation of the Photon 223
8.10 Quantum Zeno Effect for Photons 225
8.11 Observation of Photon Trajectories 227
8.12 Creation of Matter by Colliding Photons 228
8.13 Gauge Theory and Size of the Photon 230
References 231

9 Quantum Applications of the Photon 235
9.1 Entanglement-Enhanced Quantum Communication 235
9.1.1 Theoretical Foundation 235
9.1.2 Experimental Quantum Communication Using Dense Coding 235
9.2 Encrypted Quantum Communication Using Photons 238
9.2.1 Quantum Cryptography Protocols 238
9.2.2 Single-Photon Beam for Secure Quantum Communication 240
9.2.3 Ideal Properties of a Pulsed Single-Photon Light Source 240
9.2.4 Single-Photon Light Source Using Diamond Nanocrystals 241
9.2.5 Single-Photon Light Source Using a Whispering Gallery 242
9.2.6 Other Room Temperature Single-Photon Light Sources 244
9.2.7 Single-Photon Detectors for Encrypted Photons 246
9.2.8 Secure Quantum Communication Through Free-Space 248
9.3 Ultra-High Data Transmission Using Photons 249
9.3.1 Angular Momentum of a Photon 249
9.3.2 Data Transmission Through Free-Space Using OAM Photons 250
9.3.3 Ultra-Fast Data Transmission Through Optical Fibers Using Photons 251

9.4 Long-Range Communication Using Quantum Repeaters 252
9.4.1 Quantum Memory 252
9.4.2 Principles of Quantum Teleportation 258
9.4.3 Quantum Teleportation Through Optical Fibers 262
9.4.4 Quantum Teleportation Through Free-Space 265
9.4.5 Quantum Repeaters for Long-Range Optical Fiber Transmission 267
9.4.6 Photonic Quantum Networks 269
9.4.7 A Quantum Internet 270
9.5 Photonic Random Number Generation 272
9.6 Quantum Computing Using Photons 277
9.6.1 Quantum Gates Using Linear Optics 277
9.6.2 Boson Sampling 280
9.6.3 Quantum Computation Efficiency 286
9.6.4 Deutsch and Shor Quantum Algorithms 286
9.6.5 A Universal Quantum Computer Architecture 288
9.7 Timelike Curves, Time Travel, and Computing 290
References 295

10 Light in Free-Space and the Cosmos 301
10.1 The Speed of Light in the Cosmos 301
10.1.1 The Cosmic Speed of Light and Planck Length 301
10.1.2 Gamma-Ray Bursts and the Cosmic Speed of Light 302
10.1.3 The Speed of Light and the Fine Structure Constant 302
10.1.4 Isotropy of the Speed of Light in Space 303
10.1.5 Slowing the Speed of Light in Free-Space 305
10.1.6 Waves Exceeding the Speed of Light in Free-Space 307
10.1.7 Particles Exceeding the Speed of Light in Free-Space 309
10.2 The Warp Drive of Alcubierre 310
10.3 Dark Matter and Gravitational Lensing 310
10.4 The Cosmological Beginning of the Photon 311
10.5 Temperature of the Universe and Wien's Law 312
10.6 The Cosmological Fate of the Photon 313
References 314

Appendices 317

Timeline of Some Notable Achievements in Light Science 323

A Selection of Additional Readings 325

Index 327

Chapter 1
Emerging Theories of Light and Measurements of Light Speed

1.1 Theories of Light in Classical Antiquity

Prior to the emergence of Greek intellectuals in the classical antiquity era, little is recorded of any scientific or critical thinking on the nature of light or its interaction with the eye. Around 500 BC the Greek philosopher and mathematician Pythagoras of Samos (ca. 560 BC–480 BC) contemplated and theorized on the physical nature of visible light. He believed that emitted light from luminous objects could undergo reflections and be received by the human eye to produce vision. His well-known Pythagorean theorem has found use in deriving modern mathematical models for relativity and spacetime. The Greek philosopher Empedocles (ca. 490 BC–430 BC) believed that light emanating from the eye interacted with the viewed object to produce vision, and that the speed of this light was finite. The Greek philosopher Plato (ca. 427 BC–347 BC) invented a theory of vision involving three streams of light—one from the object seen, one from the eyes, and one from the illuminating source. Plato's student Aristotle (ca. 384 BC–322 BC), thought that light was nothing of substance—being incorporeal, immobile, indefinable, and not worth serious contemplation. Euclid (ca. 330 BC–260 BC) thought light reflected from an object entered the eye to produce vision, and formulated a law of reflection. The Egyptian astronomer Ptolemy (ca. 100 CE–165 CE) discovered the refraction of light and prepared several tables illustrating this effect. The Arabian scientist Ibn al Haythen (965–1040), known as Alhazen in the western world, studied reflection and refraction, and formulated laws of reflection in the period 1028–1038. He described the construction of the first pinhole camera, or *camera obscura*.

D.F. Vanderwerf, *The Story of Light Science*,
DOI 10.1007/978-3-319-64316-8_1

1.2 From Kepler Through Bartholinus

In 1604 Johannes Kepler (1571–1630) formulated laws describing the rectilinear propagation of light rays. The unpublished Snell's law of refraction was discovered in 1621 by Willebrord Snell (1580–1626). Robert Hooke (1635–1703) proposed a wave theory of light in his "Micrographia" (1665), and suggested in 1672 that the vibrations in light might be perpendicular to the direction of propagation. Polarized light was first observed by Erasmus Bartholinus in 1669, when he viewed the splitting of light rays in a calcite birefringent crystal.

1.3 Newton's Corpuscular Theory of Light

Isaac Newton (1642–1727), in addition to his remarkable mechanical description of the motion of material bodies, performed significant work in optics and the theory of light. From glass prism experiments he showed that white light could be broken up into a rainbow of colored beams by the process of refraction, and that these colored beams could then be recombined into white light. Newton stated that light was composed of non-spherical particles, with the red particles being larger and more massive that the blue particles. This was the essence of his *corpuscular theory of light*, where the propagation laws of these particles were explained by the same mechanical laws of motion he had formulated for material particles. These laws of light motion required the postulation of a propagating medium for light, which was called the *luminiferous ether*. Newton's work in light was presented in his "Optiks", published in 1704.

In particular, a law of refraction could be derived from Newton's corpuscular-based theory. Newton related the angles of incidence and refraction from one medium to another, e.g. air-to-water, to an increase in velocity of the refracted light ray in the water. He stated that the velocity component perpendicular to the interface between the air and water increased due to the "pull" of the denser water, while the velocity component tangential to the interface remained the same. The velocity of light was thereby increased in the denser medium, sometimes referred to as *emission theory*. The resulting law of refraction was stated as

$$v_{air} \sin I = v_{water} \sin I', \tag{1.1}$$

where
v_{air} is the light velocity in air,
I is the angle of incidence,
v_{water} is the light velocity in water,
I' is the angle of refraction.

In addition, the ratio v_{water}/v_{air} was constant as the angle of incidence varied. After initial acceptance of Newton's corpuscular theory of light, it gradually fell into disrepute with the emergence and success of the wave theory of light. Surprisingly, it later turned out that there was truth in both the corpuscular and wave theories.

1.4 Other Investigations on the Nature of Light

Around 1670 the French Jesuit scientist Ignace-Gaston Pardies (1636–1673), developed a wave theory of refraction that conflicted with Newton's corpuscular theory of refraction. In 1678 Christiaan Huygens further supported the wave nature of light, which was formally published as his "Treatise on Light" in 1690. Leonhard Euler (1707–1783) published a mathematical theory of light in 1746. In France, Léon Foucault (1819–1868) and Armand Hippolyte Fizeau (1819–1896) worked cooperatively, but independently, to experimentally determine whether the speed of light was increased or decreased in various stationary transparent media [1, 2]. In particular, an experiment by Foucault in 1850 definitely showed that the speed of light was reduced in water, when compared to the speed in air. Thus experimentally, Newton's theory of refraction was challenged.

1.5 Emergence of the Wave Theory of Light

The famous double-slit experiment of Thomas Young in 1801 provided solid evidence of the wave nature of light. Although Young originally used pinholes rather than slits, the experiment is conventionally described using narrow slits. As shown in Fig. 1.1, light from a source passes through a narrow slit S, where it emanates as a sequence of cylindric waves. These waves are incident on slits S_1 and S_2, which are positioned a small distance apart. These two slits act as seondary sources of cylindric waves which are in phase with each other. At the viewing screen these sources produce an interference pattern consisting of bright and dark fringes. Bright fringes appear for *constructive interference*, where the two waves arrive in phase. Dark fringes appear for *destructive interference*, where the two waves arrive 180° out of phase.

Around 1816 Augustin-Jean Fresnel formulated his Huygens-Fresnel principle, which stated that each point on a wave front could act as a secondary source for new spherical waves. This principle could explain both diffraction and interference. By the year 1821, Fresnel was also able to show by mathematical methods that polarization could be explained only if light was *entirely* transverse. Based on these investigations, in the early 19th century there was little support or evidence for a particle description of light. This was further reinforced by the formulation of the electromagnetic wave equations of James Clerk Maxwell in the 1860s. The particle

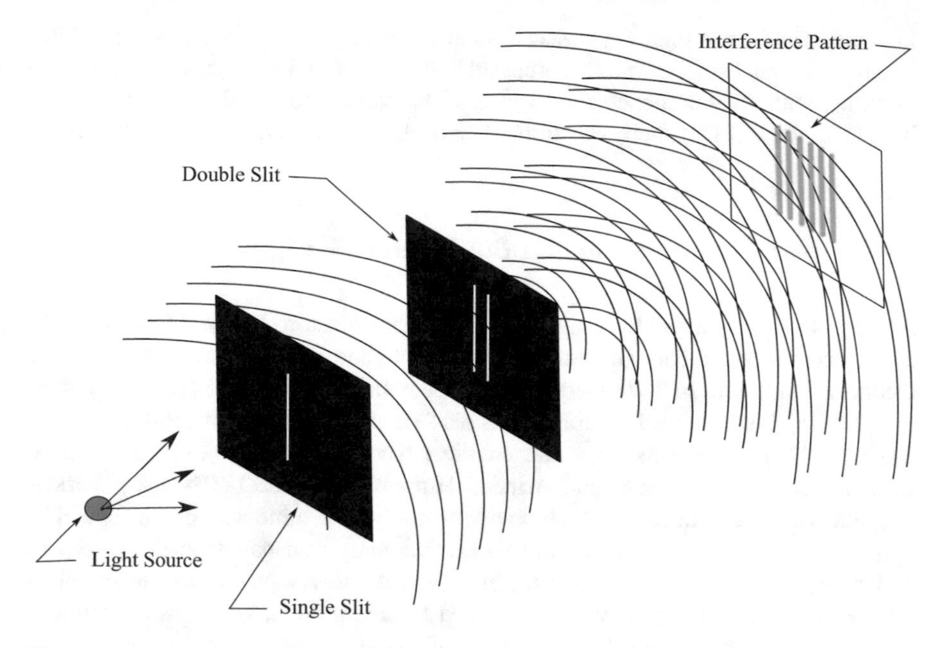

Fig. 1.1 Young's double-slit interference experiment

nature of light would not find any serious support until the discovery of the photoelectric effect in 1887.

The chronology and major contributions of many of the pioneer investigators are summarized in Table 1.1.

1.6 Experimental Speed of Light Measurements

1.6.1 Galileo and the Speed of Light

The first documented land-based attempt to measure the speed of light was made by Galileo Galilei (1564–1642). He believed that light had a finite speed, and tried to measure this speed using Galileo and an assistant, each holding a lantern at a known distance apart. This distance was varied from about one mile to ten miles. By using "synchronized" timepieces and a light shutters at each lantern, the speed of light could theoretically be calculated. However, this terrestrial-based human experiment could not detect a delay at any distance, due to the extremely high speed of light and the slow reaction times of the observers. Since the travel time of a light beam over a ten mile distance is about 5.4×10^{-6} s, the experiment was destined to fail. However, Galileo surmised that the speed of light was at least ten times faster than that of sound.

Table 1.1 Classical descriptions of thc naturc and properties of light

Investigator	Chronology	Theories and Contributions
Pythagoras	ca. 500 BC	Light emanates from a luminous object to the eye
Empedocles	ca. 450 BC	Light travels at a finite speed and emanates from the eye
Plato	ca. 400 BC	Light and vision emanate from the eye
Aristotle	ca. 300 BC	Light is nothing of substance and is indefinable
Epicurius	ca. 300 BC	Reflected light from objects travels in straight lines to the eye. Formulated a law of reflection
Euclid	ca. 300 BC	Worked with mirrors and reflection. Light travels in straight lines and speed of light may be infinite
Claudius Ptolemy	ca. 130 CE	Discovered the process of light refraction
Alhazen (Ibn al Haythen)	965–1040	Introduced concept of light ray. Described first pinhole camera (*camera obscura*). Light consisted of particles
Johannes Kepler	1571–1630	Formulated laws of rectilinear propagation of light (1604)
Willebrordus Snellius (Snel van Royen)	1580–1626	Discovered Snell's law of refraction (1621)
René Decartes	1596–1650	Described light as a pressure wave transmitted at infinite speed through a pervasive elastic medium (1637)
Francesco Grimaldi	1618–1663	Observed diffractive light wave effect. Named this phenomenon *diffraction*. Work published in 1665
Robert Hooke	1635–1703	Proposed a wave theory of light (1672)
Erasmus Bartholinus	1625–1698	Observed and described polarized light in birefringent calcite crystal (1669)
Isaac Newton	1642–1727	Postulated that light consists of tiny particles (corpuscles). Color depends on size of particle. Proposed existence of ethereal medium for light propagation (ether). "Optiks" published in 1704
Ignace-Gaston Pardies	1636–1673	Provided a wave description for the refraction of light
Christiaan Huygens	1629–1695	Proponent of the wave theory of light (1678). "Treatise on Light" published in 1690
Leonhard Euler	1707–1783	Published a mathematical theory of light (1746)
Thomas Young	1773–1829	Discovered interference of light rays. Performed double-slit interference experiment (1801)
Augustin-Jean Fresnel	1788–1827	Experimentally demonstrated interference, supporting the wave theory of light ($\sim$1817). Formulated Huygens-Fresnel principle ($\sim$1816)
Léon Foucault Armand Fizeau	1819–1868 1819–1896	Jointly demonstrated the reduction of light speed in a transparent medium, challenging Newton (1850)
James Clerk Maxwell	1831–1879	Formulated the electromagnetic wave equations in 1864. Identified light as an electromagnetic wave

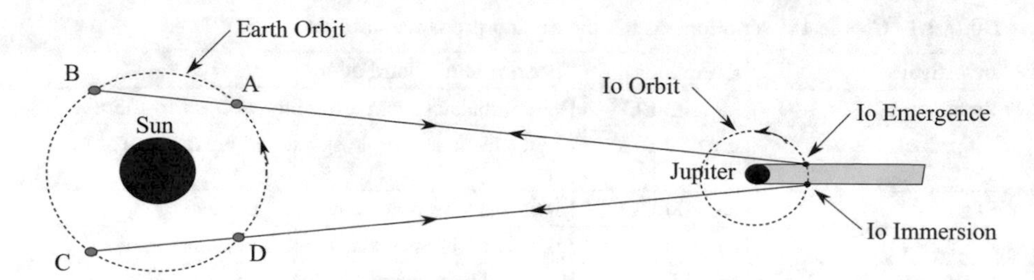

Fig. 1.2 Rømer's observation of the eclipse of Jupiter's moon Io to determine the speed of light

1.6.2 The Measurements of Rømer

Around 1676 the Danish astronomer Olaus Rømer (1644–1710) made the first astronomical-based attempt to determine the speed of light. He noticed that the observed timing of the eclipses of Jupiter's moon Io varied with the time of year. This was based on a fixed value of the period of Io being about 42.5 h. Figure 1.2 illustrates a possible arrangement during a series of observations. When Io emerges from the shadow of Jupiter, the beginning of the eclipse can be observed from the earth at position A. One Io period later, the emergence is observed at position B. Rømer reasoned that the additional time that light would take in traveling along the segment AB would result in an increase in the measured period at observing position B. Similarly, the observation of the immersion of Io into Jupiter's shadow at Earth position C would produce a decrease in the timed period of Io when measured at position D. From this data and the values of the orbital radii of Earth and Jupiter around the Sun, as known at the time, and assuming a fixed value of the Earth-Jupiter separation (later shown to be not true), Rømer concluded that the speed of light was finite and very large. He did not arrive at an actual number, but estimated the speed to be much greater than one Earth diameter per second. An acquaintance of Rømer, Christian Huygens, used Rømer's data to arrive at an "order of magnitude" value for the speed of light to be between 2.14×10^8 and 2.27×10^8 m/s.

1.6.3 Speed of Light Measurement by Bradley

During the period 1725–1728 the English astronomer James Bradley used his discovery of the aberration of starlight to calculate a value for the speed of light [3]. He measured the angular changes in position of a directly overhead star Eltanin, which is the brightest star in the constellation Draco, about 148 light-years from Earth. He attributed this change not to a change in *relative position* between the

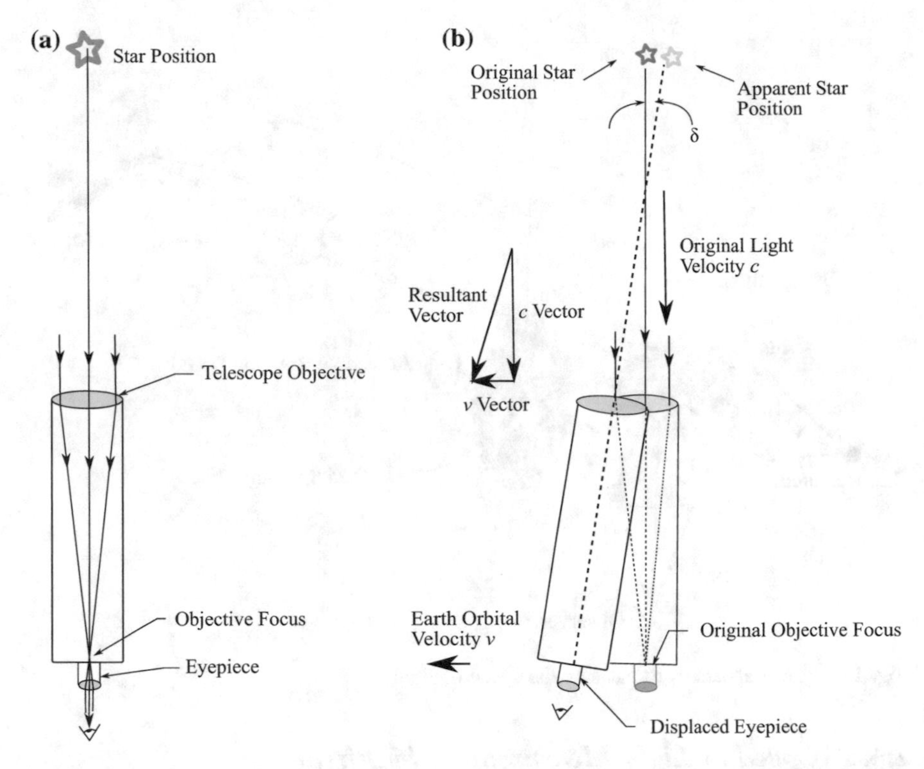

Fig. 1.3 **a** Viewed image of star with vertically mounted telescope. **b** Tilted telescope showing displacement in viewed star image

earth and star (parallax), but to a *relative motion* between the two. This is called *stellar aberration*. In Fig. 1.3a a vertically mounted telescope at rest forms an image of an overhead star at the center of the eyepiece. However, when the telescope is earth-mounted, the orbital velocity of the Earth moves the telescope relative to the star during the short time that light moves through the telescope, causing a focus displacement from the center of the eyepiece. When the telescope is tilted slightly to recenter the focus at the eyepiece as in Fig. 1.3b, the viewed star image appears displaced from the actual star position. This small *aberration angle* δ is related to the ratio of the orbital speed of Earth around the Sun to the much larger speed of light. Over a 12 month observation period Bradley measured the variation of the aberration angle, and arrived at an angular aberration value $\delta = 20.44$ arc seconds. Using a modern value for the Earth mean orbital velocity $v = 29.78$ km/s, the speed of light in space c can then be calculated from $c = \frac{v}{\tan \delta}$, yielding a value $c = 3.01 \times 10^8$ m/s. This result is surprisingly close to the modern accepted value of the speed of light $c = 2.998 \times 10^8$ m/s.

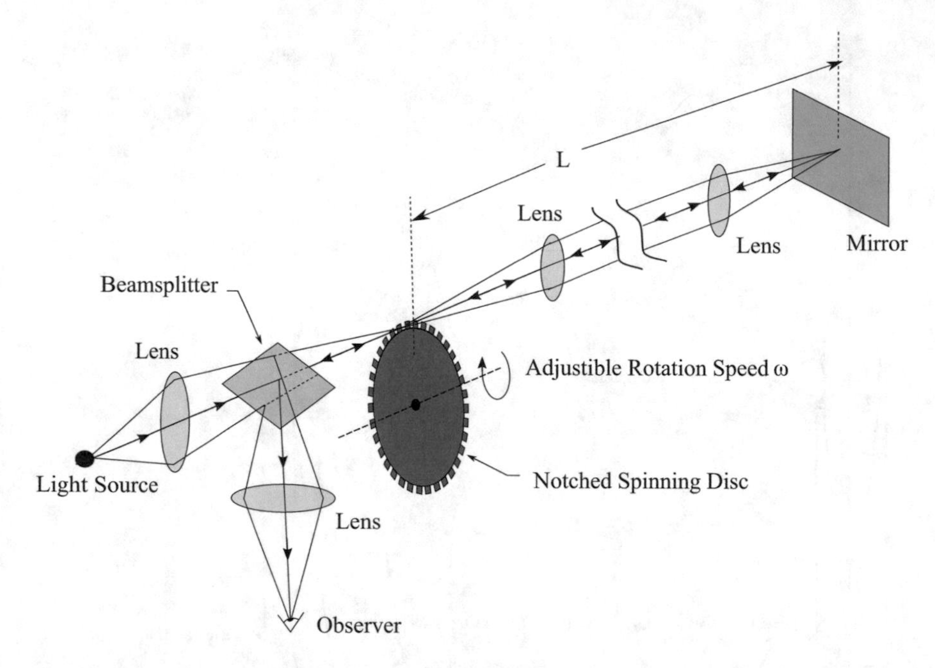

Fig. 1.4 Fizeau apparatus to measure the speed of light

1.6.4 Speed of Light Measurement by Fizeau

In 1849 Armand Hippolyte Fizeau made a successful terrestrial measurement of the speed of light [4]. The apparatus is shown in Fig. 1.4. A spinning circular disc had a series of 720 evenly spaced teeth, and 720 gaps. The apparatus was in a house at Suresnes, France, and the mirror was situated on the height of Montramartre, with a separation $L \approx 8633$ m. A light source and lenses formed a collimated beam that was chopped by the spinning disk and reflected back to the disk from the distant mirror. A beamsplitter allowed an observer to view the lightness or darkness of the returning beam. The time to travel from gap to adjacent tooth can then be related to the speed of the light c. For example, when the wheel rotates at $\omega = 10$ rev/s, the time T for a complete revolution is 0.10 s, and the time t to move between an adjacent gap and tooth is

$$t = T/(720 + 720) = 0.1\ \mathrm{s}/1440 = 6.94\ \times\ 10^{-5}\ \mathrm{s} \tag{1.2}$$

By adjusting the speed of the disc, the transition from a light to a dark condition can be observed when light leaving a gap is blocked by the adjacent tooth upon its return. This was observed at a disc speed $\omega = 12.6$ rev/s ($T = 0.079$ s) yielding $t = 5.51 \times 10^{-5}$ s. The speed of light c could then be calculated from

$$c = 2L/t = 17326\ \text{m}/5.51 \times 10^{-5}\text{s} = 3.14 \times 10^{8}\ \text{m/s} \tag{1.3}$$

Fizeau took the average of about 28 measurements. His value was a little high compared with the modern value of c, but was a significant improvement over the value of $c \approx 2.1 \times 10^8$ m/s obtained from Rømer's data.

1.6.5 Speed of Light Measurements by Foucault

Around 1850 Fizeau collaborated with Léon Foucault to construct another apparatus to investigate the speed of light in various media, using a spinning mirror instead of a toothed disc, and a 4 m path length. This was called the *Fizeau-Foucault apparatus*. The two investigators worked cooperatively, but independently, to experimentally determine whether the speed of light was increased or decreased in various stationary transparent media [1, 2].

Foucault continued improving the accuracy of the technique over the next several years, and in 1862 performed a speed of light measurement using a folded path length and an air-driven turbine drive for the spinning mirror [5, 6]. The spinning mirror frequency was accurately measured using a stroboscope and clock. Figure 1.5 illustrates the setup, where a series of lenses and concave mirrors are employed to focus the returning light from the spinning mirror to a reticle having a calibrated scale. The one-way distance L from position A at the spinning mirror to the focal point position B at the last concave mirror was $L = 10$ m, giving a round-trip path length $2L = 20$ m. The mirror was spun at an angular velocity $\omega = \frac{d\theta}{dt}$, and for the light time interval Δt for the round-trip path A-B-A, a small angular displacement $\Delta\theta$ of the mirror occurred. The resultant angular deviation of the returning light ray from the mirror is $2\Delta\theta$, which produced a measurable focus displacement on the reticle scale. Here $\Delta\theta = \omega(\Delta t) = \frac{2\omega L}{c}$, and the speed of light c could be calculated from

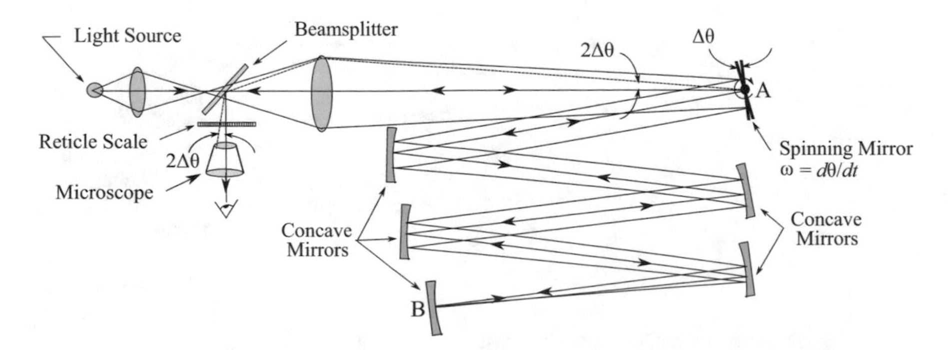

Fig. 1.5 The spinning mirror apparatus of Foucault and Fizeau to measure the speed of light

$$c = \frac{2\omega L}{\Delta\theta}, \tag{1.4}$$

where ω, L, and Δθ were determined from the geometry and operating parameters of the system. From these measurements, Foucault obtained a more accurate speed of light, $c = 2.98 \times 10^8$ m/s.

By insertion of a water tube or a glass rod in the light path, Foucault in 1862 definitely showed that the speed of light was reduced in these media by a factor ≈ 2/3 of the speed in air. Thus Newton's theory of refraction was disproved.

1.6.6 Speed of Light Measurements by Michelson

In 1879 Albert Michelson made the first of many measurements of the speed of light based on the Foucault method. He improved Foucault's method by using improved optical components, an octagonal spinning mirror, and a much longer path length. See Fig. 1.6. With a return mirror positioned on a mountain top at a distance L = 35 km from the spinning mirror, the total round-trip path length was 70 km. He determined the mirror angular velocity ω to produce a steady image in the observer's eyepiece, corresponding to a return reflection from an adjacent facet of the mirror. The time of light travel t for 1/8 revolution of the mirror was then $\frac{1}{8\omega}$. The speed of light was then calculated as

$$c = \frac{2L}{t} = 16L\omega. \tag{1.5}$$

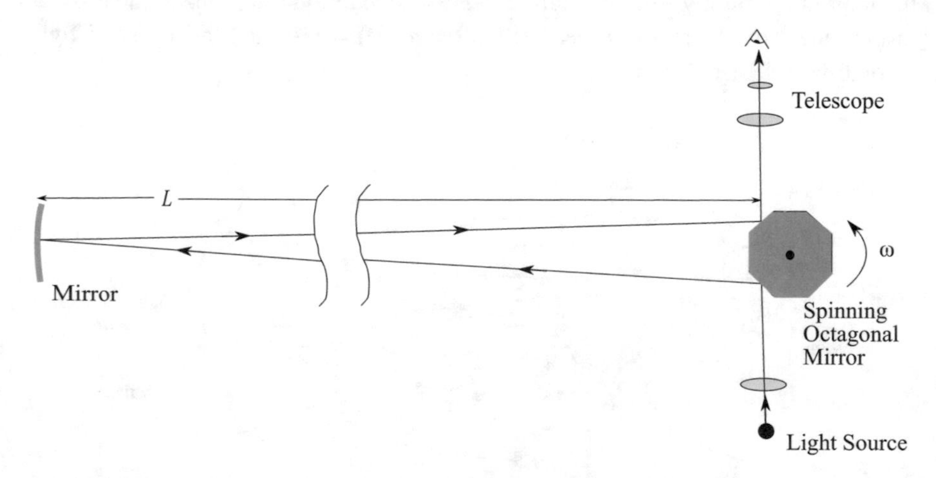

Fig. 1.6 Basic apparatus for Michelson's measurements of the speed of light

As an illustrative example, suppose that for a steady observed image, $\omega = 533.33$ rev/s and $L = 35$ km. Then from Eq. (1.5), $c \approx 2.987 \times 10^8$ m/s. Michelson actually obtained a value $c = 2.9991 \times 10^8$ m/s, and performed many speed of light measurements over a 50 year period, with increasing accuracy. He also attempted to perform a measurement in a mile-long vacuum tube with five multiple reflections to achieve a ten mile light path. In 1931 he made his final measurements at Santa Ana, California. From this data, his final calculated speed of light value was posthumously determined to be $c = 2.99774 \times 10^8$ m/s.

1.6.7 A Highly-Accurate Speed of Light Measurement

Using the fundamental relationship $c = \lambda f$, where λ is the light wavelength and f is the light frequency, the speed of light c was determined in 1972 with 100 times more certainty than the previously accepted value [7]. Hall and Barger at JILA and NBS in Colorado measured the 3.39 μm wavelength of a methane-stabilized laser using frequency-controlled interferometry as $\lambda = 3.392231376$ μm. Cooperatively, Evenson and co-workers at NBS Colorado measured the frequency of the laser by infrared frequency synthesis as $f = 88.376181627$ THz. The product of these values yielded $c = 299{,}792{,}456.2$ m/s. This value of the speed of light was accepted internationally the following year as the modern accepted value of $c = 2.997925 \times 10^8$ m/s. Table 1.2 summarizes some of the well-known light speed measurements and the determined values.

Table 1.2 Some notable speed of light experiments

Investigator	Chronology	Method and light speed result
Galileo Galilei	1638	Believed the speed of light to be faster than speed of sound. Tried unsuccessfully to measure the speed of light using Galileo as the light sender and an assistant as the receiver
Olaus Rømer	1676	Calculation of finite speed of light from eclipses of Jupiter's moon Io, giving $2.14 \times 10^8 \leq c \leq 2.27 \times 10^8$ m/s
James Bradley	1728	Discovered and measured stellar aberration. Calculated speed of light value $c = 3.01 \times 10^8$ m/s
Armand Fizeau	1849	Terrestial experiment yielded $c = 3.14 \times 10^8$ m/s
Léon Foucault	1862	Improved terrestial experiment yielded $c = 2.98 \times 10^8$ m/s
Albert Michelson	1931	Based on Foucault method, calculated $c = 2.99774 \times 10^8$ m/s
J.L. Hall R.L. Barger K.M. Evenson et al.	1972	Precise measurements of λ and f, yielding $\lambda f = c = 2.997925 \times 10^8$ m/s (the accepted international value)

References

1. L. Foucault, Méthode général pour mesurer la vitesse de la lumière dans l'aire et les milieux transparents. C. R. Acad. Sci. **30**(24), 551–560 (1850)
2. H. Fizeau, H. Breguet, Sur la vitesse comparitive de la lumière dans l'aire et dans l'eau. C. R. Acad. Sci. **30**(24), 771–774 (1850)
3. J. Bradley, Account of a new discovered motion of the fix'd stars. Philos. Trans. R. Soc. Lond. **35**, 637–660 (1729)
4. H. Fizeau, Sur une expérience relative à la vitesse de propagation de la lumière. C. R. Acad. Sci. **29**, 90–92 (1849)
5. L. Foucault, Determination expérimentale de la vitesse de la lumière: description des appareils. C. R. Acad. Sci. **55**, 792–796 (1862)
6. P. Lauginie, Measuring speed of light: Why? Speed of what? in *Fifth International Conference for History of Science in Science Education* (Keszthely, Hungary, 2004) pp. 81–82
7. K.M. Evenson et al., Speed of light from direct frequency and wavelength measurements of the methane-stabilized laser. Phys. Rev. Lett. **29**(19), 1346–1349 (1972). doi:10.1103/PhysRevLett.29.1346

Chapter 2
Light as an Electromagnetic Wave

2.1 The Development of Classical Electrodynamics

From the late 1700s through the 1800s there was intense activity and progress in defining the properties of electrostatic and electromagnetic forces, resulting in what is known as *classical electrodynamics*. These investigations would prove to be valuable prerequisites in later descriptions of the nature of light. Some of the scientists and their contributions are summarized below.

Georg Simon Ohm (1787–1854) stated that for conducting materials, the ratio of the current density and the electric field is a constant, being independent of the electric field producing the current. That is, the current density J and the electric field E are related by $J = \sigma E$, where the constant of proportionality σ is the *conductivity*. This empirical relationship is known as *Ohm's law*.

Charles Augustin de Coulomb (1736–1806) described the electrostatic force of attraction or repulsion between point charges, where the unlike charges attract and like charges repel. This was formulated around 1783 as *Coulomb's law*, where the electrostatic force F has the units of Newtons [N].

$$F = \left(\frac{1}{4\pi\varepsilon_0}\right)\frac{q_1 q_2}{r^2}, \tag{2.1}$$

where

$\left(\frac{1}{4\pi\varepsilon_0}\right)$ = Coulomb's constant = 9×10^9 N m^2/C^2,
q_1 and q_2 are the values of the charges in units of **C**,
r is the distance between the charges in meters,
ε_0 is the permittivity of free-space = 8.8542×10^{-12} C^2/N m^2.

The unit of charge **C** is defined in terms of one of the smallest known charges e, that of a single electron, where one unit of **C** corresponds to the charge carried by 6.3×10^{18} electrons.

D.F. Vanderwerf, *The Story of Light Science*,
DOI 10.1007/978-3-319-64316-8_2

Carl Friedrich Gauss (1777–1855) described basic properties of magnetic flux using the magnetic field vector **B**. This was formalized as *Gauss' law of magnetism*. It can be expressed in an integral or differential form, these forms being mathematically equivalent. We present the integral form here.

$$\int_S \mathbf{B} \cdot d\mathbf{a} = 0, \tag{2.2}$$

where **B** is the magnetic vector, $d\mathbf{a}$ is an vector perpendicular to an infinitesimal area of a closed surface S, and $\mathbf{B} \cdot d\mathbf{a}$ is the scalar product of these vectors (product of the magnitude of the vectors and the cosine of their included angle). $\int_S \mathbf{B} \cdot d\mathbf{a}$ is the net magnetic flux Φ_{mag} through this closed surface. The net magnetic flux is zero since isolated magnetic poles do not exist.

Gauss also developed a law for electric flux that is often simply called *Gauss' law*. It relates the net flux Φ through a closed surface S to the electric field vector **E** and the enclosed charge.

$$\Phi = \int_S \mathbf{E} \cdot d\mathbf{a} = \frac{q_{enc}}{\varepsilon_0}, \tag{2.3}$$

where d**A** is an vector perpendicular to an infinitesimal area of a closed surface S, $\mathbf{E} \cdot d\mathbf{a}$ is the scalar product of these vectors, q_{enc} is the net charge inside the closed gaussian surface S, and ε_0 is the permittivity of free-space.

Hans Christian Oersted (1777–1851), around 1819 was the first to discover that a compass needle was deflected when placed in the proximity of a current-carrying wire. He concluded that electric currents produce magnetic fields, and thus established a link between electricity and magnetism, called *electromagnetism*.

André-Marie Ampère (1775–1833), in 1826 formulated his circuital laws relating electric current to the magnetic field over a closed path around a steady current source. One form of *Ampère's law* states that for a symmetric conductor (eg. a straight wire) carrying a steady electric current with a closed path surrounding the conductor, the magnetic field **B** on the path and the enclosed electric current I_{enc} are related by

$$\int_C \mathbf{B} \cdot d\mathbf{l} = \mu_0 I_{enc}, \tag{2.4}$$

where $\int_C$ is the line integral of the closed path, **B** is the magnetic field vector tangent to the path, vector **dl** is an infinitesimally small section of the path, $\mathbf{B} \cdot d\mathbf{l}$ is the scalar product of **B** and **dl**, μ_0 is the permeability of free-space, and I_{enc} is the total steady current through the conductor enclosed by the path.

Michael Faraday (1791–1867) was the first to describe electromagnetic induction, where a changing magnetic field induces a current in a nearby conductor, or produces an electric field. This was discovered independently by **Joseph Henry** (1797–1878). It was formalized in *Faraday's law of induction*, where a changing magnetic field produces a proportional electromagnetic force. He also rotated the plane of polarization of linearly polarized light with a magnetic field oriented in the direction of the light travel. This is called the *Faraday effect*, and established a connection with light and electromagnetism.

Wilhelm Eduard Weber (1804–1891) was a colleague of Gauss in Germany where he worked with him in studies of magnetism. He invented the electrodynamometer, which measured electric current and voltage by the interaction of two magnetic coils. Weber formalized an electrical law which integrated Ampère's law with the laws of induction and Coulomb's law. He also collaborated in a breakthrough experiment in electromagnetism, as described in the next section.

2.2 The Experiment of Weber and Kohlrausch

In 1855 Wilhelm Eduard Weber and Rudolf Kohlrausch performed an experiment that established a fundamental relationship between electricity, magnetism, and light [1]. The measured quantity was the ratio of the absolute unit of electrostatic charge to the absolute unit of magnetic charge, and remarkably had units of velocity, although no actual velocities were involved in the experiment. They measured this constant ratio by an experiment that involved charging a Leyden jar capacitor and measuring the electrostatic force. Then the Leyden jar was discharged and the magnetic force from the discharge current was measured. The ratio of the forces had a value 3.107×10^8 m/s, which was close to known measurements of the speed of light. Weber had previously named this constant ratio the letter "c" in an 1846 paper, although this ratio was not associated with the speed of light. This measured force ratio corresponds to the value of the later well-known expression $c = \sqrt{\frac{1}{\mu_0 \varepsilon_0}}$.

2.3 Maxwell and Electromagnetism

2.3.1 Maxwell's Electromagnetic Equations

During the period 1864–1865 James Clerk Maxwell developed a remarkable set of equations describing the properties of electromagnetic radiation [2]. There is a large volume of literature describing the development and application of these equations.

In our case the concern is primarily on their influence in the development of the physics of light in free-space. Maxwell had originally formulated 20 quaternion equations describing electric and magnetic phenomenon. These were later combined into four equations that we now refer to as *Maxwell's equations*. We state them here in their integral form with a brief explanation.

The first Maxwell equation is Gauss' law for electric flux.

The First Equation:

$$\int_S \mathbf{E} \cdot d\mathbf{a} = \frac{q_{enc}}{\varepsilon_0} \tag{2.5}$$

The second Maxwell equation is Gauss' law of magnetism.

The Second Equation:

$$\int_S \mathbf{B} \cdot d\mathbf{a} = \mathbf{0} \tag{2.6}$$

The third of Maxwell's equations is Faraday's law of induction. The mathematical form is

The Third Equation:

$$\int_C \mathbf{E} \cdot d\mathbf{l} = -\frac{d\Phi_{mag}}{dt} = -\frac{d}{dt}\int_S \mathbf{B} \cdot d\mathbf{a} \tag{2.7}$$

This equation states that the line integral of the scalar product of the electric field vector $\mathbf{E}$ with an infinitesimal length vector $d\mathbf{l}$ around a closed path equals the time rate of change of the magnetic flux Φ_{mag} through that path.

The fourth of Maxwell's equations is an extension of Ampère's law that adds the effect of time-varying electric flux, and is often called the *Ampère-Maxwell law*.

The Fourth Equation:

$$\int_C \mathbf{B} \cdot d\mathbf{l} = \mu_0 I_{enc} + \varepsilon_0 \mu_0 \frac{d}{dt}\int_S \mathrm{E} \cdot \mathrm{d}\mathbf{a} \tag{2.8}$$

Here I_{enc} is the current flowing through a closed curve C, and $\int_S \mathrm{E} \cdot d\mathbf{a}$ is the electric flux Φ_{elec} through the surface S bounded by the curve C. The $\mu_0 I_{enc}$ term in

Eq. (2.8) is called the *displacement current*. Using these four Maxwell equations, a complete description of electromagnetic phenomena can be obtained.

2.3.2 The Electromagnetic Wave Equations

The crowning achievement of Maxwell was the synthesis of the wave equations for electric and magnetic fields in 1864 using the four electromagnetic equations. The wave equations can be derived by converting the integral forms of Maxwell's equations to the differential form, and in free-space are given by

$$\nabla^2 \mathbf{E} = \mu_0 \varepsilon_0 \frac{\partial^2 \mathbf{E}}{\partial t^2} \tag{2.9}$$

and

$$\nabla^2 \mathbf{B} = \mu_0 \varepsilon_0 \frac{\partial^2 \mathbf{B}}{\partial t^2} \tag{2.10}$$

where ∇^2 is the Laplacian operator, which represents the second spatial derivative of the vector following it.

The derivation of these wave equations is explained in several texts, such as that of Fleisch [3]. Since both **E** and **B** are vector quantities, each wave equation actually contains three equations, one for each of the x, y, and z spatial directions.

The three-dimensional wave equation for a vector **U** has the form $\frac{\partial^2 \mathbf{U}}{\partial t^2} = v^2 \nabla^2 \mathbf{U}$, where v is the propagating velocity. The propagating velocity v in the wave equations for **E** and **B** must then be $v = \sqrt{\frac{1}{\mu_0 \varepsilon_0}}$. When previously known values of μ_0 and ε_0 were inserted in this equation, a value of v was obtained that was very close to Foucault's measured speed of light in 1850, $c \approx 2.98 \times 10^8$ m/s.

Without any direct experimental confirmation, Maxwell proposed that light was an electromagnetic wave that propagated through space at velocity c. We anticipate this later confirmation and state that

$$c = \sqrt{\frac{1}{\mu_0 \varepsilon_0}}. \tag{2.11}$$

Another electromagnetic wave relationship that follows from Maxwell's equations is

$$c = \frac{E}{B}, \tag{2.12}$$

where E and B are the magnitudes of the **E** and **B** vectors at any instant of time.

2.4 The Experiments of Hertz

Around 1886 Heinrich Hertz was the first to experimentally produce the electromagnetic waves proposed by Maxwell. He built a transmitter of electromagnetic waves using an induction coil connected to two small spherical electrodes separated by a small gap. The receiver was a loop of wire with two small spheres separated by another small gap, with the transmitter and receiver separated by several meters. When the oscillation frequency of the receiver was adjusted to match that of the transmitter, a small high-frequency spark appeared at the receiver. These propagated waves through space became known as *radio waves*.

The continuous propagation of electromagnetic waves in free-space can be described as a supportive interaction between varying electric and magnetic fields. An initial change in the electric field induces a change in the magnetic field from Maxwell's third equation (Faraday's law), Eq. (2.7). In turn, the changing magnetic field induces a change in the electric field from Maxwell's fourth equation (Ampère-Maxwell law), Eq. (2.8), and the process continues as a sustainable continuous wave having a velocity c.

In 1888 Hertz experimentally measured the frequency f and wavelength λ of radio waves in order to calculate their velocity v from the relationship $v = \lambda f$. The frequency of the wave was first measured. He then performed the more difficult measurement of the wavelength by wave interference effects using a reflected wave from a boundary, and measuring the nodal points. In one method he transmitted guided waves on a two-wire coaxial line [4]. His calculated value for v was remarkably close to the speed of light c. Hertz concluded that light was indeed an electromagnetic wave, thereby affirming Maxwell's theory and the electromagnetic nature of light. Hertz was also the first to observe the photoelectric effect (Sect. 4.1). Ironically, the photoelectric effect later provided evidence for the particle nature of light.

2.5 The Discovery of X-Rays

In 1895 Wilhelm Roentgen was studying the high-speed movement of electrons in an evacuated glass vessel. The impact of the electrons on the glass envelope produced a type of radiation that penetrated solid materials and produced distinct shadows of interior structures, such as human bones, on developed photographic film. Because of the unknown nature of this type of radiation, it was referred to as an *X-ray*. X-rays were positively identified as another type of electromagnetic radiation when in 1912 Max von Laue demonstrated interference effects of X-rays and measured their wavelength.

2.6 The Electromagnetic Spectrum

At this point it is convenient to join together the radio waves of Hertz, the visible light perceived by the human eye, and the X-rays of Roentgen, as types of electromagnetic radiation whose primary difference is their wavelength λ and frequency f. They all propagate at a common speed, which is the speed of light in vacuum $c = \lambda f$. These are usually presented as regions of a general *electromagnetic spectrum*, as illustrated in Fig. 2.1. The visible light spectrum is the smallest of these three frequency ranges, having a frequency spread between about 4×10^{14} and 7.9×10^{14} Hz, where one Hz (Hertz) $\equiv$ one cycle/s. The radio wave portion of the spectrum is the broadest, running from about 3 to 3×10^{11} Hz. This includes the microwave region. The X-ray region of the spectrum covers the frequency range from about 3×10^{16} to 3×10^{19} Hz. The combined ultraviolet, visible, and infrared region is often called the *optical region*. The terahertz radiation region from about 3×10^{11} to 3×10^{12} Hz, is useful for communication and security screening. Other important high-frequency regions of the electromagnetic spectrum are identified as cosmic radiation and gamma radiation.

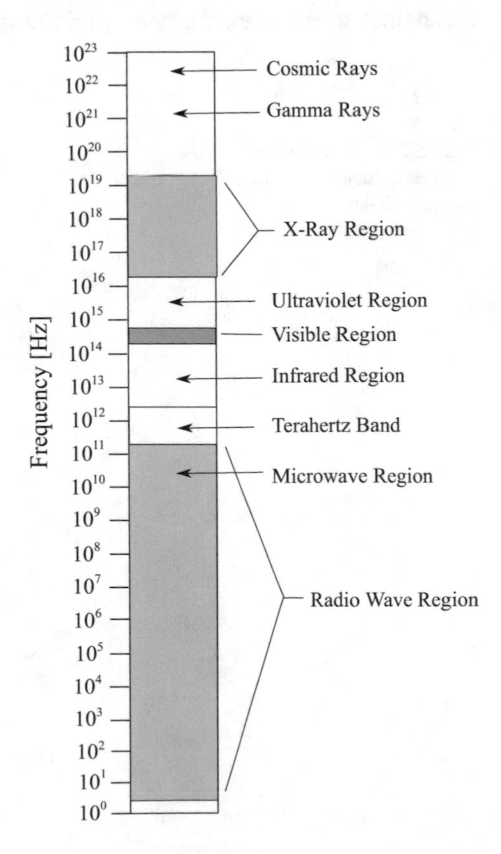

Fig. 2.1 Frequency regions of the electromagnetic spectrum

It is important to emphasize that all types of radiation in the electromagnetic spectrum are considered as types of light having similar physical properties. For example, we can speak of *visible light*, *X-ray light*, *gamma ray light*, and *radio wave light*. Thus certain properties of one type of light can be determined or verified from experiments that are specifically designed for, or more easily performed on another type of light. This technique has proved to be extremely useful in the scientific investigation of light.

2.7 Polarization of Light

Considering light as a transverse electromagnetic wave, the **E** and **B** vectors are perpendicular to each other and are perpendicular to the wave direction, as shown in Fig. 2.2a. Polarization of light is described by consideration of the motion of the **E** vector. In the splitting of light rays in a calcite birefringent crystal, the split rays are polarized, with each ray linearly polarized in orthogonal planes to each other. Figure 2.2b illustrates a linear polarized light wave, where the **E** vector remains in a plane tilted with respect to the propagation axis z. The wave theory of light provides a convincing explanation of the polarization of light.

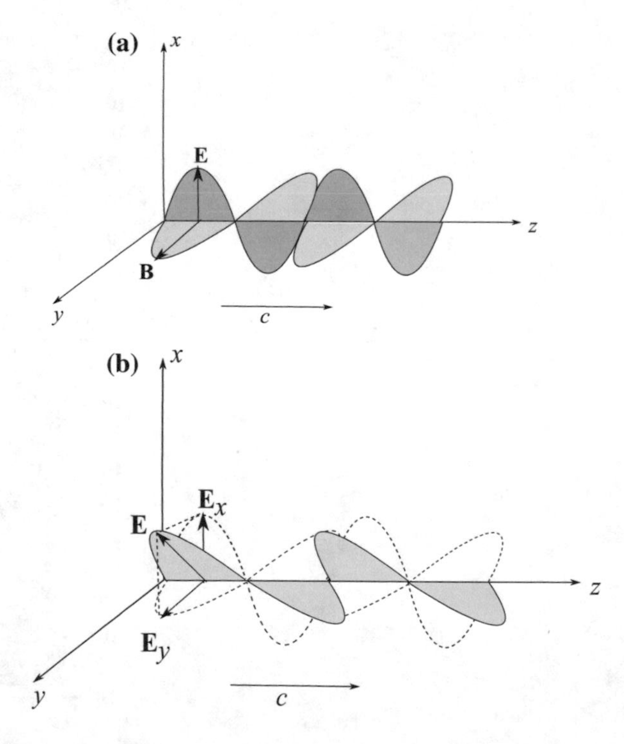

Fig. 2.2 **a** Illustration of a transverse electromagnetic wave. **b** Illustration of a linear polarized light wave

References

1. R. Kohlrausch, W. Weber, Electrodynamische maassbestimmungen insbesondere zurückführung der stromintensitäs-messungen auf mechanisches maass, in *Abhandlungen der Königl. Sächs. Gesellschaft der Wissenschaften, mathematisch-physische Klass*, vol. III, (Leipzig, 1857), pp. 221–290
2. J.C. Maxwell, A dynamical theory of the electromagnetic field. Phil. Trans. R. Soc. Lond. **155**, 459–512 (1865). doi:10.1098/rstl.1865.0008
3. D. Fleisch, *A Student's Guide to Maxwell's Equations*, 1st edn. (Cambridge University Press, New York, 2008)
4. J. Bryant, Heinrich Hertz's experiments and apparatus: his discovery of radio waves and his delineation of their properties, in *Heinrich Hertz: Classical Physicist, Modern Philosopher*, ed. by D. Baird et al. (Kluwer Academic Publishers, Dordrecht, 2010)

Chapter 3
Light and Its Application to Relativity

3.1 Light Speed in the Fizeau Moving Medium Experiment

In 1851 Fizeau made an interferometric measurement to determine how the speed of light is affected in a moving medium [1]. The chosen medium was water, flowing at a velocity v of about 700 cm/s through the U-tube configuration shown in Fig. 3.1. Each tube section had a length $\approx$150 mm, with a tube diameter $\approx$5.3 cm. Light from a source is directed to a 50/50 beamsplitter and lens, where two collimated beams A and B entered the end of each tube. The light beams exited the other ends of the tubes, where they were focused to a flat mirror by a second lens. The reflected beams were then returned through the tubes and focused to a screen where they formed interference fringes, the path lengths of each beam being the same. Light beam A moved with the velocity of the water, while light beam B moved against the velocity of the water. It was found that the speed of light v in the moving medium was affected by the velocity of the medium, according to the following equation:

$$v = \frac{c}{n} \pm \left(1 - \frac{1}{n^2}\right) w \tag{3.1}$$

where

c = speed of light in vacuum,
n = refractive index of water = 1.33,
w = velocity of the moving medium,

and the plus sign is used when the light beam and the medium velocity are moving in the same direction, and the minus sign for movement in opposite directions. During the 1885–1886 period Albert A. Michelson and Edward W. Morley

D.F. Vanderwerf, *The Story of Light Science*,
DOI 10.1007/978-3-319-64316-8_3

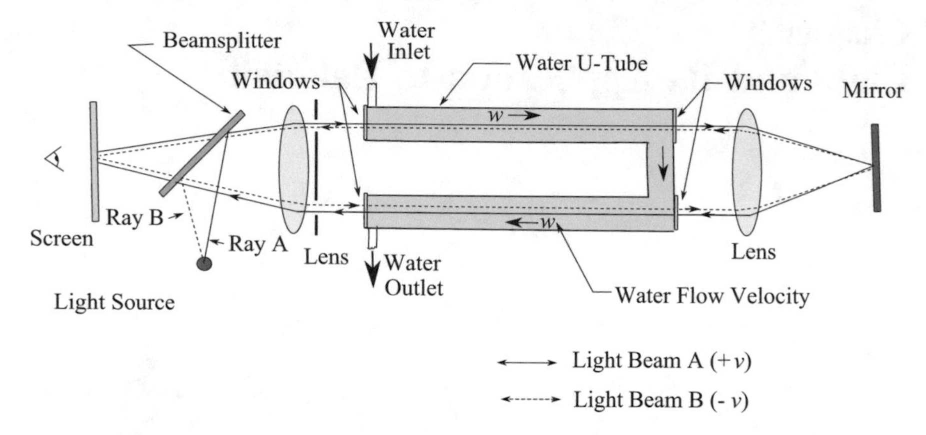

Fig. 3.1 Fizeau setup for measuring light speed in a moving medium

performed a more precise interferometric measurement of the flowing water experiment [2]. They confirmed the validity of Eq. (3.1).

3.2 The 1887 Experiment of Michelson and Morley

During the late 1880s the prevailing belief among physicists was that for light waves to propagate in space, there must be a fixed and invisible propagation medium to support the oscillations—the luminiferous ether. The Michelson-Morley experiment of 1887 is one of the most famous experiments in the history of physics, not for what it achieved, but for what it *did not* achieve. Michelson and Morley set out to accurately detect the relative motion between the revolving earth speed and the speed of light in the absolute reference frame of the ether. This was determined by a sensitive interferometer measurement in a device moving relative to the speed of light. An earlier experiment had been performed in 1881 by Albert Michelson in Germany with no detected change in the speed of light. The 1887 experiment with Edward Morley at Case Institute in Cleveland, Ohio used more accurate instrumentation with a longer interferometric path length [3].

The *Michelson interferometer*, invented in 1881 by Michelson, was a key component used in this experiment. See Fig. 3.2. Light from a sodium light source of wavelength $\lambda = 550$ nm passed through a 50/50 semi-reflecting beamsplitter plate, where the transmitted component traveled along an arm of length l_1 with a fixed mirror M_1, and the reflected component traveled along a perpendicular arm of length l_2 with a precisely moveable mirror M_2. Interference fringes were then observed through the cross-hairs of a telescope to determine slight differences in path length between the two beams in each arm of the device. A compensation plate with the same thickness as the beamsplitter plate was inserted in the transmission

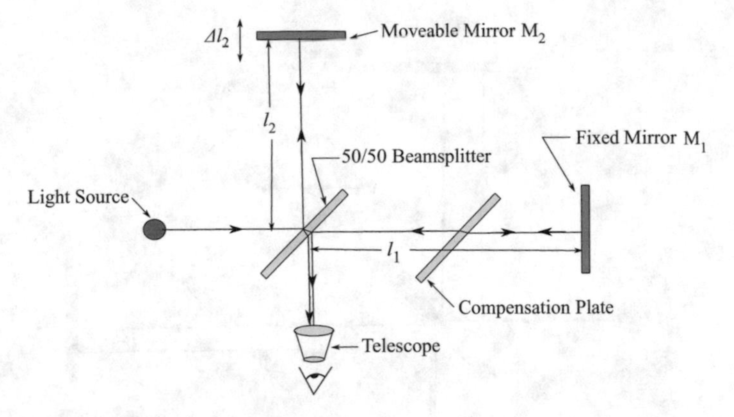

Fig. 3.2 Basic layout of the Michelson interferometer

arm to equalize the physical path lengths at the reference point position. By precise adjustment of the moveable mirror position, the distance change Δl_2 was calculated by counting the number of fringes N which passed by a reference point, where

$$\Delta l_2 = N\left(\frac{\lambda}{2}\right) \tag{3.2}$$

The Michelson interferometer provided a very precise method of measuring small changes in distance—to within a fraction of a wavelength.

The Michelson-Morley experiment of 1887 attempted to detect minute differences in light speed between two light paths, one path parallel to the mean orbital velocity v of the Earth ($v \approx 30$ km/s), and another path transverse to the Earth's orbital velocity. The existence of the fixed ether should then produce a noticeable "ether wind". The interferometer arm lengths were equal and fixed ($l_1 \approx l_2 = l$) and the interferometer was first oriented with the source S in the direction of the earth's orbital velocity v (position **A**). See Fig. 3.3a. If we assume that Galilean transformations of v and c occur, the time t_1 for the light to travel along the M_1 arm in the direction of v would differ from the time t_2 for light to travel along the M_2 path, which is transverse to the direction of v. The expected fringe shift was calculated from the expected difference between the total light travel times t_1 and t_2 for each arm. For the M_1 arm

$$t_1 = \frac{l_1}{c-v} + \frac{l_1}{c+v}. \tag{3.3}$$

or

$$t_1 = \frac{2l}{c(1 - v^2/c^2)}, \tag{3.4}$$

where $(v^2/c^2) = (9 \times 10^8)/(9 \times 10^{16}) = 1 \times 10^{-8}$.

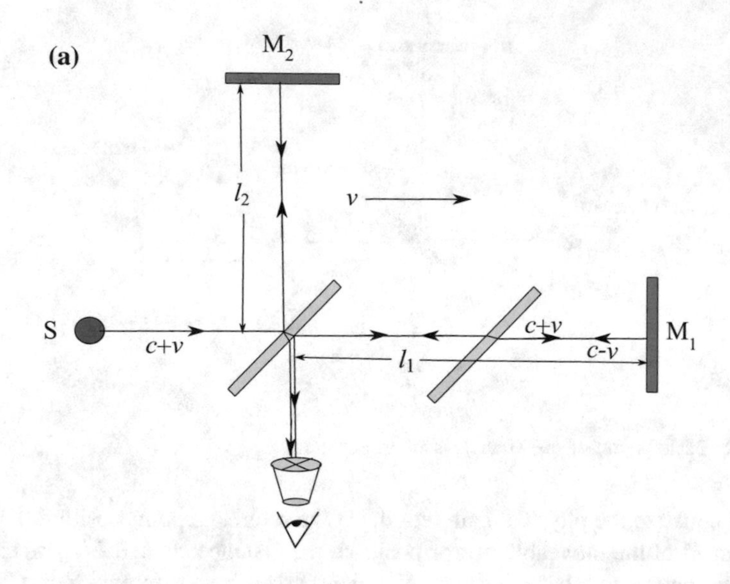

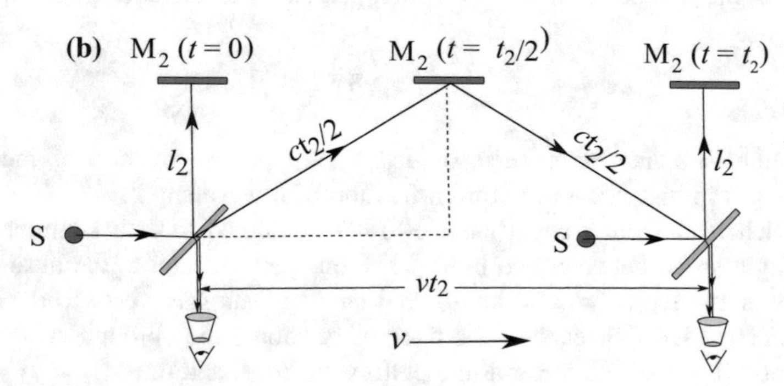

Fig. 3.3 **a** Michelson-Morley experiment—light travel along arm M_1. **b** Michelson-Morley experiment—light travel along arm M_2

For the transverse M_2 arm in Fig. 3.3b, the mirror M_2 moved a distance vt_2 over the time t_2, and the light path distance was increased over the M_1 arm. Use of the Pythagorean theorem gives

$$(ct_2)^2 = 4l^2 + (vt_2)^2. \tag{3.5}$$

Solving for t_2, the following is obtained:

$$t_2 = \frac{2l}{c\sqrt{1 - v^2/c^2}}. \tag{3.6}$$

Use the series expansion $(1+x)^{-n} = 1 - nx + \frac{n(n+1)}{2!}x^2 - \frac{n(n+1)(n+2)}{3!}x^3 + \cdots$, where $x^2 < 1$, and set $x = -(v^2/c^2)$. Since $(v^2/c^2) \approx 10^{-8}$, the first two terms of the expansion gives

$$1/(1 - v^2/c^2) \approx 1 + (v^2/c^2), \tag{3.7}$$

and

$$1/\sqrt{1 - v^2/c^2} \approx 1 + \frac{1}{2}(v^2/c^2). \tag{3.8}$$

Using these expressions Eqs. (3.4) and (3.6) become

$$t_1 = \left(\frac{2l}{c}\right)(1 + (v^2/c^2)), \tag{3.9}$$

and

$$t_2 = \left(\frac{2l}{c}\right)\left(1 + \frac{1}{2}(v^2/c^2)\right). \tag{3.10}$$

This introduces a time difference $\Delta t = t_1 - t_2$ at position **A**, where

$$\Delta t = \left(\frac{l}{c}\right)(v^2/c^2) \tag{3.11}$$

By the use of sixteen reflecting mirrors, the distance l was set at 11 m. For $\lambda = 550$ nm and $v = 3 \times 10^4$ m/s, a time difference $\Delta t \approx 3.7 \times 10^{-16}$ s should occur. The corresponding path difference Δl_1 is

$$\Delta l_1 = ct_1 = 2l[1 + (v^2/c^2)]. \tag{3.12}$$

The interferometer was then rotated 90° to position **B**, reversing the roles of mirrors M_1 and M_2, with the M_1 arm now having the longer light travel distance. The path difference Δl_2 is then

$$\Delta l_2 = ct_2 = 2l\left[1 + \frac{1}{2}(v^2/c^2)\right]. \tag{3.13}$$

The expected fringe shift is

$$\frac{\Delta l_1 - \Delta l_2}{\lambda} = \left(\frac{2l}{\lambda}\right)\left(\frac{v^2}{c^2}\right) \approx 0.41. \tag{3.14}$$

A fringe shift of this magnitude should be visually detectable with the apparatus. The experimenters fully expected to observe this fringe shift, indicating a change in the speed of light emitted from the source oriented in different directions.

The apparatus, positioned on a massive stone slab floating on a pool of mercury, was continuously and slowly rotated during the measurement process to eliminate any induced strains due to stopping or starting. Visual fringe observations were made over three month intervals, with light from the source moving in the same direction of the orbital velocity v of the Earth, and in the opposite direction six months later. Surprisingly, no fringe shift was ever detected, showing no change in the speed of light for any position. The failure of the Michelson-Morley experiment indicated the absence of a stationary ether, or perhaps that the ether was dragged along with the Earth's velocity (later disproved by Michelson and others). It implicitly showed that the speed of light c was independent of the motion of the source. In the following decades, similar experiments of other investigators have also failed to detect any fringe shift. It remained for Einstein's special theory of relativity in 1905 to formulate an explanation.

3.3 Lorentz-FitzGerald Contraction

In 1889 George FitzGerald proposed length contraction of a moving object in the direction of motion in an absolute reference frame consisting of the luminiferous ether. This was called the *FitzGerald contraction*. The physical basis of this contraction was an electromagnetic force compression between the atoms of matter in the direction of motion, and utilized Maxwell's equations. The negative result of the Michelson-Morley experiment could then be explained by a contraction of the arm of the interferometer in the direction of the Earth's motion. Around 1890 Hendrik Lorentz developed the mathematics to calculate this contraction. The contraction then became known as the *Lorentz-FitzGerald contraction* or the *Lorentz contraction*. The contracted length L' is given by the following formula:

$$L' = L\sqrt{1 - v^2/c^2}. \tag{3.15}$$

where L is the length of the object at rest, v is the velocity of the object, and c is the speed of light.

The Lorentz contraction is often expressed as $L' = L/\gamma$, where the factor $\gamma = \frac{1}{\sqrt{1-v^2/c^2}}$ is called the *Lorentz factor*. This contraction cannot be measured in a moving object reference system, since the measuring "ruler" would be similarly contracted in the direction of motion. Since this formulation was based on electromagnetic forces between the atoms in the object, it did not rely on the constancy of the speed of light in a moving reference frame, as later postulated by Einstein. There is still some controversy today over whether this contraction is a physically real attribute of matter [4]. In any event, the Lorentz factor would prove quite useful in Einstein's theory of special relativity. Appendix 3A gives a calculation for the Lorentz contraction of a modern spacecraft.

3.4 Lorentz Transformations

The Lorentz transformations were originally derived in the year 1890, before Einstein's 1905 theory of special relativity, and are an extension of the Lorentz contraction. However, Einstein found these transformations to be entirely applicable to his theory, and he adapted them along with the implied invariance of the speed of light from the Michelson-Morley experiment. The equations relate how the coordinates and times of two events are perceived by two observers, one event and observer in a fixed reference frame, and another event and observer in a reference frame moving in the x-direction at a uniform speed v with respect to the fixed frame. The derivation of these equations has been published in many sources, such as in the text of Serway [5]. The equations are stated here.

$$x' = \gamma(x - vt) \tag{3.16a}$$

$$y' = y \tag{3.16b}$$

$$z' = z \tag{3.16c}$$

$$t' = \gamma\left(t - \frac{vx}{c^2}\right) \tag{3.16d}$$

$$x = \gamma(x' + vt') \tag{3.16e}$$

$$y = y' \tag{3.16f}$$

$$z = z' \tag{3.16g}$$

$$t = \gamma\left(t' + \frac{vx'}{c^2}\right) \tag{3.16h}$$

Consider a fixed *inertial reference frame* (no acceleration allowed) **1** having local coordinates (x, y, z, t), and another inertial reference frame **2** moving at uniform velocity v with respect to reference frame **1**, as shown in Fig. 3.4. Equations (3.16a–3.16d) then give the position and time coordinates (x', y', z', t') of the moving reference frame **2**, as observed from the fixed reference frame **1**. However, an observer in the moving reference frame **2** having local position and time coordinates (x', y', z', t'), measures the position and time coordinates (x, y, z, t) of an event occuring in the fixed reference frame **1** according to Eqs. (3.16e–3.16h).

The *Lorentz transformations of velocity* use a slightly different perspective. Again, reference frame 1 is fixed, and reference frame 2 moves at uniform velocity

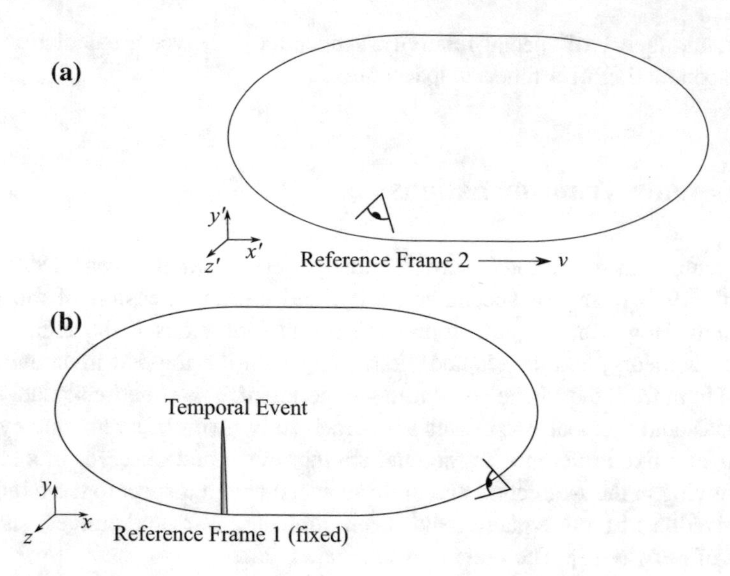

Fig. 3.4 **a** Illustration of Lorentz coordinate transformations for reference frame 1. **b** Illustration of Lorentz coordinate transformations for reference frame 2

v relative to frame **1**. The speed of an object (e.g. airplane) moving in the same direction as frame **2** is measured by an observer in frame **2** as u'. An observer in frame **1** measures the object speed as u. See Fig. 3.5. Then the measured speeds u' and u are related by the following equations:

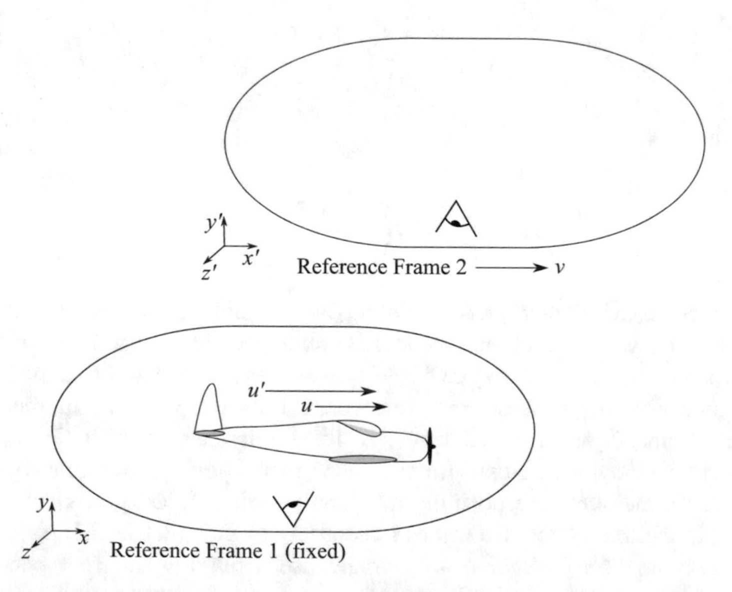

Fig. 3.5 Illustration of Lorentz velocity transformations

$$u' = \frac{u - v}{1 - \left(\frac{v}{c^2}\right)u} \tag{3.17}$$

$$u = \frac{u' + v}{1 + \left(\frac{v}{c^2}\right)u'} \tag{3.18}$$

These equations provide several interesting results. Suppose that the object speed u is measured as the speed of light c from reference frame **1**. Then for $u = c$, Eq. (3.17) reduces to $u' = c$, and for $u' = c$, Eq. (3.18) yields $u = c$. Thus the speed of light has a constant value c that does not change when measured from various moving inertial reference frames, or the speed of light is *invariant*. These equations also show that we cannot exceed the speed of light c by combining values of the velocities u' and v. The result is always c or is less than c. A speed less than c in any reference frame will always be less than c in any other relative reference frame. If $v << c$, then $\left(\frac{v}{c^2}\right) \approx 0$, and we obtain the Galilean transformations of velocity, $u' = u - v$, and $u = u' + v$. The Lorentz transformations of velocity do not predict that c is the ultimate speed of a moving object, or show that an object moving at velocity v can never reach the speed of c. To show this, we need to progress to the paradigm shift of Albert Einstein.

3.5 Einstein Special Theory of Relativity

In the year 1905 Albert Einstein, while working as a patent clerk in Zurich, published five scientific papers, two of which defined the *special theory of relativity*, which changed the foundations of physics. Central to this theory were the speed of light and the invariance of light. Influenced perhaps by the Michelson-Morley experiment and certainly by the Lorentz transformations, Einstein in his June 1905 paper [6] proposed two postulates:

1. The same laws of physics apply in all inertial reference frames that are moving at a constant velocity with respect to each other.
2. The observed speed of light is the same for *any* reference frame moving at a uniform velocity.

With these postulates Einstein formulated a theory of relativity that applied to all laws of physics, including mechanics, electrodynamics, and optics, giving rise to the *principle of relativity*. Einstein used the Lorentz velocity transformations and the Lorentz-FitzGerald length contraction to form a coherent physical theory for bodies that travel in different reference frames. He derived the Lorentz transformations using these relativistic principles. The results are commonly known as the *Lorentz-Einstein transformations*.

Einstein performed a famous *Gedankenexperiment* (thought experiment) to define the relativity of time in terms of ticking clocks in moving inertial reference frames. He concluded that there is no way to determine absolute time in space, as was assumed by Newton. Time depends on movement of the reference frame, and moving clocks run slower than clocks at rest. This is known as *time dilation* (sometimes called time "dilatation" in the literature), and is also a consequence of the Lorentz transformations. The relativistic equation for time dilation can be derived using Fig. 3.3b from the Michelson-Morley experiment. Consider the time interval Δt from $t = 0$ to $t = t_2$. In this figure, the time interval $\Delta t'$ for light to travel from the beamsplitter to M_2 and back, measured by an observer in the moving reference frame (at velocity v) is $\Delta t' = \frac{2l}{c}$. For an observer in a fixed reference frame the time interval Δt is calculated as $(c\Delta t)^2 = 4l^2 + (v\Delta t)^2$. Then substituting $l = \frac{c\Delta t'}{2}$ in this equation and solving for Δt yields

$$\Delta t = \frac{\Delta t'}{\sqrt{1 - v^2/c^2}} = \gamma \Delta t', \tag{3.19}$$

where γ is the Lorentz factor. Here t' is the time interval measured in the moving reference frame, t is the time interval measured in the fixed reference frame, and since $\gamma > 1$, Eq. (3.19) states that time runs slower in a moving reference frame, or that moving clocks run slower. Another consequence is that events that are observed as occurring simultaneously in one reference frame will be observed as occurring at different times from another reference frame moving relative to it. Thus there is no absolute simultaneity between events. However, the cosmic speed of light c in vacuum is absolute and invariant.

Since the principles of the theory do not support the concept of an absolute stationary space, the existence of the luminiferous ether is not required and is not real. Einstein also recognized that Maxwell's equations were not compatible with the Galilean transformations, and modified Maxwell's equations using relativistic principles for a reference frame moving with velocity v. He also examined the effect of motion on the optics of moving bodies in terms of light pressure.

The June 1905 paper by Einstein also introduced the concept of relativistic mass by analysis of a moving electron. In modern terminology the relativistic mass of a moving body is given by

$$m = \frac{m_0}{\sqrt{1 - v^2/c^2}} = \gamma m_0 \tag{3.20}$$

where m_0 is the rest mass of the body, and γ is the Lorentz factor. This expression for relativistic mass can also be derived by considering a grazing collision between two elastic balls, and using the relativistic conservation of momentum.

The *Doppler frequency shift* for light is another relativistic phenomenon described by Einstein in his June 1905 paper. A light source at rest emits electromagnetic waves at frequency f, and an observer a very large distance away moves at velocity v at an angle φ relative to a fixed coordinate system at the light source. Then the frequency f^1 perceived by the observer is given by

$$f' = f\frac{1 - (v/c)\cos\varphi}{\sqrt{1 - v^2/c^2}}. \tag{3.21}$$

This equation implies that there is a longitudinal component ($\varphi = 0°$) and a transverse component ($\varphi = 90°$) of Doppler shift. For the longitudinal Doppler shift, Eq. (3.21) reduces to

$$f'_{long} = f\frac{1 - v/c}{\sqrt{1 - v^2/c^2}} = f\sqrt{\frac{1 - v/c}{1 + v/c}}. \tag{3.22}$$

For the transverse Doppler shift, the following equation results:

$$f'_{trans} = f\frac{1}{\sqrt{1 - v^2/c^2}} = f\gamma \tag{3.23}$$

3.6 Minkowski Spacetime

In 1908 the mathematician Hermann Minkowski (1864–1909) combined the three spatial dimensions with the single time dimension of the special theory of relativity, into a single four-dimensional continuum known as Minkowski space or *spacetime*. It is formally defined as a four-dimensional vector space, which is non-Euclidian, and uses hyperbolic quaternions. Einstein collaborated with Minkowski to formulate a new type of spacetime that is known as *Einstein-Minkowski spacetime*, replacing the classic Galilean separation of space and time. It rejected the idea of an absolute time that could produce simultaneity of events.

3.6.1 The Light Cone

The mathematics of Minkowski four-dimensional spacetime can be quite complex and obtuse. A more visually agreeable presentation is the use of the *light cone*. The light cone illustrates the observation of events, the cosmic speed of light, the event horizon, and causality between events. For the light cone illustrated in Fig. 3.6a, an observer at position **O** is considered to be instantaneously in the "here and now". The vertical coordinate is *time* and the space coordinates (x, y, z) are represented in a horizontal plane called *space*. The sides of the cone define a boundary along which light travels at the cosmic speed c. The top region of the cone defines the

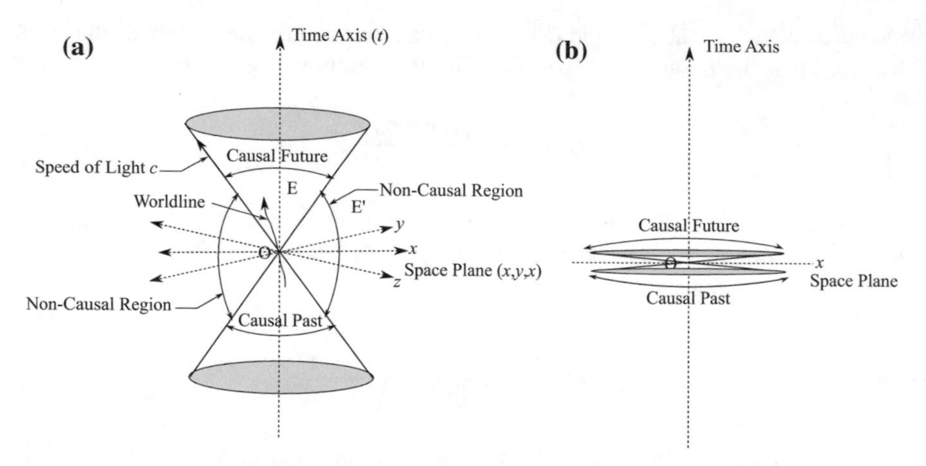

Fig. 3.6 **a** Light cone representation of Minkowski four-dimensional spacetime. **b** Light cone representation when $v << c$

region where the causal future of the observer **O** exists, and the bottom region of the cone is the region where the causal past of **O** exits. An event E occurring in a causal region can influence, or be influenced by the observer **O**. The physical law of *cause and effect* applies. Outside the cone boundaries, to the right and left of **O** are the non-causal regions, and an event **E′** in these regions cannot influence or be influenced by the observer **O**. The observer **O** moves forward in time at velocity v through the upper causal region, and traces out a path known as a *world line*.

For terrestrial events, the observation time between events is relatively short. For example, the time it takes for light to travel between two events on Earth separated by 1000 km is ≈0.03 s. This results in an extreme flattening of the light cone as shown in Fig. 3.6b. Terrestrial events occur almost in a plane of simultaneity, which fortunately makes interaction with earthly events and fellow humans appear to be instantaneous.

For events that are separated by astronomical distances, or are moving closer to the speed of light, the light cone shown in Fig. 3.7 is useful. This shows a vertical section of the light cone, where space is considered as a single dimension x. The boundaries of the light cone are set at 45° by expressing time along the vertical time line as a distance called a light-second [ct]. Then the resultant 45° light cone boundaries are defined by $x = ct$. For an event **A** on the light cone boundary to influence an observer at **O** at time t, the maximum distance away, $x = ct$, is determined by the ultimate cosmic speed of light c. An event **E′** outside the boundary would not arrive in time to be observed, since it would have to travel faster than c, and the observer has moved forward in time at velocity $v = x/t$ to position **O′**. For this reason, the non-causal regions are sometimes referred to as the unknown zones, since events in these regions cannot be observed by an observer at **O**, and the cone boundary is referred to as the *event horizon*.

Fig. 3.7 Two-dimensional planar section of light cone

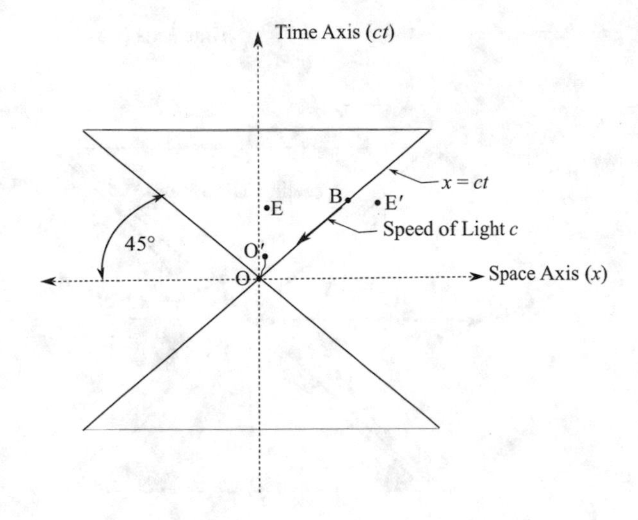

3.6.2 *The Invariance of Spacetime*

The Lorentz transformations have shown that distance x, velocity v, and time t in one reference frame are not invariant when viewed from another reference frame moving relative to it. The speed of light c is invariant, according to Einstein's second postulate. The spacetime of Minkowski produces another invariant called *spacetime distance s*, which is invariant under the Lorentz transformations. The two-dimensional planar spacetime shown in Fig. 3.8 has a single space coordinate x, and the *spacetime distance s* is defined by the hyperbolic equation

$$s^2 = (ct)^2 - x^2, \tag{3.24}$$

or

$$s = \sqrt{(ct)^2 - (vt)^2} \tag{3.25}$$

where s is the distance connecting two points in spacetime. The causal/non-causal separation of events of the light cone remains in spacetime. Although observers in moving reference frames may measure relativistic values x', v', and t' in another moving frame, and vice versa, the measured value of the spacetime s will be the same for all inertial reference frames. These hyperbolic curves can be expanded to a spacetime continuum. This simplified two-dimensional representation of spacetime can be expanded by the appropriate mathematics to the actual four dimensions, but the visualization of a four-dimensional spacetime continuum is a stretch for the human mind.

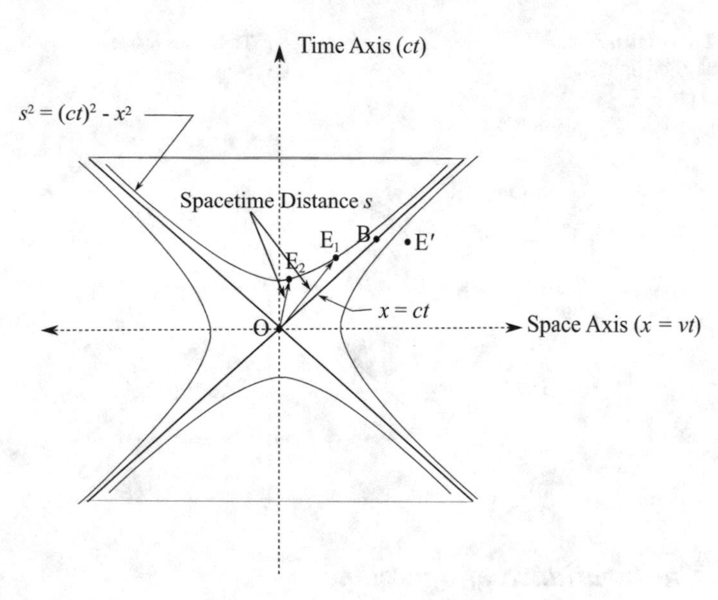

Fig. 3.8 Two-dimensional representation of Minkowski spacetime showing a hyperbolic spacetime curve

3.7 The Speed of Light and Einstein's $E = mc^2$

3.7.1 *Derivations of* $\mathbf{E = mc^2}$

The September 1905 paper by Einstein [7] introduced the concept of the conversion of mass into energy, leading to what is probably the most famous single equation in physics: $E = mc^2$. We identify the quantity c as the cosmic or maximum speed of light, where $c = 2.99792458 \times 10^8$ m/s. Thus, there is a tremendous amount of contained energy in a small mass of material. The principles proposed and developed in the two 1905 papers are referred to as Einstein's special theory of relativity. There are various approaches to deriving the equation $E = mc^2$. One method uses the relationships between energy and momentum for two photons emitted from an atom moving at constant velocity v in an inertial reference system [8]. Each photon has energy $E_{photon} = hf$ and momentum $p_{photon} = hf/c$. The atom has a momentum component $p_{atom} = mv$ in the velocity direction. Using conservation of momentum, the decrease in momentum of the atom equals the momentum of the photons, or $\Delta p_{atom} = \Delta m_{atom} v = p_{photons}$. This momentum decrease in the atom causes a decrease in its mass. The change in energy of the atom ΔE_{atom} is then related to the energy of the emitted photons $E_{photons} = 2hf$. This leads to $\Delta E = \Delta mc^2$ for the atom, which relates the change of energy in the atom for a change in its mass, and is a form of Einstein's energy-mass equation.

Another approach to derive $E = mc^2$ considers the force F acting on a mass particle moving at a relativistic speed v. The calculations are detailed in Appendix 3B.

3.7.2 *Relativistic Energy, Momentum, and Mass of a Photon Particle*

The energy of a particle moving at relativistic speed v has been shown to be $E = mc^2 = \gamma m_0 c^2$, and the momentum p of the particle is $p = mv = \gamma m_0 v$, where m is the relativistic mass, m_0 is the rest mass, and $\gamma = 1/\sqrt{1 - (v^2/c^2)}$. Then $E^2 = m^2c^4 = \frac{m_0^2 c^4}{1-(v^2/c^2)}$. Substituting $v^2 = (p/m)^2$ and rearranging terms gives $E^2 - m^2c^2(p^2/m^2) = m_0^2 c^4$, which yields the energy-momentum-mass relationship

$$E^2 = p^2c^2 + m_0^2c^4. \tag{3.26}$$

This relationship holds for all types of moving particles, and each particle type has a specific and constant rest mass. For a particle at rest, $p = 0$ and the particle has energy $E = m_0c^2$.

The cosmic speed of light c could also be considered as the *speed of a massless particle*—yet photons have both energy and momentum. Although a photon particle is never really at rest and has zero mass at any speed, a photon rest mass m_0 is defined such that $m = m_0 = 0$. Then from Eq. (3.26) the energy and momentum of the photon are related by $E_{photon} = pc$. Experimental proof that the photon mass is *exactly* zero has not been reported. However, it is experimentally possible to set upper limits on the photon rest mass. The most accurate limit values have been based on extra-terrestrial measurements. In 1994 an upper limit of the photon mass m was reported as 1×10^{-48} g, using satellite measurements to detect any effect of nonzero photon mass to the Earth's magnetic field [9]. A review paper in 2004 summarized types of experimental methods showing no conclusive proof of a finite photon mass [10].

3.8 Einstein General Theory of Relativity

In 1907 Einstein began a series of papers on the *general theory of relativity*. The 1907 paper introduced the equivalence principle, gravitational redshift, and the gravitational bending of light [11]. The general theory of relativity extended the principles of special relativity to accelerating (non-inertial) reference frames. This gave rise to the *principle of equivalence*, in which the force of gravity experienced by a stationary observer cannot be distinguished from the force experienced by an observer in an accelerating reference frame. In 1911 Einstein published a paper where he realized the need for this new theory to replace special relativity and Newton's law of gravitation [12]. In another paper published in 1915, he explained the anomalous precession of the planet Mercury, and described the bending of light by gravity [13]. Einstein developed and refined the theory of general relativity well into the 1920s.

When electromagnetic waves of light emanate from a source in a strong gravitational field, for an observer positioned in a weaker gravitational field, the light appears to be shifted toward a longer wavelength. For visible light, this is referred to as the *gravitational redshift*.

The gravitational effect of matter causes a curvature of spacetime, as opposed to the flat spacetime considered in special relativity. Curved spacetime is usually referred to as Minkowski spacetime, with a spacetime distance defined as in Eq. (3.25). In the universe, spacetime is locally warped near massive astronomical bodies. Since light follows the curvature of Minkowski spacetime, we would expect an observed bending or deflection of light in the vicinity of these bodies. In the Einstein 1911 paper he calculated the expected light deflection α near an astronomical body to be

$$\alpha = 2\frac{\mathrm{G}M}{c^2\Delta}, \tag{3.27}$$

where G is the gravitational constant, M is the mass of the body, c is the speed of light, and Δ is the distance of the light ray from the center of the body. Einstein then predicted that the light from a star viewed close to the edge of the Sun during a solar eclipse would be deflected by $\approx$0.83 arc seconds. He then challenged any astronomer to attempt to experimentally verify this result. This challenge was soon taken up.

3.9 Experimental Affirmations of Relativity

3.9.1 The Kennedy-Thorndike Experiment

The results of the 1887 Michelson-Morley experiment could be explained by Lorentz contraction. In 1932 Kennedy and Thorndike performed an experiment to directly test the validity of length and time contraction as predicted by Einstein [14]. They used a modified Michelson interferometer having one arm shorter than the other. The unequal path lengths in the interferometer would produce different light travel times in each arm, forming a circular fringe pattern.

The solar system and Earth appear to be moving in the direction of the constellation Leo at a net speed of about 375 km/s. The interferometer on the Earth's surface was not rotated, but its position was fixed pointing westward such that once a day the apparatus would be oriented in the direction of Leo. This interferometer speed would vary for other times, due to Earth's rotation, orbital revolution, and the deviation of the direction towards Leo. By introducing the variable of changing velocity, one could observe a phase shift in the pattern as the net speed of the Earth varied. In practice, no fringe shifts occurred as the velocity of the apparatus varied, implying the presence of both length contraction and time dilation, as predicted by the Lorentz-Einstein transformations.

3.9.2 Transverse Doppler Shift

The tranverse Doppler shift for light predicted by Eq. (3.23), $f'_{trans} = f\gamma$, is referred to as the *transverse Dopppler effect* (TDE), and can provide a direct value of relativistic time dilation. Using the classic Galilean transformations, the expected Doppler shifts, for $v << c$, would be

$$\lambda_{red} = \lambda_0(1 + v/c)\cos\varphi, \tag{3.28}$$

and

$$\lambda_{blue} = \lambda_0(1 - v/c)\cos\varphi, \tag{3.29}$$

where λ_{red} is the observed longer wavelength of a receding source of light, λ_{blue} is the observed shorter wavelength of an approaching source of light, λ_0 is the reference wavelength, v is the light beam velocity, and φ is the angle between the observer's line-of-sight and the light beam direction. These equations are modified for special relativity through multiplication by the Lorentz factor γ. Then

$$\lambda_{red} = \lambda_0\gamma(1 + v/c)\cos\varphi, \tag{3.30}$$

and

$$\lambda_{blue} = \lambda_0\gamma(1 - v/c)\cos\varphi. \tag{3.31}$$

Ives and Stilwell, in the time period 1938–1941, performed a series of experiments at Bell Laboratories where they were able to distinguish the small TDE from the much larger longitudinal Doppler effect [15, 16]. A direct measurement of the TDE at exactly a 90° transverse angle is very difficult, since any slight deviation from this angle introduces significant longitudinal shift. Therefore.the TDE was not meaured directly, but extracted from simutaneous longitudinal measurements in the forward (φ = 0°) and backward (φ = 180°) directions of the light. A version of the apparatus is illustrated in Fig. 3.9. Positive particles produced in a hydrogen arc from a filament cathode F were accelerated between a pair of perforated electrodes A and B. The voltage at accelerating electrode B was up to 30,000 V DC above the grounded electrode A.. This produced canal rays (positive ion particles) of uniform

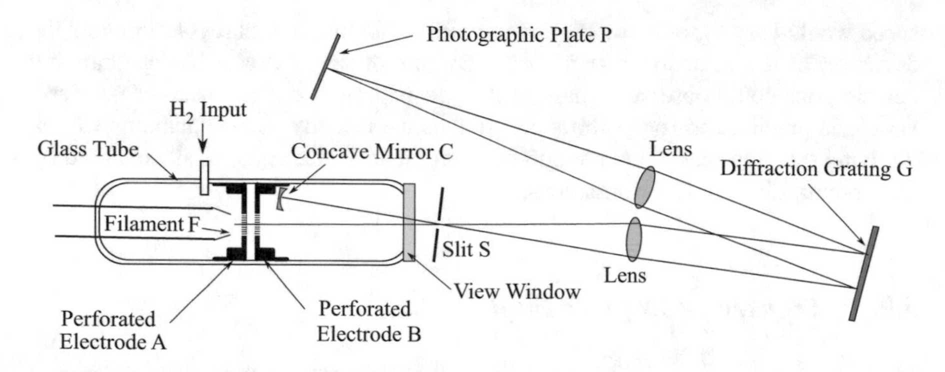

Fig. 3.9 Illustration of the Ives-Stilwell experiment to detect the transverse Doppler effect

velocity v. The spectral line studied was the blue-green H_β line (λ_0 = 486.1 nm). A small off-axis concave mirror C formed an image of the slit S back on itself, allowing a simultaneous backward view of the canal rays. The rays were dispersed by a metal-on-glass ruled diffraction grating G and imaged on a photographic plate P. The forward direction produces a maximum blue Doppler shift and the backward direction produces a maximum red Doppler shift. Since $\lambda_{trans} = \lambda_0/\gamma$ and $\Delta\lambda_{trans} = \lambda_0/\gamma - \lambda_0$, the following equations result:

$$\Delta\lambda_{red} = \lambda_{\mathrm{red}} - \lambda_0 = \gamma(v/c)\lambda_0 \cos\varphi + \Delta\lambda_{trans}, \tag{3.32}$$

$$\Delta\lambda_{blue} = \lambda_{blue} - \lambda_0 = \gamma(v/c)\lambda_0 \cos\varphi - \Delta\lambda_{trans}, \tag{3.33}$$

where $\Delta\lambda_{red}$ is the total shift of the receding light, $\Delta\lambda_{blue}$ is the total shift of the advancing light, and $\Delta\lambda_{trans}$ is the transverse Doppler shift. Subtracting Eq. (3.33) from Eq. (3.32) gives:

$$\Delta\lambda_{trans} = \frac{\Delta\lambda_{red} - \Delta\lambda_{blue}}{2}, \tag{3.34}$$

and adding Eqs. (3.33) and (3.32) with φ = 0 deg gives

$$\frac{\Delta\lambda_{blue} + \Delta\lambda_{red}}{2\lambda_0} = \gamma(v/c). \tag{3.35}$$

Thus, in theory, the transverse Doppler shift can be calculated from Eq. (3.34) using longitudinal measurements of the blue Doppler shift at $\varphi = 0°$ and the red Doppler shift at $\varphi = 180°$, avoiding any transverse measurements at $\varphi = 90°$. A value of the velocity v is not specifically required. The presence of a transverse Doppler shift would be indicated by unequal measured values of $\Delta\lambda_{blue}$ and $\Delta\lambda_{red}$ from the reference wavelength λ_0. In fact this was the case, indicating the validity of time dilation in the relativistic theory. Figure 3.10 illustrates the Doppler shifts.

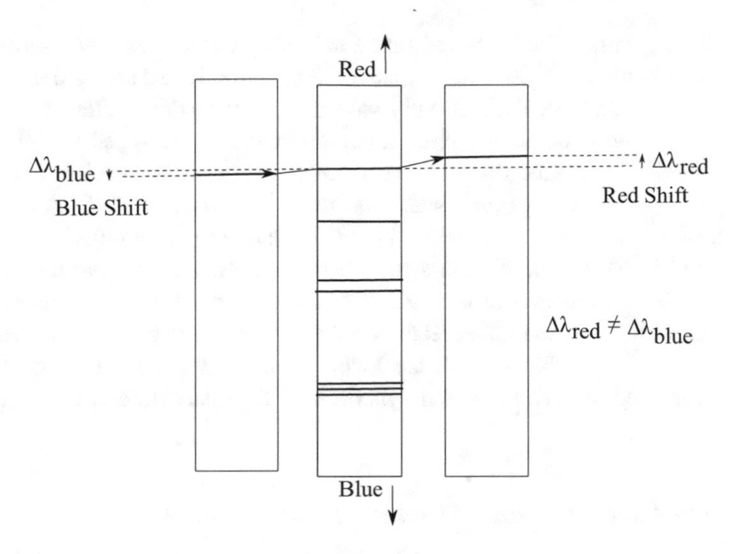

Fig. 3.10 Doppler wavelength shifts in the Ives-Stilwell experiment

In 1979 a direct measurement was made of the Doppler shift from a source moving at right angles to the line of sight [17]. The experimenters used the H_α line emitted by hydrogen atoms with a linear velocity range 2.53×10^6 to 9.28×10^6 m/s. Another experiment in 2007 measured the time dilation factor by comparing Doppler shifts at two different speeds of positive lithium ($^7Li^+$) ions in a storage ring [18]. The fractional accuracy of the measured frequency was 10^{-9}, with a deviation from the relativistic formula of 2.2×10^{-7} for speeds close to the speed of light c.

3.9.3 *Time Dilation and Muon Lifetime*

In 1941 Rossi and Hall provided an experimental confirmation of time dilation in muons by the detection of cosmic ray muons at the top (≈2000 m) and base (≈sea level) of Mount Washington, New Hampshire [19]. Muons are short-lived elementary particles having the charge of an electron but with a mass 207 times heaver. They are produced in the upper atmosphere (15–20 km up) by the collision of cosmic rays with atoms. They travel at a velocity $v \approx 0.994c \approx 2.98 \times 10^8$ m/s, and have a mean lifetime $\tau \approx 2.2$ μs at rest. Velocity and lifetime values of muons have been measured over several decades, for example, the fairly recent report of Liu and Solis [20].

Rossi and Hall measured the muon arrival rates [muons/h] at the mountain top and at sea level. Over the 2000 m Earth travel distance, very few of the short lifetime muons should have arrived at sea level. However, the significant arrival rate of muons at sea level could be predicted using muon time dilation as observed from the

Earth reference frame. The measured muon arrival rate at sea level was close to the rate predicted using relativistic time dilation. Appendix 3C describes details of the actual experiment and calculations. This experiment clearly demonstrated relativistic time dilation of muon particles having a velocity close to the speed of light c.

Atmospheric muon time dilation experiments of increased precision were performed by other investigators, such as that by Frisch and Smith at MIT (Massachusetts Institute of Technology) [21]. Time dilation of muons has also been demonstrated in particle accelerators, where the lifetimes of positive and negative muons have been measured in the CERN Muon storage Ring [22]. Here the calculated relativistic time dilation agreed with the experimental value with a fractional error $\approx 2 \times 10^{-3}$ at a 95% confidence level. Today, time dilation of particles is routinely observed in cyclotrons and synchrotons for many different particles.

3.9.4 Time Dilation and Moving Atomic Clocks

In 1971 Hafele and Keating flew four cesium-beam atomic clocks on commercial airliners around the world twice [23, 24]. One flight was in an eastward direction, in the direction of the Earth's rotation, and the other westward, opposite to the direction of the Earth's rotation. The clocks were initially synchronized with a ground-based reference atomic clock at the US Naval Observatory in Washington. The expected relativistic time dilation after each flight would be small, but measurable.

The predicted total relativistic time dilations for each direction were calculated by addition of the expected velocity time dilation from special relativity and the expected gravitational time dilation from general relativity. The gravitational effect was dependent upon the height of the airplane above the Earth surface. The results are summarized in Table 3.1.

The total measured time loss for the eastward flight direction was -59 ± 10 ns, and the measured time gain for the westward flight direction was $+273 \pm 7$ ns [24]. For the expected experiment accuracy of $\approx 10\%$, the agreement between the predicted and measured values was quite good. When the flown clocks were returned to Earth and compared to the ground-based reference clocks, they had gained about 150 ns.The published outcome of the experiment was consistent with special and general relativity. The observed time gains and losses were different from zero to a high degree of confidence, and were in agreement with relativistic predictions to within the $\sim 10\%$ precision of the experiment.

Table 3.1 Relativistic predictions from the Hafele-Keating experiment

Direction	Time Gain/Loss [ns] Special Relativity	Time Gain/Loss [ns] General Relativity	Total Gain/Loss [ns]
Eastward	-184 ± 18	$+144 \pm 14$	-40 ± 23
Westward	$+96 \pm 10$	$+179 \pm 18$	$+275 \pm 21$

In 1996, on the 25th anniversary of the Hafele-Keating experiment, a single cesium atomic clock was flown round-trip between London, England and Washington, D.C. [25]. The increased accuracy of cesium clocks allowed the experiment to be performed over this shorter trip. The reference atomic clock was at the National Physical laboratory in England. The predicted clock gain due to gravity was 53 ns for a total flight time of 14 h, with a mean altitude over 10 km. The predicted clock loss due to velocity was predicted to be 16.1 ns. The predicted net clock gain was 39.8 ns, and the measured clock gain was 39.0 ns with an uncertainty of ± 2 ns.

3.9.5 Time Dilation in GPS Satellite Clocks

In 1956 Winterberg proposed that Einstein's theories of relativity could be tested by placing atomic clocks on artificial satellites [26]. This was before the launch of the first man-made satellite—*Sputnik*. The current global positioning system, known as GPS or NAVSTAR, consists of 24 solar-powered satellites, where three or four are used for a position calculation. They orbit the earth at a height of about 19 km at a velocity of about 4 km/s. The velocity frequency shift of special relativity produces a time loss of about 7 μs/day relative to a ground observer. The gravitational frequency shift of general relativity produces a time gain of about 45.9 μs/day. Several other minor corrections are for the ellipticity of the satellite orbits, and also for the *Sagnac effect*, which causes a signal delay between satellite and ground due to Earth rotation. The net result is about a 38 μs/day gain in the satellite clock, relative to a ground-based observer. This correction is applied to the clock satellites prior to launch, since uncorrected satellite clocks would give about a 10 km/day error in computed earth distances.

3.9.6 Detection of Lorentz Length Contraction

According to special relativity, length contraction of a moving object in a reference frame can only be measured from another reference frame that is moving or stationary with respect to it. However, direct measurement of the length contraction of a moving object by this method is difficult, due to its small magnitude. An early attempt was the Trouton–Rankine experiment, where the other "proper" reference frame was envisioned as the luminiferous ether [27]. They tried to measure the electrical resistance of a coil as it was oriented in the direction of motion, and perpendicular to the direction of motion. No measureable contraction was detected. In accordance with the prediction of special relativity, no change would be expected for a measurement made or observed in the same reference frame. However, there are several indirect ways to detect Lorentz contraction in the laboratory. In the Relativistic Heavy Ion Collider (RHIC) at Brookhaven National Laboratory, two oppositely directed heavy gold ion beams are each accelerated to a velocity $v \approx 0.99995c$. At this velocity, the ions take the shape of spheres flattened by a factor $\approx$100 in the direction of motion. The beams are then collided head-on. The results obtained for these gold ion beam collisions can only be explained when the increased nucleon density due to Lorentz contraction is considered. For an electrically charged particle in motion, there is a Lorentz contraction of the surrounding Coulomb field. This contraction produces a stronger electric field perpendicular to the direction of motion. The result is an increased ionization of charged particles in motion. This increased ionization is routinely observed in high-speed particle accelerators.

In the muon experiment (Sect. 3.9.3), the 15 km travel distance through the atmosphere appears to be only 1641 m in the muon reference frame. The measured survival rate of the muons on Earth then indicates that a real Lorentz length contraction has occurred, as observed from the ground-based reference frame. Sometimes a distinction in wording is made between Lorentz length contraction and Einstein length contraction, the former being considered as *real*, and the latter as *relative* [28].

3.9.7 Bending of Light by the Sun

Einstein's theory of general relativity (Sect. 3.8) predicted a gravitational bending of light rays passing near large astronomical bodies, and challenged astronomers to detect it. Perhaps due to this challenge, the astronomer Arthur Eddington and several colleagues waited for the perfect observation of distant stars near the Sun's periphery during a solar eclipse [29]. Such an eclipse occurred on the island of Principe, off the west coast of central Africa, on May 19, 1919. The observed star cluster was Hyades, at about 151 light years distance from Earth, and consisting of about 300 to 400 similar stars within a few light years apart. Backup measurements were made by colleagues at Sobral, Brazil. The measured light deflection was close to Einstein's predicted 1.75 arc second, a corrected calculation that doubled the 0.83 arc second value originally stated in his 1911 paper. Accepted by most scientists, these experimental measurements were a stellar confirmation of Einstein's theory, making Einstein famous and bringing Eddington much publicity.

This gravitational effect on starlight was confirmed by the Hipparcos satellite launched in 1989 by the European Space Agency. The three-and-a-half year mission charted over a hundred thousand stars with an accuracy of about 0.001 arc second. This accuracy allowed the gravitationl bending to be perceived by stars in the night sky, without the need of a solar eclipse. The uncertainty in Einstein's prediction was then determined to be $\approx$0.1%.

3.10 Source Motion and the Constancy of the Speed of Light

Several more experiments involving Einstein's second postulate have been performed to examine whether the speed of light is independent of the velocity of the light producing source. In 1964 a laboratory experiment was performed using the CERN proton synchrotron at Geneva [30]. The collision of an accelerated high-speed proton with a proton in a beryllium target nucleus produced two protons and a neutral pion (π^0). The pion, acting as a source of light moving at a calculated speed $\approx 0.99975c$, quickly decayed into two gamma rays. The speed of the gamma rays relative to the fast-moving pion light source was determined from the measured travel time between two gamma ray detectors spaced 31 m apart. This speed was 2.9979×10^8 m/s, which is very close to the value $c = 2.997925 \times 10^8$ m/s. Thus it was again shown that the speed of the light photons is independent of the speed of the source.

The so-called *emission theory* or *ballistic theory* of Ritz in 1908 predicted that the speed of light c' emitted from a source moving at velocity v would be $c' = c + v$. Based on this theory, an objection to the Michelson-Morley experiment was proposed stating that the speed of light c after transmission through a glass plate could be affected by the velocity v of the plate. In 1964 a terrestrial experiment was performed by Babcock and Bergman at the Michelson Laboratory in China Lake, California using a modified Michelson-Morley apparatus [31]. If the emission theory was correct, the speed of light exiting the plate could have a speed $c' = c + \kappa v$, where $\kappa \approx 1$. Einstein's special relativity theory predicted that $\kappa = 0$. Considering another factor, the presence of air might have affected the results. To address these issues, a very sensitive interferometer was constructed as in Fig. 3.11, and was enclosed in a vacuum chamber that could be evacuated to a pressure of 0.02 torr. Two flat glass windows of thickness 0.34 cm and refractive index 1.5 were mounted to a bar, which could be continuously rotated in both clockwise and counter-clockwise directions. With a stroboscopic light source incident upon a beamsplitter, interference fringes were observed through a telescope focused at infinity for both rotation directions. The stroboscopic light source of wavelength 474 nm was triggered only when the windows were positioned normal to the light path, the triggering being controlled by an auxiliary collimated light source and electronics. The distance L between the fixed mirrors M_1 and M_2 was 276 cm and the distance r between the rotation axis of the bar and the window centers was 13.3 cm. The average rotational speed of the bar was 45 revolutions per second. For these parameters a value $\kappa = 1$ would predict a fringe shift of 2.9 fringes, while for $\kappa = 0$, the predicted fringe shift would be 0.0036 fringes. A series of total fringe shift determinations was made both visually and on a photographic plate, with the bar rotating in both directions. A total fringe shift <0.02 fringes was obtained, supporting the special relativity theory over the emission theory.

In 1979 Kenneth Brecher at MIT made astronomical measurements from strong (≈ 70 keV) X-ray binary pulsars to further test the validity of Einstein's second

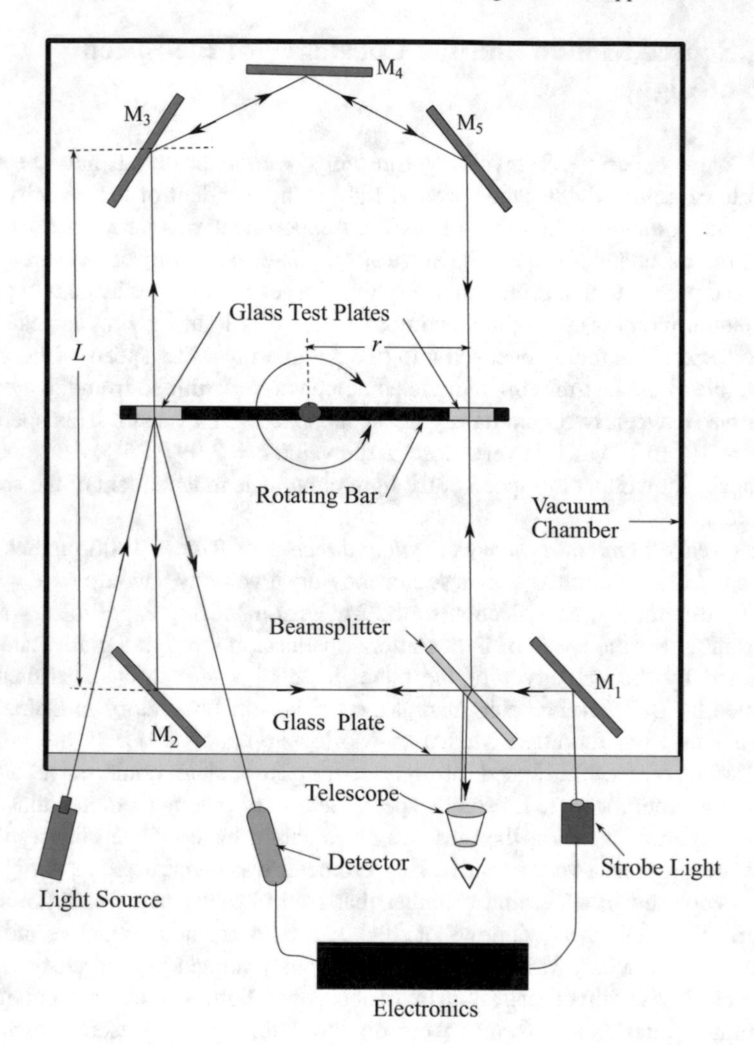

Fig. 3.11 Apparatus of Babcock and Bergman to determine the effect of motion on the speed of light (Adapted from Ref. [31] with permission from the Optical Society of America)

postulate [32]. As in the Babcock and Bergman experiment, Brecher expressed the pulsar light velocity $c' = c + \kappa v$, where $\kappa \approx 1$ for the emission theory, and $\kappa \approx 0$ for special relativity theory. He examined the arrival time of pulses from the Hercules X-1 (pulse rate 1.237 s) and the Small Magellanic Cloud SMC X-1 (pulse rate 0.717 s) X-ray binary pulsars in order to determine some upper limits on the value of κ. He found that $\kappa < 2 \times 10^{-9}$, concluding that the speed of light c was independent of the velocity v of the source. Brecher in the year 2000 made another series of astronomical measurements by examining exploding star gamma ray bursts having durations Δt as short as one millisecond [33]. These bursts originated near the outer edge of the universe, at a distance $D \approx 10^{26}$ m ($\approx 10^{10}$ light years) from the Earth. If the emission of light from these exploding stars depended on their internal velocity v, and assuming that $v > 0.1c$, he calculated that κ was $<10^{-20}$, giving a very precise confirmation of the second postulate of Einstein's special relativity.

References

1. M.H. Fizeau, Sur les hypothèses relatives à l'éther lumineux. Annales de Chimie et de Physique III **lvii**, 385–404 (1859)
2. A. Michelson, E. Morley, Influence of motion of the medium on the velocity of light. Am. J. Sci. **31**(185), 377–386 (1886). doi:10.2475/ajs.s3-31.185.377
3. A. Michelson, E. Morley, On the relative motion of the Earth and the luminiferous ether. Am. J. Sci. **234**(203), 333–345 (1887). doi:10.2475/ajs.s3.203.333
4. N. Hamdan, Can the Lorentz-FitzGerald contraction hypothesis be real? Proc. Pak. Acad. Sci. **44**(2), 121–128 (2007)
5. R. Serway, *Physics for scientists and engineers/with modern physics* (Holt, Rinehart and Winston, New York, 1983), pp. 839–841
6. A. Einstein, On the electrodynamics of moving bodies. Ann. Phys. **17**, 891–921 (1905). doi:10.1002/andp.19053221004
7. A. Einstein, Does the inertia of a body depend on its energy content? Ann. Phys. **17**, 639–641 (1905). doi:10.1002/andp.19052231314
8. R. Baierlein, E = mc^2, in *Newton to Einstein: The Trail of Light* (Cambridge University Press, Cambridge, United Kingdom, 2002)
9. E. Fischbach et al., New geomagnetic limits on the photon mass and on long-range forces coexisting with electromagnetism. Phys. Rev. Lett. **73**(4), 514–517 (1974). doi:10.1103/physRevLett.73.514
10. L.-C. Tu, J. Luo, G.T. Gillies, The mass of the photon. Rep. Prog. Phys. **68**(1), 77–130 (2004). doi:10.1088/0034-4885/68/R02
11. A. Einstein, On the relativity principle and the conclusions drawn from it. Jahrbuch der Radioaktivität **4**, 411–462 (1907)
12. A. Einstein, On the influence of gravitation on the propagation of light. Ann. Phys. **35**, 898–908 (1911)
13. A. Einstein, Explanation of the perihelion motion of mercury from the general theory of relativity. Preussische Akademie der Wissenschaften, Sitzungsberichte, Part 2, 831–839 (1915)
14. R.J. Kennedy, E.M. Thorndike, Experimental establishment of the relativity of time. Phys. Rev. **42**(3), 400–418 (1932). doi:10.1103/PhysRev.42.400
15. H. Ives, G. Stilwell, An experimental study of the rate of a moving atomic clock. J. Opt. Soc. Am. **28**, 215–226 (1938). doi:10.1364/JOSA.28.000215
16. H. Ives, G. Stilwell, An experimental study of the rate of a moving atomic clock II. J. Opt. Soc. Am. **31**, 369–374 (1941). doi:10.1364/JOSA.31.000369
17. D. Hasselkamp, E. Mondry, A. Scharmann, Direct observation of the transversal Doppler-shift. Zeitschrift für Physik A **289**(2), 151–155 (1979). doi:10.1007/BF1435932
18. S. Reinhardt et al., Test of relativistic time dilation with fast optical atomic clocks at different velocities. Nat. Phys. **3**, 861–864 (2007). doi:10.1038/nphys778
19. B. Rossi, D.B. Hall, Variation of the rate of decay of mesotrons with momentum. Phys. Rev. **59**(3), 223–228 (1941). doi:10.1103/PhysRev.59.223
20. L. Liu, P. Solis, The speed and lifetime of cosmic ray muons. MIT Undergraduate Report, 18 Nov 2007

21. D.H. Frisch, J.H. Smith, Measurement of the relativistic time dilation using μ-mesons. Am. J. Phys. **31**(5), 342–355 (1963). doi:10.1119/1.1969508
22. H. Bailey et al., Measurements of relativistic time dilatation for positive and negative muons in a circular orbit. Nature **268**, 301–305 (1977). doi:10.1038/268301a0
23. J.C. Hafele, R.E. Keating, Around-the-world atomic clocks: predicted relativistic time gains. Science **177**(4044), 166–168 (1972). doi:10.1126/science.177.4044.166
24. J.C. Hafele, R.E. Keating, Around-the-world atomic clocks: observed relativistic time gains. Science **177**(4044), 168–170 (1972)
25. News from the National Physical Laboratory, *Metronia*, Issue 18, United Kingdom, Winter (2005)
26. F. Winterberg, Relativistische zeitdilatation eines künstlichen satelliten. Astronautica Acta **2**(1), 25–29 (1956)
27. F.T. Trouton, A. Rankine, On the electrical resistance of moving matter. Proc. R. Soc. **80**, 420 (1908). doi:10.1098/rspa.1908.0037
28. C. Sherwin, New experimental test of Lorentz's theory of relativity. Phys. Rev. A **35**(9), 3650–3654 (1987). doi:10.1103/PhysRevA.35.3650
29. F.W. Dyson, A.S. Eddington, C. Davidson, A determination of the deflection of light by the Sun's gravitational field, from observations made at the total eclipse of 29 May 1919. Philos. Trans. R. Soc. Lond. **220A**, 291–333 (1920). doi:10.1098/rsta.1920.0009
30. T. Alväger et al., Test of the second postulate of special relativity in the GeV region. Phys. Lett. **12**(3), 260–262 (1964). doi:10.1016/0031-9163(64)91095-9
31. G.C. Babcock, T.G. Bergman, Determination of the constancy of the speed of light. J. Opt. Soc. Am. **54**(2), 147–150 (1964). doi:10.1364/JOSA.54.000147
32. K. Brecher, Is the speed of light independent of the velocity of the source? Phys. Rev. Lett. **39** (17), 1051–1054 (1977). doi:10.1103/PhysRevLett.39.1051
33. K. Brecher, Precision test of special relativity using gamma ray bursts. Bull. Am. Phys. Soc. **45**(2), No. 34, May 2000 Meeting, Long Beach, California

Chapter 4
The Quantum Nature of Light

4.1 The Photoelectric Effect

The *photoelectric effect* is a classic example of light interacting with matter. It was observed and documented by Heinrich Hertz in 1887 while he was experimenting with electromagnetic waves [1]. When ultraviolet light from a mercury lamp illuminated a spark gap, the amount of voltage required to produce a spark was reduced. The required voltage also decreased when the light intensity was increased. Insertion of a glass plate between the electrodes decreased this effect, while the effect remained when a quartz plate was inserted. This is often referred to as the *Hertz effect*.

An assistant of Hertz, Phillip Lenard further investigated the effect by constructing the apparatus shown in Fig. 4.1 [2]. An evacuated glass tube contained a negative potential metal plate (cathode) and a positive potential metal plate (anode). Ultraviolet light entered the tube as shown and was incident upon the cathode plate. A variable voltage source provided the potential between the metal plates. An almost instantaneous flow of current in the circuit was measured. Lenard identified the emitted particles from the cathode metal plate as electrons, which were being collected by the anode metal plate. This produced a *photoelectric current* and the emitted electrons were called *photoelectrons*. However, a remarkable conclusion of these experiments was that the emission was highly wavelength dependent, with a specific light wavelength below which there was no photoemission. When expressed as a light frequency f, there was a threshold frequency f_0 below which no electron emission occurred, regardless of the light intensity. This f_0 value varied with the type of metal, but for $f > f_0$ the number of emitted electrons was proportional to the intensity of the incident light for any wavelength that produced electron emission. However, the kinetic energy or speed of the emitted electrons varied only with frequency, increasing as the incident light frequency increased. For his experimental work on the photoelectric effect, Phillip Lenard received the Nobel Prize in Physics in the year 1905. Although Lenard probably suspected that the

D.F. Vanderwerf, *The Story of Light Science*,
DOI 10.1007/978-3-319-64316-8_4

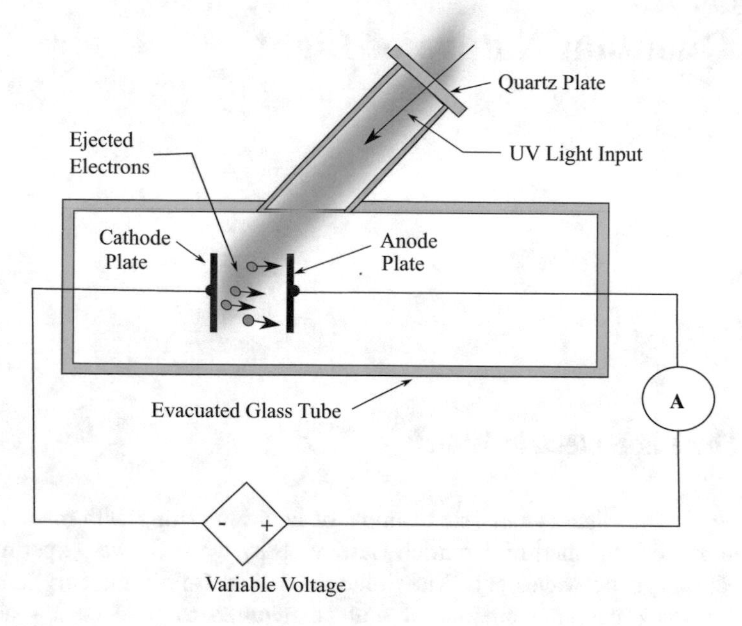

Fig. 4.1 The photoelectric measurement apparatus of Lenard

photoelectric effect involved an interaction between light and the electrons in atoms, it was Einstein in 1905 that provided the theoretical framework to explain the effect when he considered light as a quantized particle.

4.2 Quantum Blackbody Radiation

4.2.1 Wien's Displacement Law

The birth of quantum theory arose from the attempts to mathematically describe the radiation emitted from a blackbody cavity. An ideal blackbody radiator emits energy from the extreme ultraviolet region (≈15–80 nm) to the far infrared region ($\approx 5 \times 10^4$–1×10^6 nm), with a small intermediate band of visible light.

In 1893 Wilhelm Wien, by thermodynamic analysis, formulated a relationship between the wavelength λ_{max} [nm] of the highest emitted intensity and the corresponding temperature T [K] of a blackbody radiator as

$$\lambda_{max} T = \text{constant} = 2.8978 \times 10^6 \text{nm K}. \quad (4.1)$$

This relationship is known as *Wien's displacement law*, and besides having many important applications by itself, provided a starting point for significant advances on the theory of the distribution of radiation from a blackbody [3].

4.2.2 Development of the Radiation Distribution Laws

In 1896 Wilhelm Wien developed a blackbody radiation distribution formula, assuming that radiation emitted by heated atomic oscillators followed a continuous Maxwell velocity distribution, and arrived at a (λ^5) wavelength dependence [4]. Max Planck, a colleague of Wien, was impressed with Wien's analysis and produced a form of this law known as *Wien's radiation law*:

$$I(\lambda, T) = \frac{2bc^2}{\lambda^5} e^{ac/\lambda T}, \tag{4.2}$$

where

$I(\lambda, T)$ is the emitted energy in watts per steradian per cubic meter,
λ is the wavelength [m],
T is the temperature [K],
c is the speed of light = 2.998×10^8 m/s,
a is a constant = 4.818×10^{-11} K s,
b is a constant = 6.885×10^{-34} J s.

The fit to experimental data was good, except that significant deviation occurred at longer wavelengths.

In response to this deviation, thinking that Wien's approach was incorrect, Lord Rayleigh (a.k.a. John Strutt) in 1901 set out to develop another blackbody radiation formula. He assumed a group of atomic oscillators in the walls of a radiating cavity that emitted a continuous spectrum of wavelengths at temperature T. The formula was further developed with James Jeans and emerged in 1905 as the *Rayleigh-Jeans radiation formula,* or *Rayleigh-Jeans law*.

$$I(\lambda, T) = \frac{2ckT}{\lambda^4}, \tag{4.3}$$

where Boltzmann's constant $k = 1.381 \times 10^{-23}$ J/K. This formula fitted well with experimental data at the longer wavelengths, but failed miserably at the shorter wavelengths. This failure is known as the *ultraviolet catastrophe*, a term coined by Paul Ehrenfest in 1911.

Building on the work of Wien, Max Planck in 1900 derived a formula for blackbody radiation based on a concept of quantized emitted energy [5]. This formula is called *Planck's radiation law* or simply *Planck's law*, and was based on the radical idea that radiation from oscillating molecules is not continuous, but was emitted in discrete bundles of energy called *quanta.* Planck also thought that the excitation of these quanta at high frequencies would exceed the average thermal energy of the blackbody radiator. Planck's radiation law was given by

$$I(\lambda, T) = \frac{2hc^2}{\lambda^5} \frac{h}{e^{hc/\lambda kT} - 1}, \tag{4.4}$$

where $h = 6.626 \times 10^{-34}$ J s was a constant designated as a "quantum of action". Shortly thereafter, Planck's radiation law was confirmed to fit an experimental blackbody radiation curve at all wavelengths. Planck's radiation law is considered to be one of the most important breakthroughs in light science. Figure 4.2 plots the predicted intensity I [J m^{-3} s^{-1}] from a 5800 K blackbody radiator as a function of wavelength λ [nm] from the radiation laws of Wien, Rayleigh-Jeans, and Planck.

The energy of the *quantum states* of the atomic oscillators in a medium are given by discrete energy bundles $E_n = nhf$, where n is the quantum number, and f is the frequency of the atomic oscillator. Moreover, these oscillators emit or absorb these discrete energy bundles by jumping from one quantum state to another. When n changes by unity (adjacent quantum states), the energy absorbed or emitted is given by

$$E = hf \tag{4.5}$$

and the constant h is identified with the constant h in Eq. (4.4) [6]. This constant became known as *Planck's constant h.*

It is interesting to note that the constant b in Wien's radiation law of Eq. (4.2) is close to the value of Planck's constant h, while the constant a is close to the value

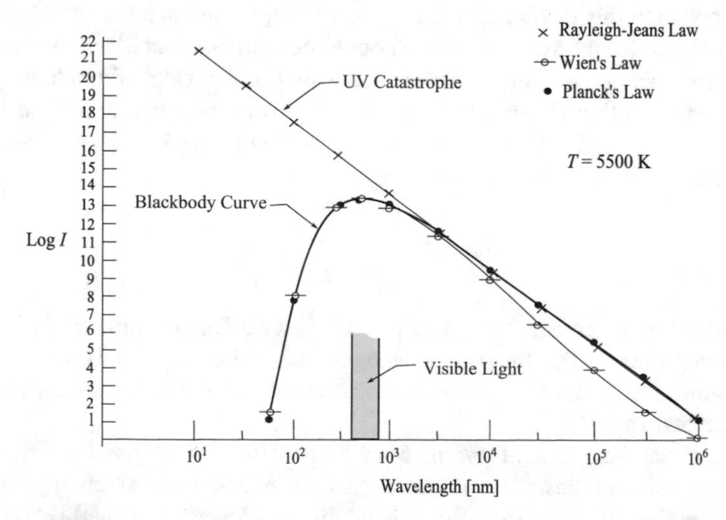

Fig. 4.2 Intensity versus wavelength for blackbody radiator predicted by Wien, Rayleigh-Jeans, and Planck radiation laws

(h/k), since $h/k = 6.626 \times 10^{-34}/1.381 \times 10^{-23} = 4.80 \times 10^{-11} \approx a$. In fact, they are numerically equivalent and Wien's radiation law is usually given in the form:

$$I(\lambda, T) = \frac{2hc^2}{\lambda^5} e^{hc/\lambda kT} \tag{4.6}$$

Planck's radiation law was later derived theoretically from a photon gas using quantum theory and statistical mechanics. Although Planck's constant was unknown in 1893 when Wien's displacement law was developed, the displacement law can be related to Planck's constant by the formula:

$$\lambda_{max} T = \frac{hc}{4.9651k} = 2.8978 \times 10^6 \text{ nm K}. \tag{4.7}$$

Moreover, Wien's displacement law of Eq. (4.1) can be directly derived from the Planck's radiation formula of Eq. (4.4). Appendix 4A gives a mathematical calculation for this derivation. For their pioneering work in radiation theory, Wien and Planck were awarded the 1911 and 1918 Nobel Prizes in Physics.

4.3 Light as a Quantized Particle

4.3.1 The Photon as a Quantized Particle

In 1905 Einstein published a paper in which he proposed, from a heuristic viewpoint, that light was a quantized particle [7]. Additionally, he identified the energy bundle in Eq. (4.5) as the energy of a quantum particle of light. He envisioned light of any electromagnetic frequency f to be composed of a stream of quantum particles, each light particle having an energy $E = hf$, where h is Planck's constant. In 1921 the American chemist Gilbert N. Lewis coined the word *photon* for Einstein's light quantum particle. A 1909 paper by Einstein stated that light should also be considered as a particle which carried momentum, but stated that light could also exhibit wavelike properties [8]

4.3.2 Explanation of the Photoelectric Effect

In the previously cited 1905 paper, Einstein explained the photoelectric effect in terms of the energy quantum states proposed by Planck. When light photons having energy $E = hf$ are incident on the surface of a metal, electrons which absorb the incident light acquire kinetic energy. To eject these electrons from the metal, additional energy is required to overcome the intrinsic potential of the metallic surface. The maximum kinetic energy K_{max} of the ejected electron is proportional to the light frequency and can be defined by the straight-line equation

$$K_{max} = hf - \varphi, \tag{4.5}$$

where $\varphi = hf_0$ is the work function, or the minimum energy to eject an electron from the surface of the metal at the threshhold frequency f_0. Then

$$K_{max} = h(f - f_0). \tag{4.6}$$

To achieve photoemission ($K_{max} > 0$), the frequency of the emitted light must exceed the threshold frequency of the metal. Although zinc was used in the defining experiments, other metals have different work functions, and the incident light frequency could be over a wide range—from visible to gamma rays [9, 10]. Table 4.1 lists the threshold frequencies and threshold wavelengths for several metals having different work functions. Figure 4.3 illustrates the dependence of maximum kinetic energy on frequency for these metals. Also, it is possible to observe the photoelectric

Table 4.1 Threshold frequencies and wavelengths for several photoelectric metals using empirical published values of their work functions (h = 4.141 × 10^{-23} eV s)

Photoelectric metal	Empirical work function φ [eV]	Threshold frequency f_0 [Hz]	Threshold wavelength λ_0 [nm]
Cesium (Cs)	2.1	5.07 × 10^{14}	591.3
Potassium (K)	2.3	5.55 × 10^{14}	540.2
Magnesium (Mg)	3.68	8.89 × 10^{14}	337.2
Zinc (Zn)	4.3	1.04 × 10^{15}	288.3
Platinum (Pt)	6.35	1.53 × 10^{15}	195.9

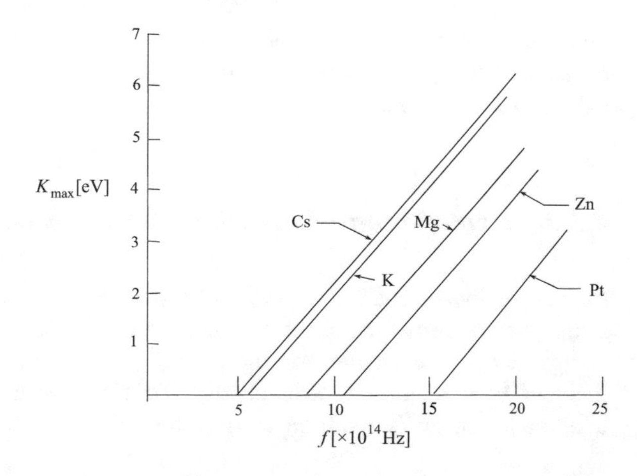

Fig. 4.3 Maximum kinetic energy versus of photoemitted electrons versus frequency for various metals

effect in certain liquids, gases, or crystals, and the emission of other types of charged particles such as ions. By providing this explanation of the photoelectric effect, Einstein received the 1921 Nobel Prize in Physics.

4.4 Stimulated Emission of Einstein

Among several other significant papers published in the period 1916–1917, Einstein published two papers describing the photonic processes of *stimulated absorption*, *spontaneous emission*, and *stimulated emission* [11, 12]. Stimulated emission occurs when energy is released from a single atom, if an electron is stimulated by an incident photon to drop from one atomic energy level to a lower atomic energy level. A quantum particle of energy $E = hf$ is then released, where E is the energy difference between the two levels. However, stimulated emission only occurs when an incident photon has a specific frequency which matches the energy difference $E = hf$ between the two levels. During the transition an additional photon is created, which has the same phase, frequency, polarization, and propagation direction as the incident photon. The original photon and the newly created photon can then cause additional stimulated emissions in the material, resulting in a *cascade effect* where a significant amount of coherent light can be emitted from the material.

In the photoelectric effect we can consider the interaction of the incident photons with the electrons as a type of collision where each particle has close to the same energy, allowing an energy exchange to take place. Calculations dealing with the resultant momentum transfer, using a corpuscular model as described in Einstein's 1917 paper, have been carried out more recently by Friedberg at Columbia University [13]. As viewed from the particle reference frame, a "drag" in the material retards the motion of the particles during impacts and recoils, which affects the momentum balance calculation. The result of this momentum calculation showed that the emitted light was coherent with the incident light and was consistent with Einstein's quantum analysis for stimulated emission.

4.5 The Experiment of Compton and the Compton Effect

Related to Einstein's 1909 paper, a 1916 paper of Einstein stated that photons must have momentum to be consistent with Planck's law [14]. In the early 1920s Arthur Compton (1892–1962) at Washington University proposed a quantum theory on the scattering of X-rays and performed a groundbreaking experiment showing the interaction of light with matter. This experiment confirmed Einstein's conclusion and changed the scientific perception of the structure of light [15, 16]. The basic experiment is illustrated in Fig. 4.4. A 1.5 kW X-ray tube emitted the

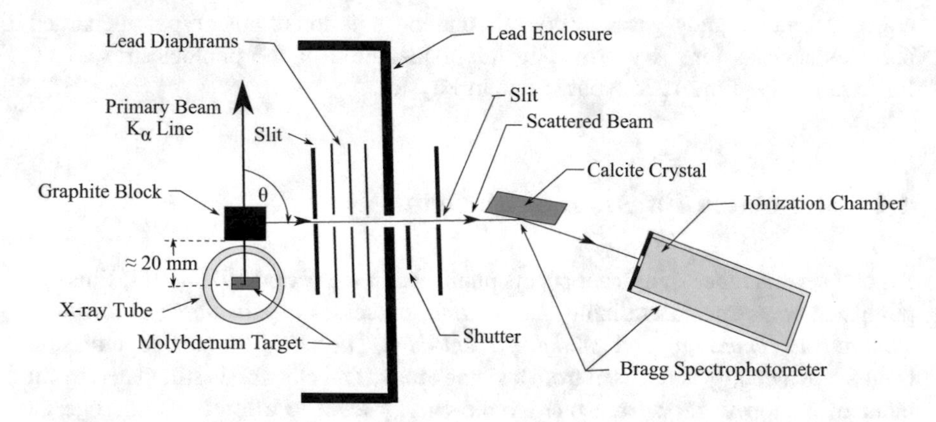

Fig. 4.4 Experiment apparatus to measure the Compton shift, shown for a scattering angle θ = 90°. (This Figure adapted from Ref. [16]. Copyrighted by the American Physical Society)

0.0708 nm K_α line from a molybdenum target. Incident X-rays of wavelength λ_{inc} collided with outer electrons in a block of graphite ≈20 mm from the target, and were scattered at velocity v at various angles θ from the primary incident beam. Scattering from the graphite block was observed at θ = 45°, 90°, and 135°, and the scattered wavelength λ_{scat} was measured by means of a Bragg spectrometer consisting of a calcite crystal and an ionization chamber.

There was a measured increase in the wavelength λ_{scat}, when compared to the wavelength λ_{inc}. This wavelength shift of the scattered light is known as the *Compton shift* or *Compton effect*, and is given by the *Compton shift equation*:

$$\Delta\lambda = \lambda_{scat} - \lambda_{inc} = \frac{h}{mc}(1 - \cos\theta), \tag{4.7}$$

where m is the electron mass. For example, at θ = 90°, the scattered wavelength λ_{scat} was 0.0731 nm, giving Δλ = 0.0023 nm. The scattered X-rays can be considered as photons with energy $E = pc = hf$, and momentum $p = h/\lambda_{scat}$. Assuming conservation of energy and momentum in the collisions, the wavelength increase of the scattered photons is caused by a transfer of energy from the incident photons to the recoiled outer electrons in the material. A recoiling electron could attain relativistic speed, with a relativistic momentum $p_{elec} = \gamma mv$ (see Sect. 3.7.2). The Compton effect could not be explained by the electromagnetic field in Maxwell's equations, but was uniquely explained by relativistic collisions between small particles, as described by Einstein. The Compton effect showed that the light scattering was a quantum phenomenon and provided solid evidence of the quantum nature of light. For these accomplishments, Compton received the 1927 Nobel Prize in Physics.

4.6 The de Broglie Wavelength

In 1927 Louis de Broglie proposed that moving particles have an associated wave [17]. For moving particles where the momentum $p = h/\lambda = h/mv$, the associated *de Broglie wavelength* λ_{dB} is defined as

$$\lambda_{dB} = \frac{h}{p} = \frac{h}{mv} \tag{4.8}$$

This important equation related the wavelength nature of light to the particle nature of light.

The wave characteristic becomes more noticeable at near-relativistic or relativistic velocities for particles of low or negligible mass (momentum), e.g. electrons or photon particles. The de Broglie wavelength for near-relativistic particles can be obtained by equating the Planck and Einstein energy formulas: $E = hf = mc^2$. Let $v \approx c$.Then $\frac{hv}{\lambda} = mv^2 = pv$, where m is the relativistic mass of Eq. (3.20). Thus we arrive at the de Broglie wavelength $\lambda_{dB} = \frac{h}{p}$.

For his development of the wave properties of fast-moving particles, Louis de Broglie received the 1929 Nobel Prize in Physics.

4.7 Wave-Particle Duality

There are several interpretations of the connection between wave and particle properties of light:

1. Light acts like an electromagnetic wave when it is propagating in free-space, but acts like a quantized particle, e.g. photon, when it is emitted or absorbed by a material.
2. Light can act like a wave or a particle, depending on how it is measured or observed.
3. All optical properties of light can be explained by considering light as a particle, as described by quantum electrodynamics.
4. Light is always quantized, whether represented as a wave or a particle.
5. Light can appear as a wave or a particle, dependent on the conscious perception of the observer.
6. Polarization and interference phenomena are rationally described for light as an electromagnetic wave, but not by quantum theory.

At this point in time, there is no reason to reject any of these interpretations, or accept only one as completely valid.

References

1. H. Hertz, Über einen einfluss des ultravioletten lichtes auf die electrische entladung. Ann. Phys. **267**(8), 983–1000 (1887). doi:10.1002/andp.18872670827
2. P. Lenard, Über die lichtelektrische wirkung. *Annalen der Physik* **313**(4), 149–198 (1902). doi:10.1002/andp.1902313510
3. W. Wien, Eine neue beziehung der strahlung schwarzer körper zum zweiten hauptsatz der wärmetheorie. *Sitzungsberichte der preussischer akademie*, 55–62 (1893)
4. W. Wien, Über die energievertheilung im emissionspectrum eines schwarzen körpers. Annalen der Physik und Chemie **294**(8), 662–669 (1896)
5. M. Planck, Zur theorie des gesetzes der energieverteilung im normalspecterum. Verhandlung der deutschen physikalischen gesellshaft **2**(17), 237–252 (1900)
6. M. Planck, Über das gesetz der energieverteilung im normalspectrum. Ann. Phys. **309**(3), 553–563 (1901). doi:10.1002/andp.19013090310
7. A. Einstein, On a heuristic point of view concerning the production and transformation of light. Ann. Phys. **17**, 132–148 (1905)
8. A. Einstein, On the development of our views concerning the nature and constitution of radiation. Physikalische Zeitschrift **10**, 817–825 (1909)
9. R. Millikan, A direct photoelectric determination of Planck's '*h*'. *Phys. Rev.* **7**(3) (1916). doi:10.1103/PhysRev.7.355
10. W.H. Souder, The normal photoelectric effect of lithium, sodium and potassium as a function of wavelength and incident energy. Phys. Rev. **8**(3), 310–319 (1916). doi:10.1103/PhysRev.8.310
11. A. Einstein, Emission and absorption of radiation in quantum theory. Verh. Dtsch. Phys. Ges. **18**, 318–323 (1916)
12. A. Einstein, On the quantum theory of radiation. Physikalische Zeitschrift **18**, 121–128 (1917)
13. R. Friedberg, Einstein and stimulated emission: s completely corpuscular treatment of momentum balance. Am. J. Phys. **62**(1), 26–32 (1994). doi:10.1119/1.17737
14. A. Einstein, On the quantum theory of radiation. Mitteilungen der Physikalischen Gesellschaft **16**, 47–62 (1916)
15. A.H. Compton, A quantum theory of the scattering of X-rays by light elements. Phys. Rev. **21** (5), 483–502 (1923). doi:10.1103/PhysRev.21.483
16. A.H. Compton, The spectrum of scattered X-rays. Phys. Rev. **22**(5), 409–413 (1923). doi:10.1103/PhysRev.22.409
17. L. de Broglie, Recherches sur la théorie des quanta. Annales de Physique **3**, 22–128 (1925)

Chapter 5
Natural and Artificial Sources of Light

5.1 The Sun

Since the dominant source of light for mankind is the Sun, it is worthwhile to discuss some of its properties. The Sun has played a major role in the development of life and civilization by providing the proper levels of temperature and visibility during Earth's "daylight" hours. In cosmological terms, the Sun is classified as a yellow dwarf star and is about 4.7 billion years old. The source of energy is a central core, having a radius about 20–25% of the Sun's radius. The energy is produced from nuclear fusion of helium from hydrogen, at a core temperature of about 15 million K and a pressure of about 250 billion atmospheres. At the Sun's core about 620 million metric tons of hydrogen are converted to helium each second from a proton-proton chain reaction, with about a 70% mass-to energy conversion efficiency. Gamma ray photons and X-ray photons are emitted toward the surface from these reactions. The Sun is currently about halfway through its fusion resources or main sequence star cycle, and its luminosity is increasing about ten percent every billion years.

The surface temperature of the Sun at the semi-transparent *photosphere* is currently accepted to be 5778 K, and results from radiation absorption, emission, and convection of the core energy as it travels the 695,000 km distance to the surface. The photosphere defines the observed spherical shape of the Sun, consisting of a semi-opaque layer of incandescent gases $\approx$300 km thick, where visible light results from the reaction of electrons with hydrogen, producing hydrogen ions. Other trace elements in the photosphere are helium, neon, argon, magnesium, and iron [1]. Gamma ray photons and X-ray photons produced in the core are also radiated through this region, after an arduous journey of thousands of years from core to photosphere. Above the photosphere lies the hotter *chromosphere* region, where the temperature can rise to about 100,000 K. The outermost layer is the *corona*, with even hotter temperatures approaching one million K.

D.F. Vanderwerf, *The Story of Light Science*,
DOI 10.1007/978-3-319-64316-8_5

If we consider the Sun as a blackbody radiator, Wien's displacement law (Eq. 4.1) can determine the prominent radiated wavelength. For a solar surface temperature $T = 5778$ K, $\lambda_{max} = 501.6$ nm, which is in the blue-green region (475–510 nm) of the visible spectrum. However, Earth sunlight could have a yellowish-white appearance due to atmospheric scattering of the blue light component [2, 3]. In fact, the retinal cones of the human eye probably evolved to produce a maximum sensitivity at about 550 nm (yellow-green), with visible perception of light over the 400–700 nm range.

The intensity of solar radiation at the Earth's surface is usually expressed in terms of the *solar constant* S, which is the integrated spectral irradiance [W/m^2] at a mean Earth-Sun distance of one *astronomical unit* (AU), sometimes called *total solar irradiance* (TSA). Figure 5.1 shows the solar irradiance spectrum outside the atmosphere of the Earth (Air Mass Zero) from World Meteorological Association (WMO) measurements, along with the blackbody spectrum at a temperature of 5800 K. Modern solar spectral irradiance measurements are usually obtained from high-altitude aircraft, spacecraft, and satellites. A currently accepted value of the solar constant is S = 1367.28 ± 0.02 W/m^2 from WMO measurements in 1985. Another Air Mass Zero value of S = 1366 ± 0.37% W/m^2 was specified in the year 2000 by the American Society of Testing and Materials (ASTM E-490) from integrated spectral irradiance data using various ground-based and space-based measurements.

The Sun, being a dynamic celestial object, is not an ultra-stable light source, resulting in a variation of the value of S. A major cause of short-term variation is the rise and decay of sunspot activity. The elliptical orbit of the Earth around the

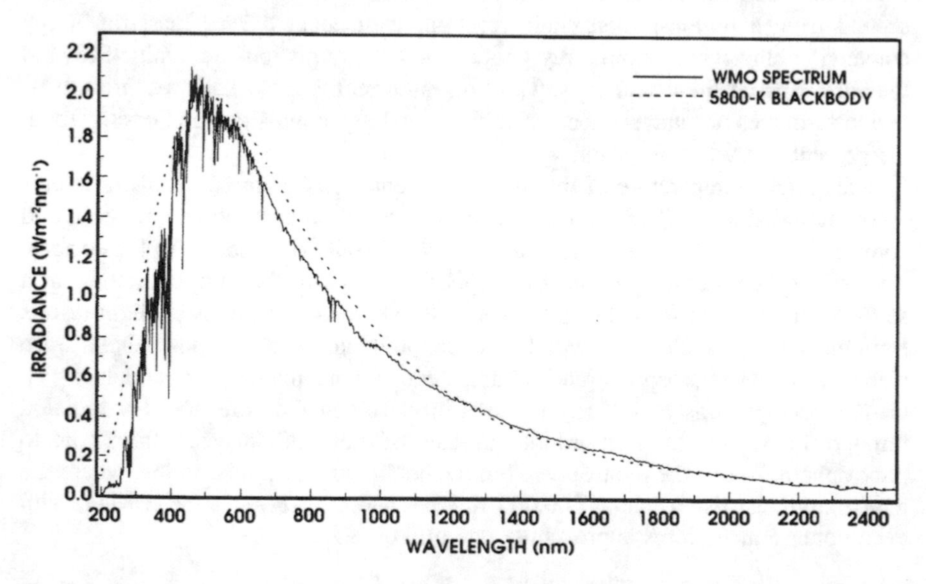

Fig. 5.1 Solar irradiance WMO spectrum outside the Earth's atmosphere, and the irradiance of a 5800 K blackbody. (Image courtesy of Newport Corporation)

Sun also causes a periodic variance of S. It is believed that the magnitudes of these fluctuations do not significantly affect the Earth's climate, or that any significant effect has yet to be identified [4, 5]. For the longer term, the steady consumption of hydrogen at the Sun's core will produce an increasing solar irradiance, and this is of more concern.

Most scientific solar measurements today are taken from satellite observations. The Solar and Heliospheric Observatory (SOHO) satellite was launched in December 1995 to study solar flares, the solar wind, sunspot activity, and other solar processes. Among the specialized instruments on this satellite are the Extreme ultraviolet Imaging Telescope (EIT), the Large-Angle and Spectrometric Coronagraph (LASCO), the Coronal Diagnostic Spectrometer (CDS), and the Solar Ultraviolet Measurements of Emitted Radiation (SUMER) instrumentation [6]. Another solar measurement satellite, the Solar Dynamics Observatory (SDO), was launched in February 2010 and contains more precision instrumentation.

5.2 Electrically-Driven Light Sources

5.2.1 Arc-Discharge Lamps

In order to facilitate human activity beyond normal sunlight hours, earthbound artificial light sources have been developed. Although combustion-based lamps were in use for thousands of years, they could not achieve the level of illumination possible with electrically-driven light sources. See Appendix 5A. The first successful electrically-driven light source was the *high-intensity arc-discharge lamp*. It produced light by the creation of an electric arc between two high-voltage electrodes surrounded by an ionizing gas. The arc-discharge lamp was developed by Sir Humphry Davy in the early 1800s. It operated in air and consisted of two horizontal charcoal rods separated by about 100 cm and connected to a series of batteries to produce the arc between the rod electrodes. The carbon rod electrodes burned away and the rods needed to be manually or automatically shifted inward, and eventually replaced. In the 1870s the young inventor Charles Brush developed a short-arc lamp where the gap between the carbon rods was automatically held constant by electromagnetic and mechanical control. Brush envisioned the use of these arc-lamps for street lighting, and in 1879 a group of twelve Brush arc-lamps was demonstrated in the lighting of Monument Park in Cleveland. The amazing brightness of these lamps signaled the beginning of the end for the existing gas streetlights, which appeared dim in comparison.

A short-arc lamp using tungsten electrodes is a more stable and robust gas-discharge lamp, with the arc in a pressurized gas environment contained in a fused silica glass envelope. Figure 5.2 shows a typical construction. With an arc gap usually less than 12 mm, the short-arc lamp provides a high-intensity point source of light. These lamps are available commercially with input wattages

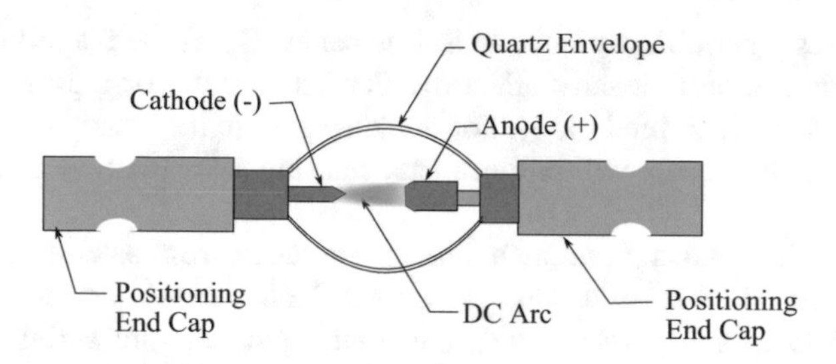

Fig. 5.2 Typical construction of a high-intensity short-arc discharge lamp

between 50 and 30,000 W, and the fill gas is usually a noble gas such as xenon or argon [7]. Short-arc lamps are used in modern motion picture projectors. For example, the IMAX projector uses a 15 kW xenon short-arc lamp. A very high-power 400 kW argon short-arc lamp has been designed and constructed for use in solar simulation [8]. Using collimating off-axis optics, an irradiance of about one solar constant S was achieved over a 4.6 m by 9.2 m area.

Long-arc lamps can be of the low or high-pressure type, and usually consist of a tubular quartz envelope where the anode and cathode are separated by a distance considerably longer than the tube diameter. For example, there are low-pressure xenon long-arc lamps, high-pressure mercury long-arc lamps, and low-pressure mercury fluorescent long-arc lamps. The metal-halide lamp is a more recent type of high-pressure gas-discharge lamp, having been developed in the 1960s and improved in the 1980s as the ceramic metal-halide lamp. Typically, the thoriated tungsten electrodes are contained in a sealed quartz tube containing a mixture of mercury vapor and metallic iodides. A second outer bulb is usually employed to filter the intense ultraviolet light emitted.

Higher powers can be achieved with a long-arc lamp by water vortex stabilization of a flowing discharge gas. This type of lamp is illustrated in Fig. 5.3. Recirculated inputs of flowing deionized water and argon gas form a rapidly moving spiral of gas and water that stabilizes the arc path through an open-ended quartz tube. Both anode and cathode are water-cooled. The argon gas vortex creates a small low-pressure region near the tube midpoint, such that the arc discharge is confined to this region. Input power capability is from 50 kW up to about 100,000 kW, making it one of the most powerful lamps of the gas-discharge type.

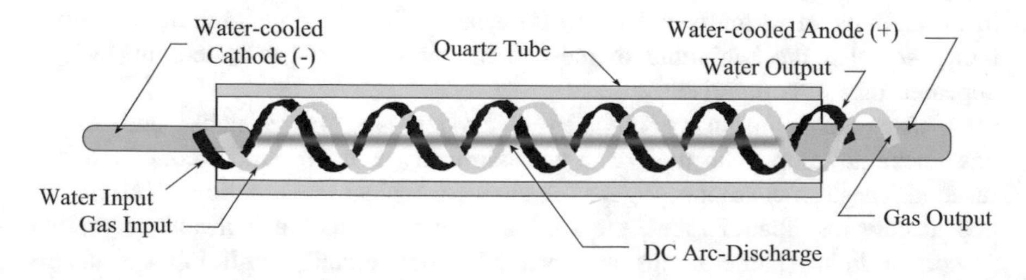

Fig. 5.3 Operation of a gas-water vortex-stabilized arc discharge lamp

Another type of lamp under development is the electrodeless gas discharge lamp. One type is the magnetic induction lamp, and another is the plasma lamp. These lamps attempt to avoid the problem of the electrode erosion of arc lamps by eliminating the electrodes entirely. The excited gas or plasma inside the glass tube emits light by an external infusion of energy by magnetic induction or radio frequency generation. A successful implementation of an electrodeless lamp is a plasma lamp, where microwaves are transported by a ceramic dielectric waveguide into a bulb filled with a mixture of argon and metal halides [9]. Electrodeless discharge lamps have operating lifetimes up to 100,000 h.

5.2.2 Incandescent Lamps

Around 1850 the British physicist and chemist Joseph Swan started investigating the transfer of an electric current through a carbon filament in a partially evacuated glass envelope. The high-temperature filament radiated visible light approximating blackbody radiation, and in 1860 he demonstrated a working incandescent lamp. He obtained a patent on this device, although the filament burned up in a short time. In the period 1875–1881, using a higher bulb vacuum and a carbonized thread filament, Swan built commercially feasible lamps. He installed these lamps in several private houses, public buildings and streets in Britain, and started his own manufacturing company—the Swan Electric Lamp Company. In 1874 Canadians Henry Woodward and Matthew Evans patented a light bulb that used an incandescent piece of shaped carbon in a nitrogen-filled envelope. See Fig. 5.4. An equivalent U. S. patent was issued in 1876.

The American inventor Thomas Edison was also working on incandescent lamps, following the design principles of Swan. Edison also purchased the patents of Woodward and Evans. In 1879 he demonstrated a light bulb that lasted about 13 h, using a carbonized thread filament in an evacuated envelope. He patented this lamp in 1879 and produced a bulb under higher vacuum with a high-resistance carbonized filament that could burn for hundreds of hours. He concentrated on commercializing the lamp, and supervised the construction of the first central electric power station in New York. The first tungsten filament light bulb was invented in 1904 by Sándor Just of Hungary and Franjo Hanaman of Croatia, with a patent assigned to the Hungarian company Tungstram, who first commercialized this type of light bulb. The tungsten filament incandescent bulb soon replaced the more fragile carbon filament light bulb and became the standard type.

The use of an inert gas such as nitrogen in an incandescent lamp envelope was shown by Irving Langmuir in 1913 to reduce filament evaporation and increase brightness. He also introduced the coiled filament geometry, where a sheath of stationary gas forms around the coils to diminish filament cooling by the natural gas flow in the envelope. This produced a hotter, and therefore brighter filament. This filament cooling reduction was further developed by the use of a coiled-coil geometry, which was first demonstrated in 1924. In 1959 Elmer Fridrich and

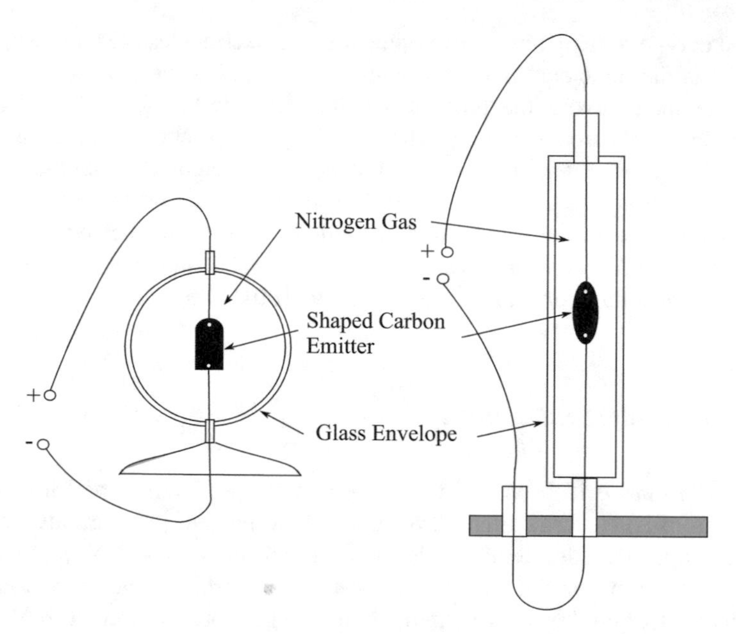

Fig. 5.4 Some early incandescent lamps built and patented by Woodward and Evans, ca. 1874

Emmett Wiley patented the first tungsten-halogen incandescent lamp. The addition of several halogen gases allowed the redeposit of tungsten that evaporated from the filament back onto the filament, instead of the interior of the glass envelope. This type of lamp, using the *tungsten-halogen cycle*, held its brightness much longer and was improved over the following years by General Electric engineers.

The era of widespread use of the common incandescent tungsten filament light bulb may be coming to an end. The *luminous efficacy* (ratio of light output in lumens to electrical input in watts) is of the order of 15%, with most of the input energy being radiated as heat. This inefficiency and required power consumption has prompted several European and Far East countries to consider banning its use. It could be replaced by more efficient tungsten-halogen lamps, compact fluorescent lamps (CFLs), and modern solid-state light sources. However, some Edison-type lamps close to the original designs are being remanufactured today, mainly for nostalgic décor use, not for operational efficiency. Also, at MIT new design techniques for incandescent lamps are being investigated that could approach the efficiency of modern light-emitting diodes [10]. This efficiency gain is accomplished by recycling wasted infrared energy from the tungsten filament. A nanocrystal-based reflective coating deposited on a larger surface area tungsten filament reflects emits heat back to the filament where it is reemitted as visible light.

5.2.3 Light-Emitting Diodes

In the early days of crystal radios, Henry Round documented the first observation of *electroluminescence*, or solid-state light emission. When a current flowed through a carborundum (SiC) crystal in a certain direction, light emission was observed in the region between two points of contact on the crystal [11]. Observed colors varied among yellow, green, orange, and blue. In the 1920s, using a zinc oxide and silicon carbide radio receiver crystal diode, Oleg Losev observed light emission and attributed it to a type of reverse photoelectric effect, publishing his results in 1927. This crystal diode light source was the beginning of a continuous series of advancements in the field of solid-state lighting. Losev is considered by many to be the discoverer of the light-emitting diode (LED) [12].

The spectacular progress in the field of solid-state physics in the 1940s through the 1960s focused on the field of electronics. Especially noteworthy was the invention of the transistor in 1948 at Bell Laboratories. However, the theory and manufacturing processes developed for semiconductor materials provided the basis for a whole new generation of solid-state lighting devices. In 1961 James Biard and Gary Pittman constructed and patented a semiconductor LED made of gallium arsenide (GaAs) which emitted infrared light [13]. The first practical visible light (red) LED was fabricated by Nick Holonyak Jr. in 1962 using synthesized gallium arsenide phosphide (GaAsP) crystals [14].

The basic principle of this type of LED is as follows: An *n-type* solid material (one containing free or conducting electrons) is interfaced with a *p-type* solid material (one containing electron or valence band holes). The materials are infused with a controlled amount of impurities (*doping*) to make them semiconducting, and form an active neutral region at the *p-n* interface. The electrons and holes are at different energy levels, forming a band or energy gap. When a negative voltage is connected to the *n*-type material (cathode), and a positive voltage is connected to the *p*-type material (anode), the LED is switched on, or *forward biased*. Electrons and electron holes then flow into the active region, where the holes are filled by the electrons. The transition of the electrons to the lower energy level of the electron holes produces light radiation as photons in all directions from the active region by spontaneous emission. The emitted light has an effective wavelength λ given by the relationship $\lambda = \frac{hc}{E}$, where E is the released energy in eV, h is Planck's constant, and c is the speed of light. The light emitted in the forward direction from every point of a bare LED chip in air is confined to a defined cone, due to total internal reflection at the air interface of the high refractive index semiconducting material. See Fig. 5.5. The emitted color of wavelength λ is determined by careful choice of materials, structure, and fabrication method of the LED.

Although red and green LEDs were developed for commercial use in the 1960s, the creation of a blue LED to complete the color triad eluded researchers for about three more decades. In the 1990s the fabrication and operation of efficient blue light-emitting diodes was reported [15, 16]. For the invention of an efficient blue

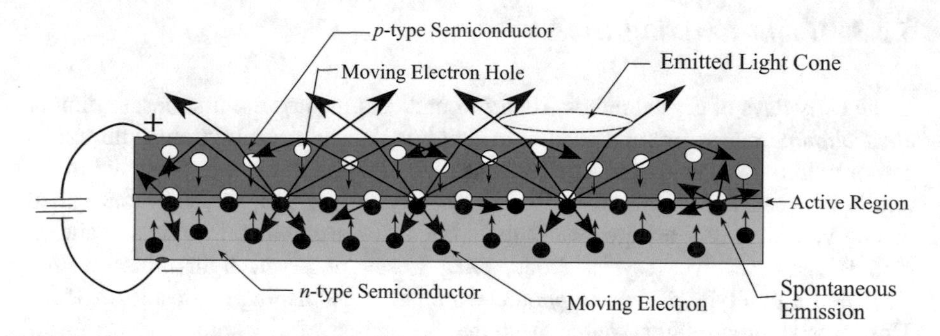

Fig. 5.5 Operating principle of the semiconductor light-emitting diode

LED, Isamu Akasaki, Hiroshi Amano, and Shuji Nakamura were jointly awarded the 2014 Nobel Prize in Physics.

Light-emitting diodes are available today in wavelength emission from infrared to ultraviolet, some having a luminous efficacy greater than 150 lumens per watt. White light LEDs can be in principle be achieved by color mixing of conjoined red, blue, and green LEDs, but the electronics is complicated for color blending and this technique is seldom used. A preferred method uses a blue LED, such as indium gallium nitride, that is coated with a broadband multi-phosphor coating to produce white light [17].

The construction of an LED lamp can take various forms. The most ubiquitous type is illustrated in Fig. 5.6. The *p-n* layers are deposited on a substrate to form an LED chip, typically about 1 mm^2 in area. Each point of the internal planar active region of the chip emits light over a wide cone angle into the *p*-type material. The chip is encased in a dome-shaped epoxy, usually of the same color as the LED light emission. The epoxy reduces the reflective losses at the epoxy-chip interface

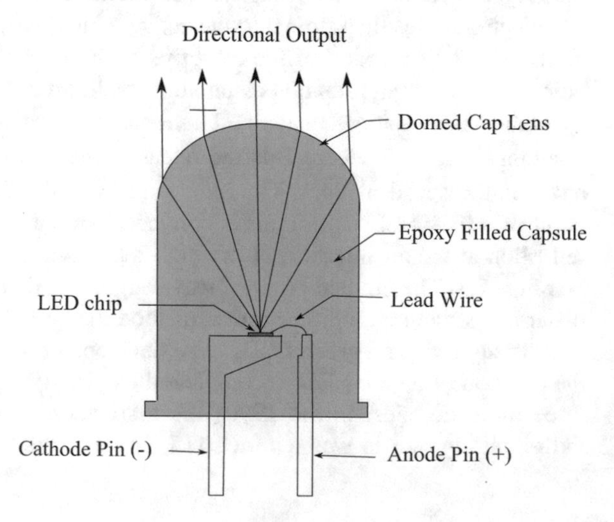

Fig. 5.6 A common epoxy dome construction for an LCD lamp

compared to an air-chip interface, and reduces total internal reflections at the external chip surface, allowing a wider cone emission angle. Also, the epoxy dome shape provides a lens effect to direct more light in the direction perpendicular to the planar active region.

The visible light LED has significant advantages over most incandescent and fluorescent sources, such as higher luminous efficacy, longer lifetime, lower operating temperature, longer on-off cycling lifetime, and smaller size. LEDs are used today in architectural and street lighting, dynamic signage, automobile and aircraft lighting, flat-panel TV backlighting, traffic signals, and indoor plant growth, with many new applications under development.

5.2.4 Organic Light-Emitting Diodes

The development of the organic light emitting diode (OLED) began with the observation of electroluminescence in a single anthracene crystal in the 1960s by Martin Pope, where light emission occurred by thermal electrons combining with electron holes. Using two electron ejecting electrodes, Helfrich and Schneider in 1965 produced electroluminescence by electron-hole combination in an anthracene crystal, called double injection recombination. The discovery of semiconductor organic polymers as light emitting materials in the 1970s provided much improved performance compared to anthracene and other organic materials. Figure 5.7 illustrates one of the first organic polymer double injection light-emitting diodes, as developed by Roger Partridge [18].

The general operation of a double injection polymer-based OLED is shown in Fig. 5.8. It consists of an upper emissive layer and a lower amorphous polymer layer. When a current is sent though the device electrons are injected into the upper layer and electrons are withdrawn from the lower layer, creating electron holes in the lower layer. The electrons and electron holes are attracted to each other and the combined pair is called an *exciton*. Light is subsequently emitted at frequency

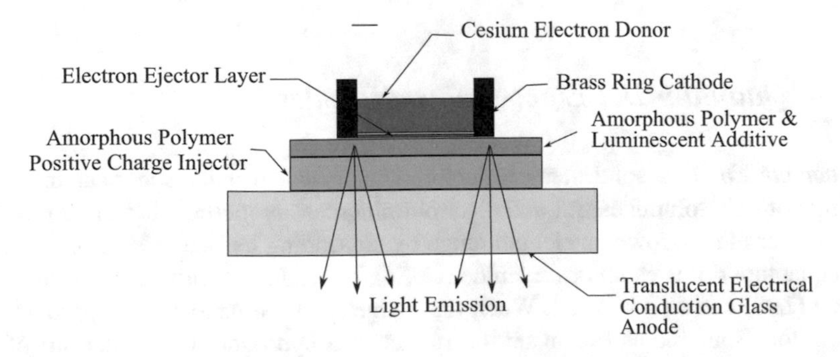

Fig. 5.7 One of the first organic polymer double injection light-emitting diodes built by Partridge

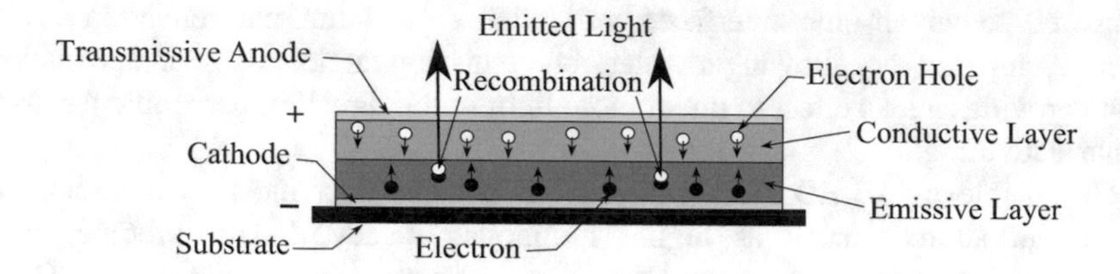

Fig. 5.8 General construction of a double injection type organic light-emitting diode

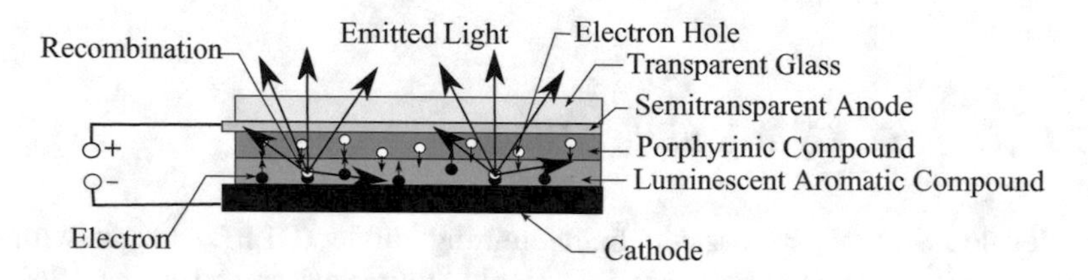

Fig. 5.9 Operation of a two-layer double injection organic light-emitting diode by Tang

$f = E/h$, where E is the energy difference [eV] between the highest energy occupied molecular orbit and the lowest energy unoccupied molecular orbit. The first practical double-injection OLED used organic molecular compounds, and was demonstrated in the 1980s by Tang and Van Slyke at Kodak [19]. A two-layer OLED is illustrated in Fig. 5.9 [20].

The advantage of the OLED over the conventional LED is a wider spectrum of available colors and the potential of being economically manufactured on thin flexible plastic substrates with high-efficiency light emission. OLEDs are currently being used as pixel elements on display screens for smaller handheld devices, where they do not require the separate backlight required for liquid crystal displays. They are also being scaled up for large screen TV applications. Large-area white illumination lighting is another application of the OLED, where development and manufacturing are currently being pursued [21, 22].

5.2.5 *Quantum Dot Light-Emitting Diodes*

A *quantum dot* is a solid semiconductor nanocrystal, usually spherical in shape, having both photoluminescent and electroluminescent properties. It can be tuned to emit a color in a narrow wavelength range by controlling its size. A suitable material for a quantum dot is cadmium selenide (CdSe), but cadmium sulfide (CdS) and zinc sulfide (ZnS) can also be used. When the diameter of the nanocrystal approximates the exciton Bohr radius (mean orbital radius of a hydrogen atom electron) of the unconfined bulk material, a phenomenon called *quantum confinement* occurs, where

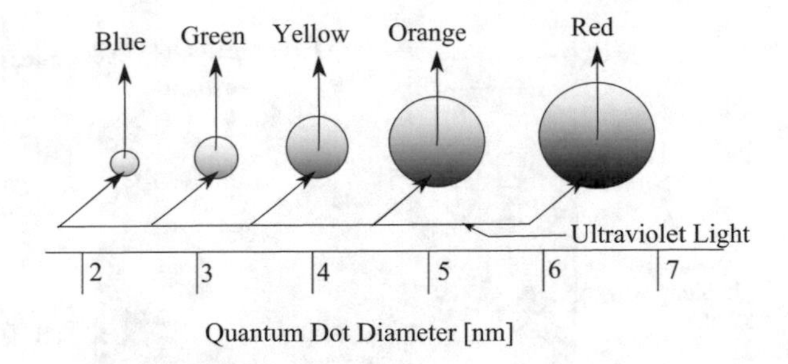

Fig. 5.10 Emitted wavelengths of photoluminescent CdSe quantum dots of various sizes, when irradiated by ultraviolet light

the electrons are confined to a volume with a specific bandgap energy [23]. This occurs at a dot diameter of ≈5.6 nm for CdSe. Then a nanocrystal can be fabricated to emit light of a specific wavelength by control of its size [24]. As the size increases, the quantum confinement decreases, and the emission wavelength shift increases. Figure 5.10 illustrates the dependence of the fluorescent wavelength of a CdSe quantum dot on its size, when illuminated by shorter wavelength ultraviolet light.

An electroluminescent quantum dot LED (QD-LED) can be fabricated using a two-dimensional monolayer of quantum dots, where the dots have a slight separation to transfer the current more efficiently through the dot layer. A typical configuration is shown in Fig. 5.11, where an inorganic semiconductor quantum dot layer is sandwiched between organic hole-injecting and electron-injecting layers. The recombination occurs in the quantum dot layer, and light is emitted through the indium tin oxide anode, with the color dependent on the dot size [25, 26].

A quantum dot film can be combined with an array of conventional light-emitting diodes to produce a high-quality white light source with a high blackbody correlated color temperature and high luminous efficacy and brightness. For example, by overlaying an array of blue-green LEDs with a sheet of aligned 627 nm red quantum dots, the resultant light absorbed and reemitted by the quantum dots had exceptional white light characteristics.

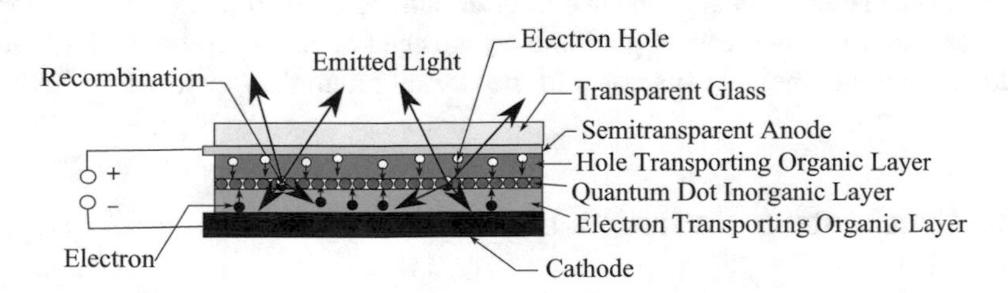

Fig. 5.11 Operating principle of a quantum dot light-emitting diode

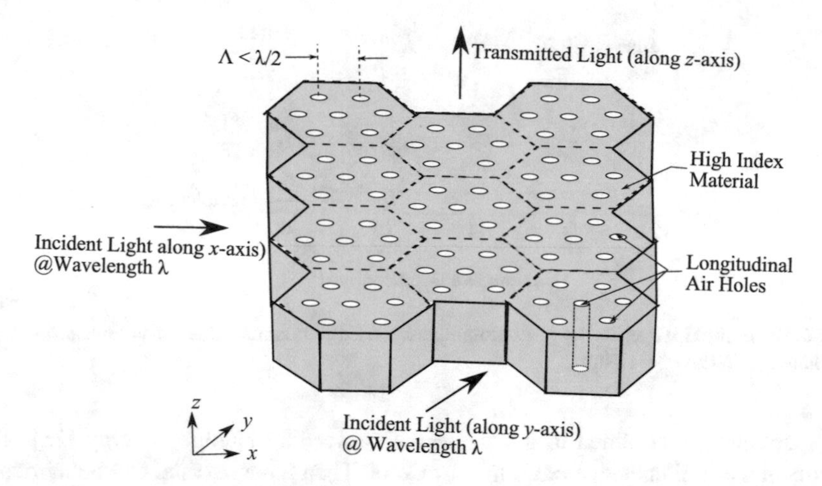

Fig. 5.12 A two-dimensional photonic crystal with allowable light transmission only in the direction of the z-axis

5.2.6 *Light Emission Using Photonic Crystals*

The emerging technology of *photonic crystals* can produce more efficient light extraction from an LED. A photonic crystal is a repeating hollow nanostructure in a dielectric material that produces a periodicity in the dielectric constant. This periodicity creates a photonic band gap, or quantum well, that prevents emission of photons whose energy lies within in this band gap. The period Λ of this repeating nanostructure is less than half the wavelength λ of the incident light. A section of a two-dimensional photonic crystal is illustrated in Fig. 5.12, consisting of a series of longitudinal air holes fabricated in a hexagonal pattern in a high refractive index material. Light can be propagated only in the direction of the z-axis. Three-dimensional photonic crystals are probably most useful for omnidirectional light control, but are much more difficult to fabricate. The extraction efficiency would then approach the quantum efficiency of the internal spontaneous emission [27, 28].

Photonic crystals can also be used as photoluminescent structures. For example, luminescent three-dimensional photonic crystals can be produced by the infiltration of high index luminescent materials in the crystal photonic gap structure [29].

5.3 Synchrotron Radiation Light Source

The last type of light source to be discussed uses *synchrotron radiation*. Synchrotron radiation was first observed as a spurious by-product in particle accelerators. It is an intense beam of radiation produced in a *synchrotron* as a result of the acceleration of electrons moving at close to the speed of light. Once the remarkable properties of

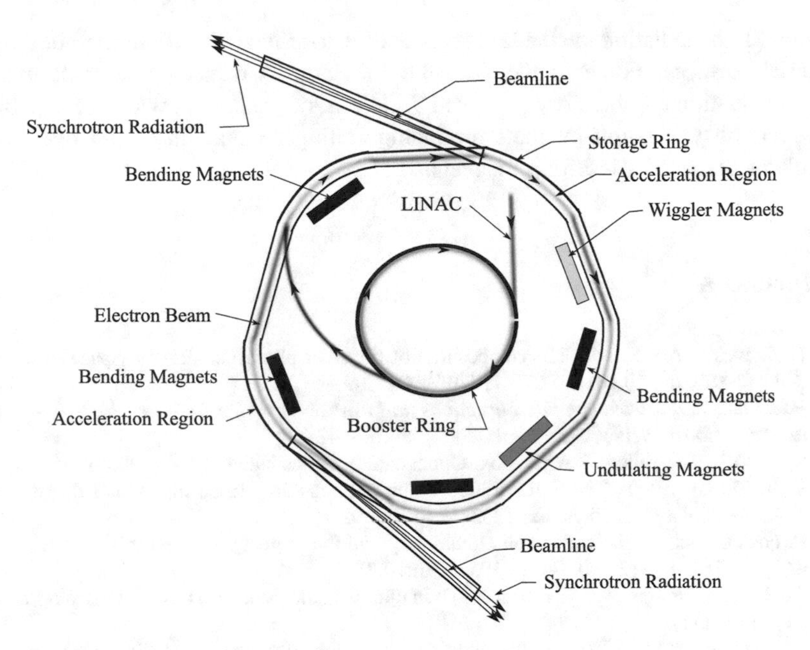

Fig. 5.13 Operation of a synchrotron to produce synchrotron radiation

synchrotron radiation were realized, dedicated synchrotron light sources were designed and built [30]. The basic operation of a synchrotron light source is illustrated in Fig. 5.13. Thermally emitted electrons enter a linear accelerator (LINAC) and attain an energy ≈90 meV, where they are injected into a *booster ring* to attain an energy ≈100 meV. These electrons are then injected into a *storage ring* where they are contained up to 48 h at a relativistic speed close to light speed. The storage ring consists of a series of curved and straight sections. Bending magnets at the curved sections produce a relativistic radial (centripetal) acceleration a toward the ring center by the changing direction of the electron velocity vector $\boldsymbol{v}$, such that

$$a = \gamma^2(v^2/r) \tag{5.1}$$

where γ is the relativistic Lorentz factor, v is the electron speed, and r is the radius of the storage ring.

Due to the electron acceleration in the curved sections and the high relativistic speed of the electrons, light is emitted in the opposite direction of the electron beam motion over a small cone angle. This produces multiple exiting *beam lines* consisting of the actual synchrotron radiation (only two exit ports are shown in Fig. 5.13). The wavelength of the emitted radiation is determined by weak *undulator* magnets inserted in the flat sections of the storage ring which transfer their periodicity to the circulating electron beam. The emitted light is usually in the ultraviolet and soft *X-ray* regions and consists of repeating pulses of a few picoseconds duration. Synchrotron radiation is one of the most intense sources of artificial light radiation available, and is useful for the study of atomic bonding at material surfaces. For example, in 1996 the large Advanced Photon Source at Argonne National Laboratory produced 7 GeV X-ray radiation, with size capability for over thirty bending magnets producing multiple beam lines, and over thirty undulator or wiggler insertion magnets.

References

1. H. Grevesse, A.J. Sauval, The composition of the solar photosphere. Adv. Space Res. **40**(1), 3–11 (2002). doi:10.1016/S0273-1177(02)00170-9
2. P.C. Plait, *Bad astronomy: misconceptions and misuses revealed from astrology to the moon landing 'Hoax'* (Wiley, New York, 2002), pp. 43–47
3. S.R. Wilk, The yellow Sun paradox. Optics & Photonics News, 12–13 (2009)
4. W.K. Tobiska et al., The SOLAR2000 empirical solar irradiance model and forecast tool. J. Atmos. Solar Terr. Phys. **62**, 1233–1250 (2000)
5. P. Foukal et al., Variations in solar luminosity and their effect on the Earth's climate. Nature **443**(14), 161–166 (2006). doi:10.1038/nature05072
6. G. Overton, Photonics plays a critical role in understanding our sun. Laser Focus World **47**(3), 61–65 (2011)
7. M. Kramer et al., Xenon and additive short arc illuminators: their capabilities and applications. SPIE Proc. **692**, 181 (1986)
8. A.J. Decker, J.L. Pollack, "*400 kilowatt argon arc lamp for solar simulation*", *NASA-TM-X-68042* (NASA John H. Glenn Research Center, Cleveland, OH, 1972)
9. F.M. Espiau, Y. Chang, Microwave energized plasma lamp with dielectric waveguide, U.S. Patent 7,518,315 (2009)
10. I. Ognjen et al., Tailoring high-temperature radiation and the resurrection of the incandescent source. Nat. Nanotechnol. **11**, 320–324 (2015). doi:10.1038/nnano.2015.309
11. H.J. Round, A note on carborundum. Electrical World **49**, 308 (1907)
12. N. Zheludev, The life and times of the LED—a 100 year history. Nat. Photonics **1**(4), 189–192 (2007). doi:10.1038/nphoton.2007.34
13. J.R. Biard, G.E. Pittman, Semiconductor radiant diode, U.S. Patent 3,293,513 (1966)
14. N. Holonyak Jr., S.F. Bevacqua,Coherent (visible) light emission from Ga(As1-xPx) junctions, Appl. Phys. Lett. **1**(4), 82–83 (1962). [doi:10.1063/1.753706]
15. I. Akasaki, H. Amano, High-efficiency UV and blue emitting devices prepared by MOVPE and low-energy electron beam irradiation treatment. Proc. SPIE **1361**, 138–149 (1991). doi:10.1117/12.24289
16. S. Nakamura, T. Mukai, M. Senoh, Candela-class high-brightness InGaN/AlGaN double-heterostructure blue-light-emitting diodes. Appl. Phys. Lett. **64**, 1687–1689 (1994). doi:10.1063/1.111832
17. S. Tanabe et al., YAG glass-ceramic phosphor for white LED (II): luminescence characteristics. Proc. SPIE **5941**(12–1), 12–16 (2005). doi:10.1117/12.61468
18. R.H. Partridge, Radiation Sources, U.S. Patent 3,995,299 (1976)
19. C.W. Tang, S.A. Van Slyke, Organic electroluminescent diodes. Appl. Phys. Lett. **51**(12), 913–915 (1987). doi:10.1063/1.98799
20. C.W. Tang, Organic electroluminescent cell, U.S. Patent 4,356,429 (1982)
21. M. Eritt et al., OLED manufacturing for large area lighting applications. Thin Solid Films **518** (11), 3042–3045 (2009). doi:10.1016/j.tsf.2009.09.188
22. J.W. Park et al., Large-area OLED lightings and their applications, Semicond. Sci. Technol. **26**(3), 0340-02-10 (2011). doi:10.1088/0268-1242/26/3/034002

23. M. Nirmal, L. Brus, Luminescence photophysics in semiconductor nanocrystals. Acc. Chem. Res. **32**(5), 407–414 (1999). doi:10.1021/ar9700320
24. D.J. Norris, M.G. Bawendi, Measurement and assignment of the size-dependent optical spectrum in CdSe quantum dots. Phys. Rev. B **53**, 16338–16346 (1996). doi:10.1103/PhysRevB.53.16338
25. S. Coe et al., Electroluminescence from single monolayers of nanocrystals in molecular organic devices. Nature **420**(6916), 800–803 (2002). doi:10.1038/nature01217
26. P.O. Anikeeva et al., Quantum dot light-emitting devices with electroluminescence tunable over the entire visible spectrum. Nano Lett. **9**(7), 2532–2536 (2009)
27. E. Yablonovich, Photonic band-gap structures. JOSA B **10**(2), 283–295 (1993). doi:10.1364/JOSAB.10.000283
28. J.D. Joannopoulous, P.R. Villeneuve, S. Fan, Photonic crystals: putting a new twist on light. Nature **386**, 143–149 (1997). doi:10.1038/386143a0
29. C.J. Summers et al., Luminescent and tunable 3D photonic crystal structures. J. Nonlinear Opt. Phys. Mater. **15**(2), 203–218 (2006)
30. M.L. Perlman, E.M. Rowe, R.E. Watson, Synchrotron Radiation—Light Fantastic. Phys. Today **27**(7), 30 (1974)

Chapter 6
Laser Light

6.1 Theoretical Foundations

6.1.1 Stimulated Emission and Population Inversion

Probably the most important light source invented and developed in the twentieth century has been the laser, with its many variations and properties. There is a vast amount of material written on laser types and their properties, accompanied by ongoing research and a multitude of new applications. In Sect. 4.4 the concepts proposed by Einstein for stimulated absorption, spontaneous emission, and stimulated emission were introduced, and set the stage for laser light production.

In *spontaneous absorption* a beam of incident photons with various energies and frequencies is incident on a material. When one of these photons has an energy $E = E_2 - E_1$, a transition of atoms from a low energy level E_1 to a higher energy level E_2 occurs. See Fig. 6.1a. The atom remains in the higher energy level for only a few nanoseconds, when it drops to a lower energy level, resulting in the *spontaneous emission* of a photon having no particular direction. Alternately, the dots can be envisioned as orbital electrons in a single atom. This process is illustrated in Fig. 6.1b. In *stimulated emission*, as shown in Fig. 6.1c, the energy $E = hf$ of any incident photon is exactly the difference between two energy levels, such that $E = E_2 - E_1$. If we envision light as a gas comprised of indistinguishable photon particles, the statistical theory developed by Satyendra Nath Bose and Albert Einstein in the 1920s predicts that photons have a tendency to travel together in pairs with identical properties. During a transition to the lower energy level, an additional photon is created that moves in the same direction as the incident photon, with identical properties of the original photon.

Normally, there are more atoms in the ground state than in the higher states. If during absorption of light, more atoms are transitioned to higher energy states than remain in lower energy states, then a *population inversion* occurs, as illustrated in

D.F. Vanderwerf, *The Story of Light Science*,
DOI 10.1007/978-3-319-64316-8_6

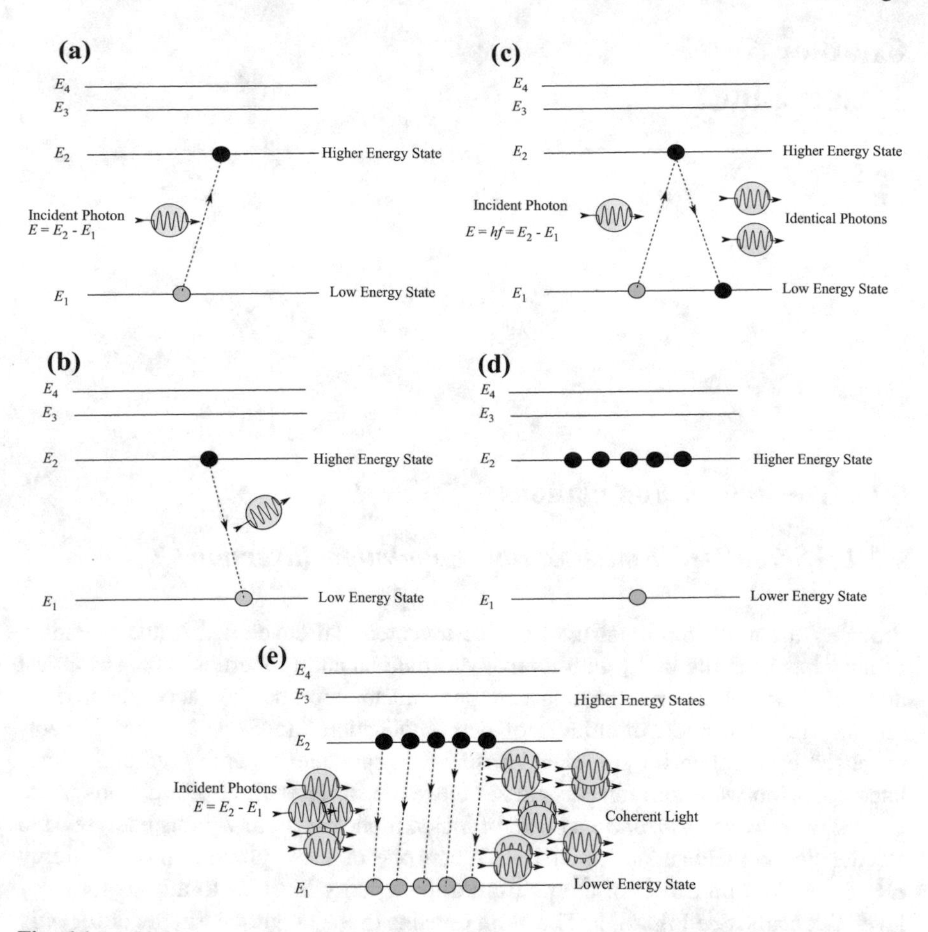

Fig. 6.1 **a** *Spontaneous absorption*: A group of photons of various energies are incident on a material. A photon with $E = E_2 - E_1$ moves an atomic electron from the energy level E_1 to a higher energy level stage E_2. **b** *Spontaneous emission*: The electron spontaneously drops from the E_2 level to the E_1 level, and a photon with random direction is emitted. **c** *Stimulated emission*: The electron at the E_2 level is stimulated to drop to the E_1 level, emitting two identical photons. **d** *Population inversion*: More electrons reside in a higher energy level E_2 than in the ground state E_1. **e** Emission of coherent light by spontaneous emission from a population inversion

Fig. 6.1d. A population inversion is necessary for stimulated emission. When these atoms transition to lower states by incident radiation of specific frequencies, stimulated emission produces significant emission of additional photons, all being perfectly in phase and moving in the same direction. Figure 6.1e illustrates this for a transition from energy level E_2 to E_1, although this effect can occur between other excited states, e.g. between E_3 and E_2, or E_4 and E_3. Stimulated emission must be initiated by incident energy and must occur before any spontaneous emission.

Population inversion and the subsequent stimulated emission are necessary and important operating principles of the laser. This can occur in certain gases, liquids, or solids, which are referred to as the *lasing medium*. To produce population inversion in the lasing medium, an external broadband source of radiation is required which contains light frequencies corresponding to energy differences between atomic energy levels. By spontaneous absorption this *optical pumping* creates a broad region of high energy states with a population inversion. Some of the excited atoms in this region revert to the ground state E_1 by spontaneous emission. However, most of the atoms undergo a radiation-free transition to a single metastable or relaxed energy level E_2, where they remain for about 10^{-3} s. A "triggering" source of radiation with energy corresponding to $E_2 - E_1$ then produces spontaneous emission from level 2 to level 1, where the emitted radiation is quasi-monochromatic and coherent.

6.1.2 Construction of a Laser

To sustain the population inversion for emission of significant coherent light in a laser, there are two basic requirements. First, an external source of energy must maintain population inversion by transition of atoms in the ground state of the lasing material to higher energy states by optical pumping. The pumping source can be optical, electrical, or even chemical in nature. Secondly, the emitted light from atoms must move other atoms to higher energy states where stimulated emission can occur. This is accomplished by bouncing the emitted light back-and-forth between two mirrors in a long resonant cavity of length L, such that the multiple passes increase the intensity of the emitted light along the longitudinal axis of the cavity. This is called *cascading* or *optical amplification*. When the mirrors are planar, the cavity acts as a Fabry-Perot etalon.

Resonance occurs within the laser cavity by the production of standing waves with a node at each mirror. Figure 6.2 illustrates a longitudinal resonance mode in a

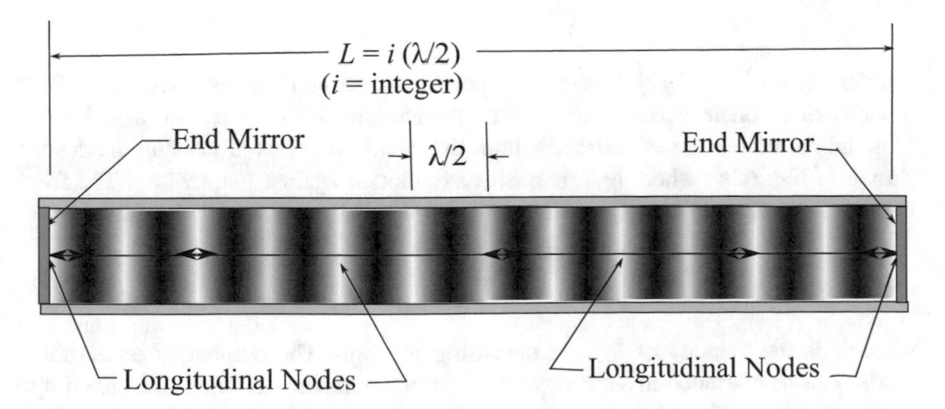

Fig. 6.2 Illustration of longitudinal nodes in a resonant laser cavity

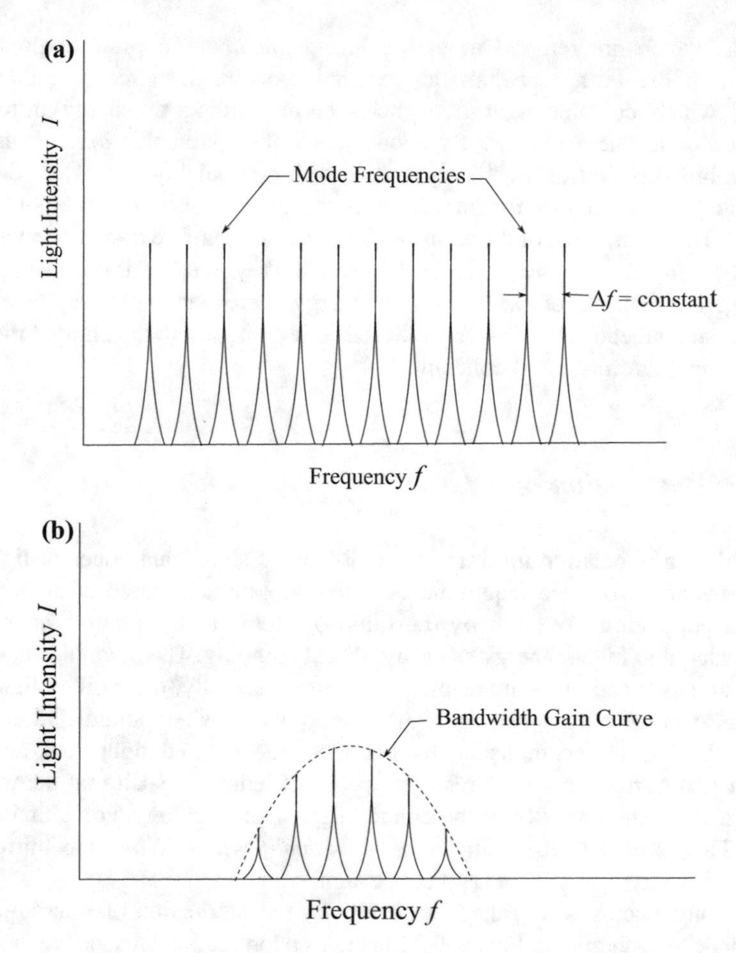

Fig. 6.3 **a** A series of longitudinal modes generated in a resonant laser cavity having a fixed frequency separation. **b** Restriction of observable longitudinal modes by the bandwidth gain of atomic transitions

laser cavity for $L = i(\lambda/2)$, where i is a specific integer and λ is the wavelength. For a series of resonate wavelengths and frequencies in the laser cavity, a series of longitudinal modes are established within the cavity. A series of possible modes is shown in Fig. 6.3a, where light intensity I is plotted against frequency f. The frequency separation Δf between consecutive modes has a constant value given by

$$\Delta f = \frac{v}{2L}, \tag{6.1}$$

where v is the velocity of light in the lasing medium. The number of observable modes can be restricted in several ways. First, the number is naturally restricted by the bandwidth profile of the atomic transitions, as shown in Fig. 6.3(b). Secondly,

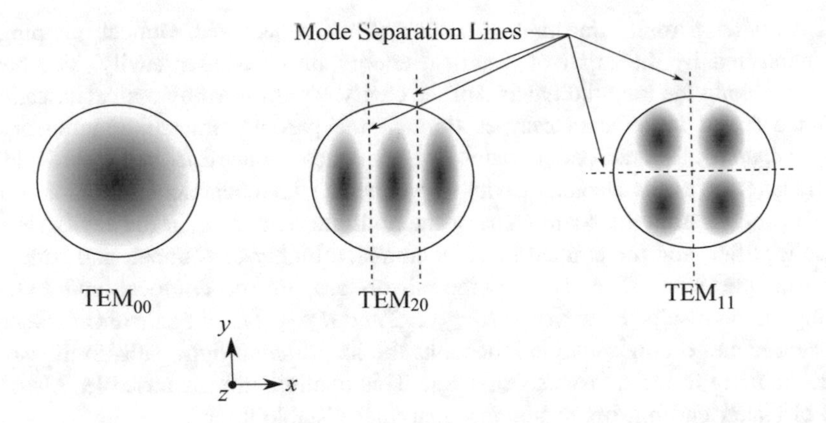

Fig. 6.4 Illustration of TEM$_{00}$ transverse mode having Gaussian profile and TEM$_{20}$ and TEM$_{11}$ transverse modes

the mirrors can be coated to reflect over a specified wavelength bandwidth. This is called *multimode operation*. When emission is restricted to one mode, *single-mode operation* occurs, and the laser beam is close to monochromatic. Transverse electromagnetic (TEM) modes are also observed in the cross-section of the cavity beam. This is expressed as TEM$_{mn}$, where m is the number of node separation lines in the x-direction, and n is the number of node separation lines in the y-direction. See Fig. 6.4. For most applications the single transverse mode TEM$_{00}$ is desired, which has a Gaussian profile and is spatially coherent.

Figure 6.5 shows a prototypical construction of a single mode laser using two concave mirrors at each end of a resonant cavity, one having ≈100% reflectance,

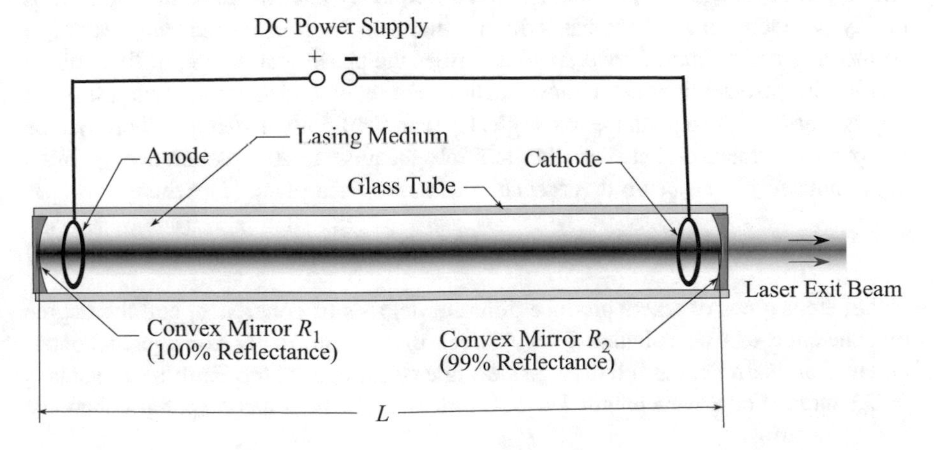

Fig. 6.5 Prototypical construction of a single-mode laser using two concave end mirrors

and the other partially transmitting having ≈99% reflectance. Optical pumping is accomplished by the input of electrical energy into the laser cavity. The beam intensity along the longitudinal axis of the cavity is reinforced by optical cascading, and the actual laser beam escapes through the partially transmitting mirror, or *output coupler*. The radii of curvature of each concave mirror R_1 and R_2 are related to the length L of the resonant cavity to produce a stable beam configuration. For a stable configuration the beam paths of the multiple reflections in the cavity remain close together, and the emitted beam is compact, highly directional, and coherent. For example, when $R_1 = R_2 \geq L$, the mirrors are close to confocal, and a stable configuration results. However, if $R_1 \leq L/2$ and $R_2 \leq L/2$, the mirrors are close to concentric and during multiple reflections the longitudinal light paths in the cavity begin moving in the transverse direction. This configuration is unstable. Also, the use of planar end mirrors results in a marginally stable laser.

6.1.3 Temporal and Spatial Coherence

Photons emitted by a laser produce coherent light, due to the fact that they are all in phase. There are two types of coherence that are relevant in the discussion of laser light. *Temporal coherence* describes the propagation time over which the electric fields in light beams can be phase correlated, and arises from the finite bandwidth of the source. It can also be specified by the *coherence length* L_{coh} of the beam. The coherence length is the distance over which the photon components of the beam remain in phase, and

$$L_{coh} = c/\Delta f, \tag{6.2}$$

where c is the speed of light, and Δf is the frequency bandwidth of the light. Thus highly monochromatic light has a high coherence length. *Spatial coherence* l_{coh} (sometimes called *lateral coherence*) describes the phase correlation, or fixed phase relationship, over a light beam cross-section. For an extended source with radiating points separated by a distance much greater than the light wavelength, there will be no phase correlation between the multiple beams. Spatial coherence is often approximated by the beam divergence φ at a reference plane [1], where

$$l_{coh} \approx 1.22\lambda/\varphi. \tag{6.3}$$

Different types of lasers produce different degrees of coherence, and the degree of coherence can be calculated for sources that are normally considered noncoherent. For example, the full divergence angle of sunlight at the Earth's surface is $\varphi \approx 9.3$ mrad. For a wavelength $\lambda = 550$ nm, sunlight then has a spatial coherence $l_{coh} \approx 72$ μm.

6.2 Fundamental Laser Types

6.2.1 The Pulsed Ruby Laser

The first operational laser was demonstrated by Theodore Maiman in 1960 and was a pulsed ruby laser [2, 3]. The basic construction is shown in Fig. 6.6, where the lasing material is a synthetic ruby crystal rod (Al_2O_3) with highly polished ends, and doped with chromium (Cr_2O_3) atoms. The broadband input energy comes from a helical gas-filled flashtube which emits a high intensity white light pulse of a few milliseconds duration. One end of the rod is perfectly reflecting and the beam exit end has a partially reflective coating. A highly reflective cylinder surrounds the flashtube to pump more energy into the cavity. This produces lasing action in the ruby rod and a coherent red beam of wavelength $\lambda = 694.3$ nm is emitted as a pulse of about 0.5 ms duration. The output energy content for a pulsed laser is usually expressed in Joules [J], and a pulsed ruby laser can deliver pulse energy greater than 20 J.

In Fig. 6.7 the three energy levels and transitions are indicated. First, spontaneous absorption of green and blue wavelengths from the flashtube transitions chromium ions in the ruby rod from the ground state E_1 to a broadband level E_3 of excited states. A fast radiationless transition from this broadband level forms a metastable state E_2 as shown. In the fluorescence lifetime of about 3 ms at this metastable level, stimulated emission occurs from energy level E_2 to the ground level E_1. Optical amplification then occurs through repeated reflections in the laser cavity and an intense red pulse is emitted. In operation, a series of pulses is emitted, determined by the flash rate of the lamp.

The intensity of an emitted pulse for a ruby laser can be considerably increased by means of *Q-switching*, where "Q" stands for "quality factor". Q-switching halts or frustrates the spontaneous emission in a laser cavity by interruption of the

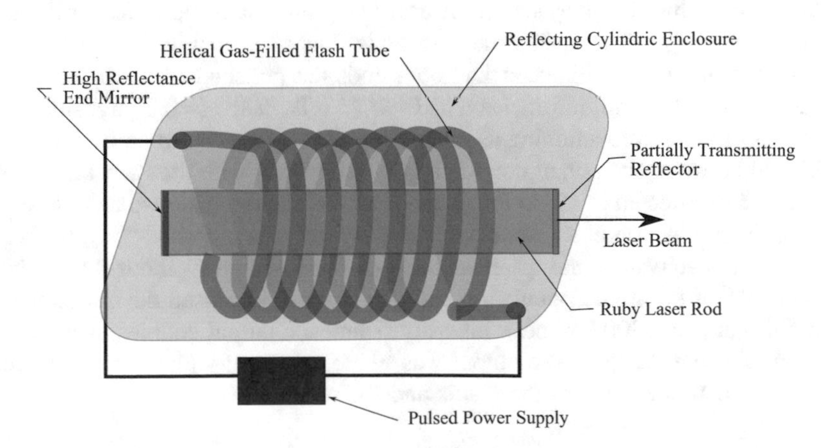

Fig. 6.6 General construction of the pulsed ruby laser of Maiman

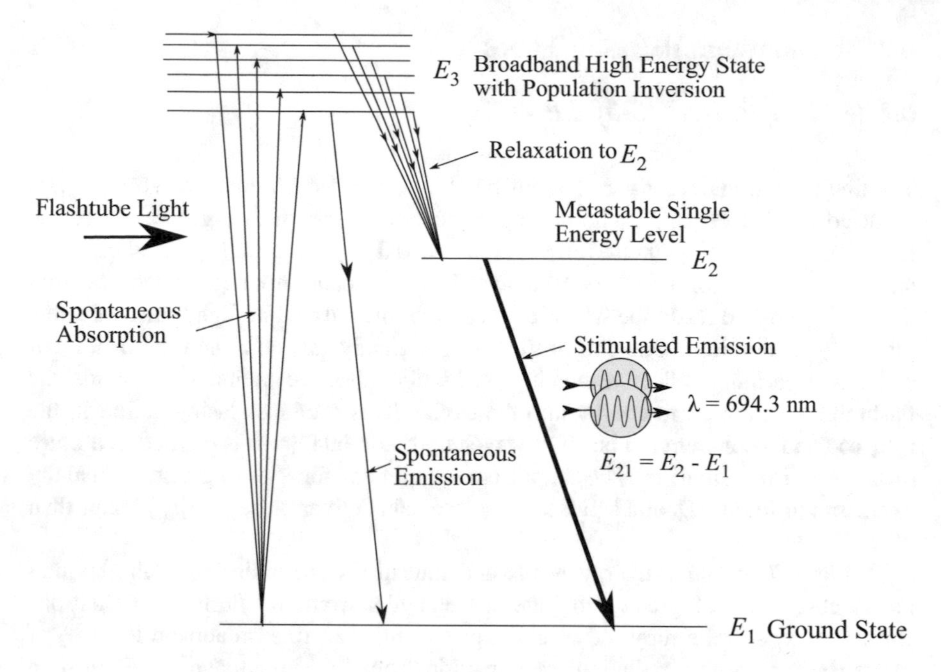

Fig. 6.7 Energy levels and lasing transitions for the ruby laser

multi-reflections between the end mirrors. During this low-Q period the population inversion of the lasing atoms is considerably increased, and *gain saturation* occurs. The lasing activity resumes by quickly changing the switch to produce high-Q in the laser cavity, and a single pulse of very high energy is emitted. *Active Q-switching* uses an external device whose transmission is actively controlled. *Passive Q-switching* uses a saturable absorber material such as a Cr:YAG crystal, where the transmission increases at a certain light level. The *giant pulse laser* was demonstrated in 1962 by McClung and Hellwarth using an electro-optic Kerr cell as an active Q-switch [4]. The basic arrangement is illustrated in Fig. 6.8. A ruby rod ≈3 cm long and 0.7 cm diameter had both ends flat polished and parallel to each other. The dielectric end mirrors each had ≈0.75 reflectance (≈0.25 transmittance) at $\lambda \approx 694$ nm, both functioning as output couplers. A nitrobenzene Kerr cell was positioned between one output coupler and the ruby rod, with the Kerr cell electric field "on" axis oriented at 45° to the plane of the crystallographic *c*-axis of the ruby rod. For the production of a giant pulse, the Kerr cell was switched "on", the pump flashlamp was activated, and the Kerr cell was switched "off" about 0.5 ms later. The collapse of the stored population energies in the rod caused the emission of a giant pulse at about 300 kW peak intensity from each output coupler, with a pulse duration of about 0.12 μs. The output was plane polarized with its electric vector perpendicular to the plane of the *c*-axis and the output axis.

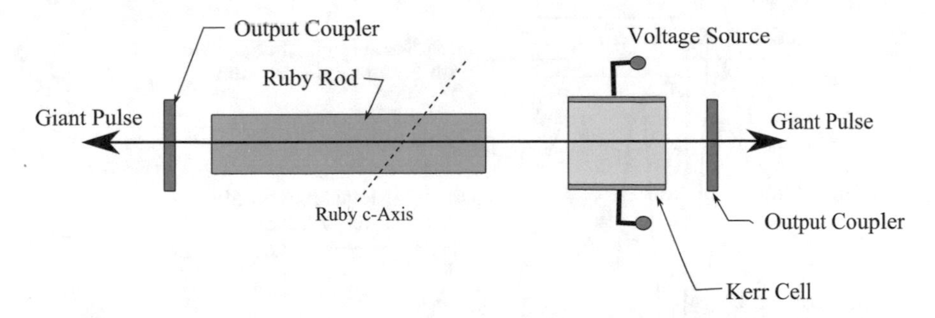

Fig. 6.8 Illustration of the giant pulsed ruby laser of McClung and Hellwarth (pump flash lamp not shown)

6.2.2 *The Nd:YAG Laser*

The first continuous wave neodymium-based solid-state laser was developed at Bell Labs in 1961 to produce coherent light at $\lambda \approx 1.1$ μm [5]. Another crystal–based laser also developed at Bell Labs around 1964 is the Nd:YAG laser, which uses the cubic garnet crystal Nd:$Y_3Al_5O_{12}$, grown in the laboratory and doped at about a 1% level with the neodymium ion [6]. This has proven to be one of the most popular and useful of the solid-state laser types. The emitted fundamental wavelength is $\lambda = 1064$ nm with a linewidth $\Delta\lambda \approx 0.5$ nm, and can be operated in a continuous wave (CW) mode or in a pulsed mode using Q-switching.

Figure 6.9 shows a typical construction of a typical pulsed operation Nd:YAG laser with side optical pumping using an elliptical reflector and a linear flashlamp. The lasing rod and the linear flashlamp are placed at the foci of the reflector. An electro-optics cell is used for active Q-switching. The four energy levels and the transition for the 1064 nm wavelength are illustrated in Fig. 6.10. Other lasing transitions can occur at 940, 1120, 1320, and 1440 nm. Frequency doubling by the insertion of nonlinear materials can also produce emitted

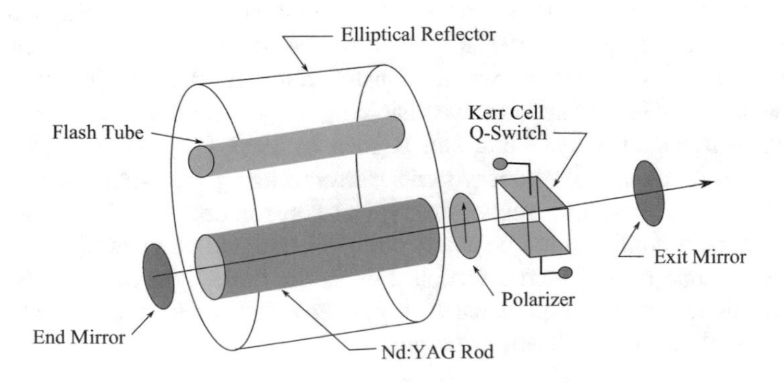

Fig. 6.9 Nd:YAG laser operating with linear lamp side optical pumping and a Kerr cell Q-switch

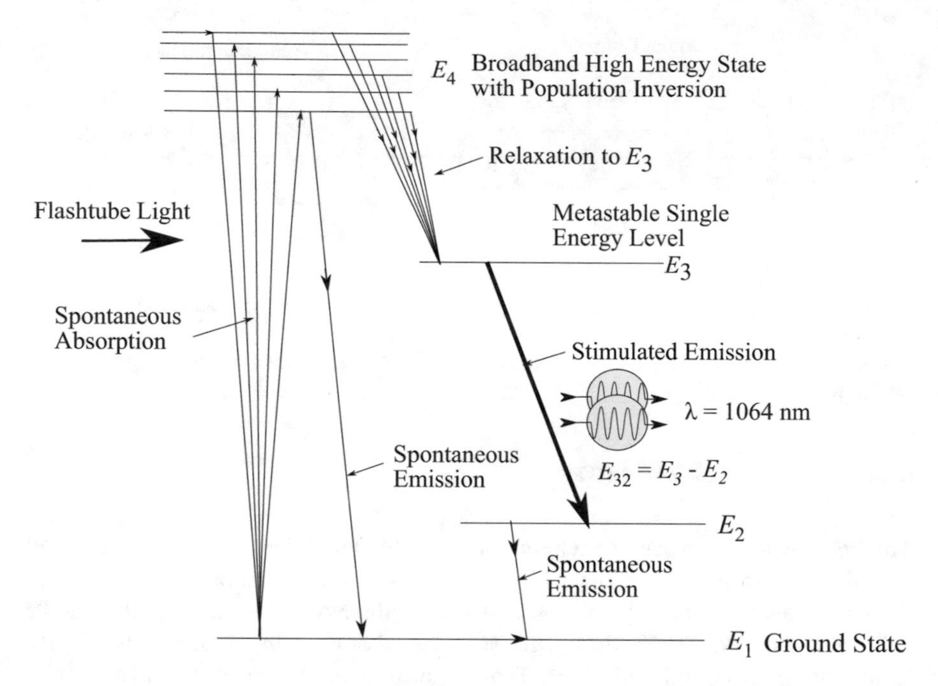

Fig. 6.10 Energy levels and transitions of Nd:YAG laser to produce lasing light at λ = 1064 nm

wavelengths of 532, 355, and 266 nm. The use of Q-switching in the Nd:YAG laser can produce pulses of several nanoseconds duration, and a peak pulsed power output has been achieved at gigawatt levels.

6.2.3 The Helium-Neon Laser

The first lasers using a gas as a lasing medium were developed in the 1960s by Arthur Schawlow and Charles Townes [7], although Gordon Gould was finally awarded the patent rights in 1977 [8]. In 1961 the use of a mixture of helium and neon gases as a lasing medium was demonstrated to emit the 1.15 μm line of neon [9]. The first He–Ne continuous wave visible light laser emitting the 632.8 nm neon line was demonstrated by White and Rigden in 1962 [10]. They used a 7 mm diameter laser cavity with concave end mirrors having maximum reflectance at λ = 632.8 nm. Figure 6.11 shows the typical construction of a HeNe laser with concave mirrors configured to be nearly confocal, and a tilted glass plate inserted at Brewster's angle to the beam direction. For the multiple light passes between the mirrors, this is optically equivalent to a gas-separated stack of plates which produces a p-polarized laser beam as output.

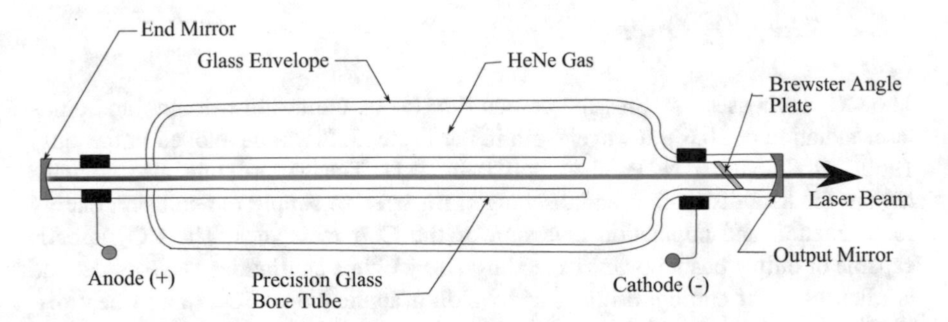

Fig. 6.11 Basic construction of a He–Ne laser, with a precision glass bore cavity, confocal mirror configuration, and a Brewster polarizing plate

Several emitted laser wavelengths can be produced by the He–Ne laser. The energy diagram of Fig. 6.12 shows how the 543.5 nm (green), the (red) and the 1523 nm (infrared) can be produced from stimulated emission between upper energy bands in neon. These neon levels are produced by collisions from excited helium ions in metastable states that have been produced by an electric discharge. The single-mode 632.8 nm laser line has made the He–Ne laser extremely popular over the last 50 years. The coherence length L_{coh} can be greater than one meter, wavelength stability is $\approx$0.01 nm, and He–Ne lasers with power outputs of about 50 mw have been built.

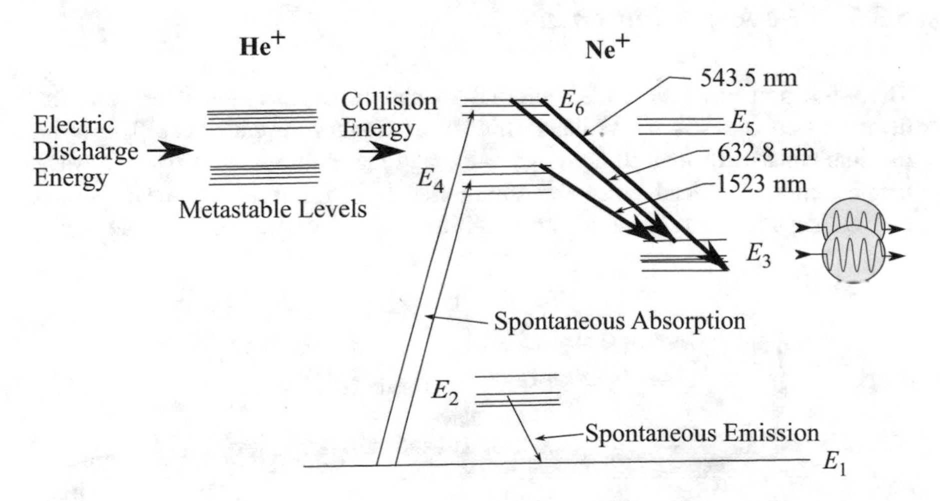

Fig. 6.12 Energy levels and transitions of He–Ne laser to produce laser light at λ = 543.5 nm (*green*), 632.8 nm (*red*), and 1523 nm (infrared)

6.2.4 The CO_2 Laser

The CO_2 laser uses low pressure carbon dioxide as the lasing medium, and emits laser radiation of 10.6 μm wavelength in the infrared. It was developed in the early 1960s by C. Kumar N. Patel at Bell Labs [11]. The gas mixture also contains helium for heat dissipation and discharged nitrogen to supply the collision energy for excitation and population inversion of the CO_2 molecules. The CO_2 laser is capable of output beam powers exceeding 100 kW in a continuous wave mode, and is often used for cutting, drilling, and welding applications. To assist in heat dissipation and maintain efficiency in the higher power CO_2 lasers, a continuous flow of carbon dioxide is created in the containment tube. An additional water-cooled cylinder enclosure is used for high-power devices. The gas flow should be turbulent to maintain stabilization of the discharge. The laser can also be Q-switched to yield pulse power at the gigawatt level.

When the flows rapidly along the axis of the laser beam, a *fast axial-flow laser* is produced. A *cross-flow laser* configuration occurs when the CO_2 flows more slowly in a direction perpendicular to the laser beam. The nitrogen gas discharge can be produced by DC direct current electrodes, AC alternating current electrodes, or RF radio frequency external source. Figure 6.13 illustrates the general operation of a DC energized fast axial-flow CO_2 laser. The mirrors are usually vacuum coated or electroplated with gold for high reflectivity in the infrared.

6.2.5 The Argon Ion Laser

The argon ion gas laser (often referred to simply as an argon laser) was first demonstrated in 1964 by William Bridges at Hughes Aircraft [12]. The lasing medium consists of ionized argon gas Ar^+, and the multiplicity of discrete wavelengths emitted extend from the ultraviolet (≈275 nm) to the near infrared (≈1092 nm). It can be used in multiline mode, where all possible lasing

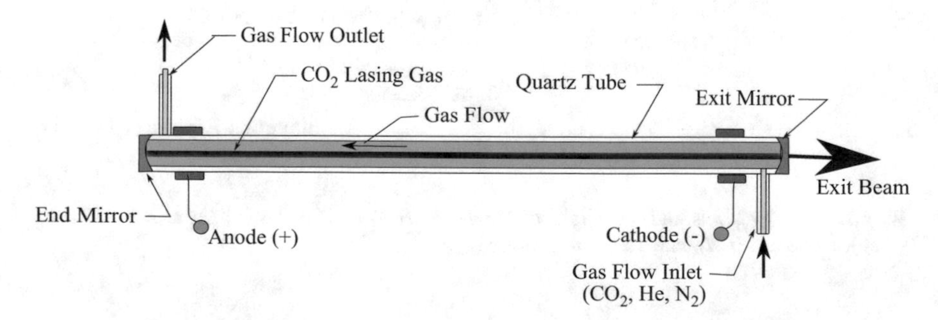

Fig. 6.13 Illustration of the general operation of a DC energized fast axial-flow CO_2 laser

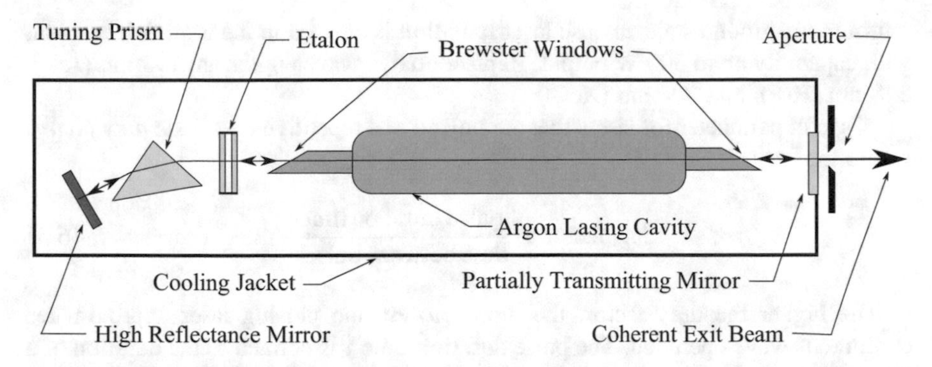

Fig. 6.14 Basic construction of an argon ion laser having single wavelength output using Brewster windows, a tuning prism, and an additional enclosed etalon

wavelengths are emitted simultaneously producing a whitish light, or a specific wavelength can be selected. Argon is one of a family of noble gases that can be used for an ionized gas laser. Several other suitable gases are krypton, xenon, or a mixture of argon and krypton.

The argon ion laser is also capable of a wide range of continuous output power, ranging from a few milliwatts to about 25 W for visible multiline operation. The input direct current required to achieve the production of argons ions is high, up to 65 amperes, and the laser cavity requires a water-cooled jacket at the highest output powers. Forced-air cooling is usually used at lower output powers. Figure 6.14 illustrates the basic construction of a wavelength selectable argon ion laser, with a tuning prism to select a single laser wavelength. An additional enclosed etalon inside the laser etalon cavity can further reduce the linewidth of single mode emitted light to a few MHz, producing a coherence length of over 100 m. Frequency doubling by the insertion of a beta barium borate (BBO) crystal can extend the emitted light into the deep UV region.

6.2.6 *The Excimer Laser*

The excimer laser is a gas laser that naturally produces short pulses of ultraviolet light at high power without the need of any frequency doubling devices. It was invented and demonstrated in 1970 at the Lebedev Physical Institute in Russia. The laser cavity is initially filled with a mixture of an inert gas (Ar, Xe, Kr) and a noble halogen or halide gas (F, Br, Cl) The cavity pressure is increased by pumping in another inert gas such as neon or helium. By application of an electric discharge, types of excited molecules are formed, such as Ar_2, Kr_2, Xe_2 (*excimer* or *excited dimmer*), or ArF, XeBr, KrF, KrCl (*exciplex* or *excited complex*). These transient-state molecules are usually formed in excited states by a pair of linear discharge electrodes running the length of the laser cavity. When these molecules return to the ground state, intense laser radiation is emitted in the ultraviolet region, with capability up to 300 W output. Representative wavelengths are 193 nm (ArF), 248 nm (KrF), and 308 nm (XeCl).

A useful parameter for lasers that are pulsed at a repetitive rate is the *duty factor*, where the

$$\text{duty factor} \equiv \frac{\text{pulse duration time}}{\text{time between pulses}}. \tag{6.5}$$

The higher the duty factor, the more closely the pulsing laser approximates continuous wave operation. The pulse duration time τ is defined as the duration of a laser pulse at half of maximum power, often called the full-width at half-maximum (*FWHM*). The time between pulses is the round-trip travel time of the pulse in the cavity, and the pulse repetition rate *PRR* is the reciprocal of the time between pulses. The total energy E in a single pulse is related to the average pulse power P_{av} and the maximum or peak pulse power P_{max} by

$$E = \frac{P_{av}}{PRR} = (\tau)(P_{max}) \tag{6.6}$$

A contained stationary gas in a large diameter cavity would require about a second to return to the thermal ground state, enabling a new exited state to form and initiate a new pulse. This long pulse repetition time reduces the duty factor but can increase the pulse energy E. To reduce this thermal delay, the lasing gas is often pumped at high speed through the laser cavity. Another method to reduce the pulse repetition time is the use of a small diameter lasing cavity, where a stationary gas can be rapidly returned to thermal equilibrium [13]. Figure 6.15 illustrates this type of excimer laser construction, using a 0.5 mm diameter capillary bore waveguide lasing cavity in a 6.2 mm diameter, 20 cm long quartz rod. For a lasing gas of KrF in a He fill, and a high voltage capacitive discharge, the emitted 248 nm radiation has an output energy of ≈50 μJ. This type of excimer construction can produce a pulse duration time ≈30 ns and a pulse repetition time $\approx 10^5$ ns, resulting in a duty factor ≈0.03%.

Excimer lasers are commercially available today that emit at discrete wavelengths over the range $\lambda = 157$ nm to $\lambda = 351$ nm. Pulse repetition rates of 2 kHz (pulse repetition time $= 5 \times 10^5$ ns) are available with pulsed output energies up to 1000 mJ. Laboratory excimer lasers have achieved pulse duration times exceeding 300 ns, pulse repetition rates over 10 kHz, with the resultant duty factor approaching 1%.

Applications for the deep UV light of excimer lasers cover many fields. It can be focused to a smaller spot than longer wavelength laser light and can perform heatless ablation over tightly defined surface areas of materials such as organic polymers, biological structures, and inorganic substances, without destructive heating effects on the surrounding areas. Excimer light at 248 nm is used in photolithography for precision micromachining of individual and patterned holes with

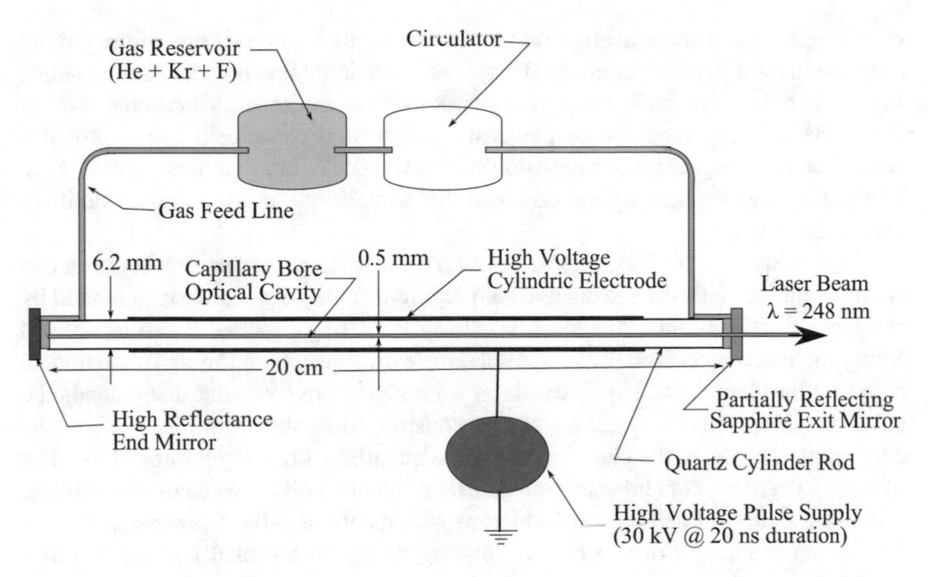

Fig. 6.15 Layout of an excimer laser, with stationary gas in a capillary cavity and high-voltage capacitive discharge to achieve excitation and lasing action

diameters in the 30 μm to 100 μm range for ink jet nozzles and semiconductor integrated circuits. In the medical field, using 193 nm excimer laser light, photorefractive keratectomy (LASIK surgery) can reshape the cornea to correct certain common vision problems. Focused 308 nm excimer laser light can treat certain local dermatological skin conditions, such as psoriasis, and 308 nm light is also used in laser angioplasty.

6.2.7 The Dye Laser

The basic dye laser uses an organic molecular dye as the lasing medium, which is dispersed in a liquid solvent. A typical dye molecule is rhodamine dispersed in methanol solvent. Various combinations of dye molecules and solvents can produce lasers that emit radiation wavelengths from about 320 nm (near UV) to 1200 nm (near IR). The emission occurs from fluorescent transitions in the dye, optically pumped by a flashlamp close to the desired dye transition wavelength. One of the first dye lasers was constructed in 1966 at IBM Labs, pumped by a ruby laser, and emitted light in the near infrared. The conventional dye laser uses an optical cavity with end mirrors, similar to other types of laser configurations. The dye laser was also discovered independently in 1966 by Schäfer and colleagues in Germany.

Figure 6.16 illustrates two energy states of a molecular dye with vibrational levels of a singlet excited state E_2 and possible stimulated emission wavelengths to the vibrational levels of the ground state E_1. With optical pumping at the proper wavelength, spontaneous absorption from the lowest vibrational level of the ground state produces a population inverted series of vibrational levels in an excited single state E_2. These vibrational levels naturally convert to the lowest vibrational level of the E_2 state. The ground state E_1, consisting of several vibrational levels, provides several paths for spontaneous emission of fluorescent light at various wavelengths. This makes the dye laser a good candidate for wavelength tuning using an auxiliary grating or prism.

The dye laser usually operates in a pulsed mode with pulse duration between one and two milliseconds for a static dye gain medium. This pulse duration is limited by chemical reactions in the dye material during the lasing process. It can be considerably increased by a flow of the dye solvent, with appropriate timing of the optical pumping flashlamp [14, 15]. A dye laser with a static dye solvent that extends the pulse duration greater than 20 ms can be constructed as shown in Fig. 6.17, where contiguous regions of the gain material are sequentially optically pumped [16]. The dye and solvent are contained in a quadrant section of a hollow-wall circular cylinder cell, and a series of linear xenon flashlamps with parabolic reflectors are sequentially fired along the length of this cell. A rotating disc with attached mirrors acts as a Q-switch and synchronously redirects the emitted light from each region to the central axis of a laser cavity between two mirrors.

To achieve continuous operation (CW), dye lasers are usually constructed in a circular configuration, the so-called *ring dye laser*. The dye cell length is often very short. The ring laser is stable and tunable to a single frequency with high-power

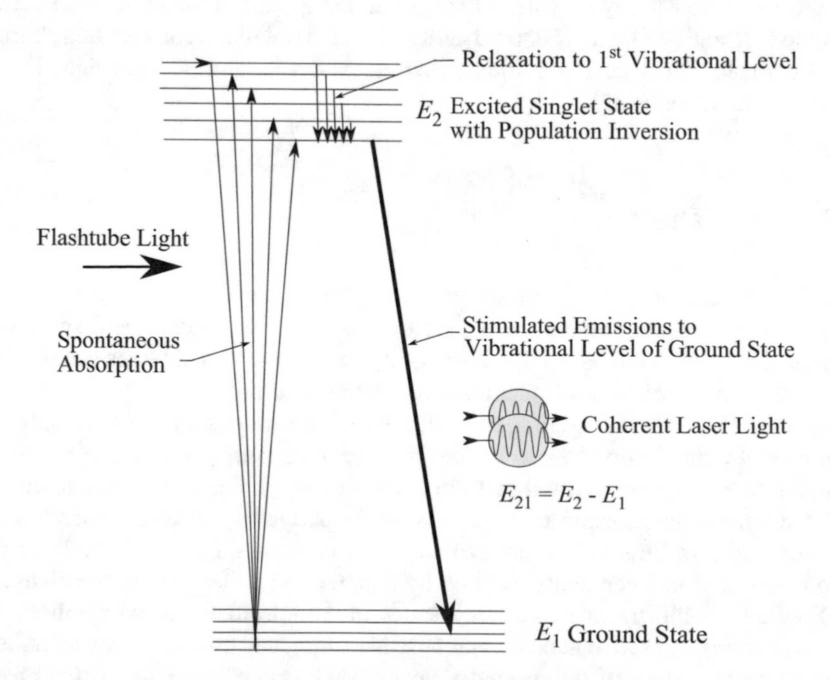

Fig. 6.16 Illustration of possible energy states in a molecular dye laser

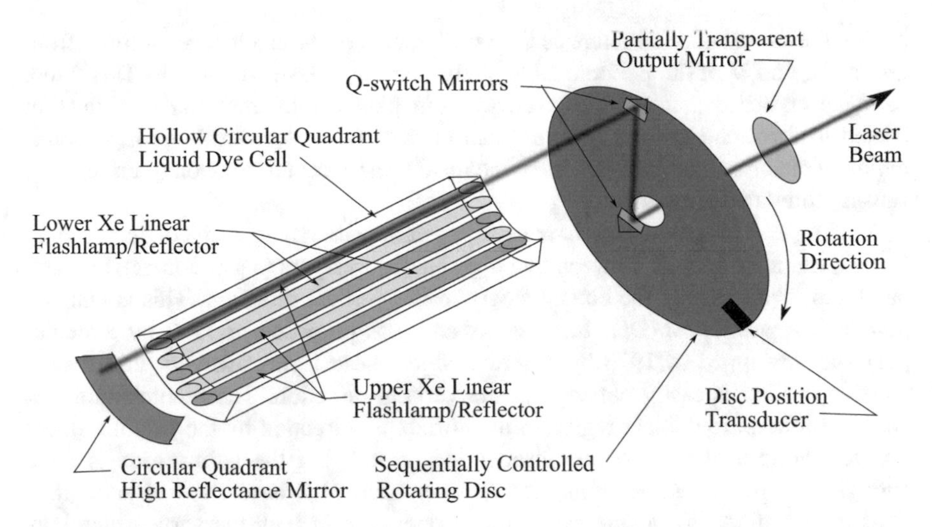

Fig. 6.17 Illustration of a dye laser with sequential excitation of regions in the dye cell to achieve longer pulse durations (Power supply, control circuits and disc servo motor are not shown)

output [17]. In addition to liquid types, solid-state dye lasers can be constructed using solid polymeric matrices doped with fluorescent dye molecules [18].

Liquid dye lasers are often used in the removal of vascular and pigmented lesions. Also, recent investigations by Sun and Fan at the University of Michigan have developed a miniature optofluidic laser consisting of a dye laser and microfluidics. Using an optofluidic ring resonator, it is possible to clearly distinguish between healthy and mutated DNA strands [19]. This could be used to identify genes that might initiate cancer and other genetic diseases.

6.3 Semiconductor Laser Diodes

6.3.1 Edge-Emitting and Quantum Well Laser Diodes

The operation of the laser diode is derived from the light emitting diode (LED) described in Sect. 5.2.3. In the LED, recombination of the electrons and electron holes at a *p-n* junction results in spontaneous emission of photons, with no amplified or coherent properties. In a laser diode, two other attributes are required: (1) a direct bandgap material where stimulated emission of photons can occur directly between conduction and valence bands, and (2) a resonant cavity for optical amplification of these photons. Certain semiconducting compounds, such as GaAs and InP, satisfy the first requirement. Simulated emission in a *p-n* junction GaAs crystal was first demonstrated by a research team at General Electric Research Lab in 1962 [20]. Laser diodes are constructed by the crystal growth of layers on a substrate such as GaAs or InP. For the GaAs based laser diode, the layers are alloys having the same lattice structure as GaAs. Elements of these alloys are chosen from groups III and V of the periodic table, including Ga, Al, In, As, and P. This direct bandgap crystal can then be cleaved to form planar sides that can be reflection coated to form end mirrors of a resonant cavity. When forward-biased, recombination of electrons and electron holes occurs at the *p-n* center junction (gain region), causes stimulated emission.

When the active *p-n* region consists of a low bandgap semiconductor (e.g. GaAs), and is positioned between two high band gap layers (e.g. AlGaAs), the high band gap layers restrict the optical cavity in the vertical direction. This is called a *double heterostructure* (DH) laser diode, and was proposed in 1963 by Kroemer [21] and developed in 1970 by Alferov. Since these layers are very thin, single mode operation is easily obtained in the vertical direction. The recombination is confined to a narrow linear region in the horizontal direction by the use of a linear contact anode at the top of the laser diode crystal, but the gain region is wide enough to support transverse multimode operation. This is referred to as a *gain-guided* laser diode. To assure single mode operation in both the vertical and horizontal directions, the optical cavity can be further constrained by surrounding the lower refractive index active region with higher index material. The gain region can then be typically restricted to about 3 μm in the horizontal direction and 0.1 μm in the vertical direction. This is an *index-guided* laser diode, and operates at lower powers than the gain-guided type. Most commercial laser diodes operate in single mode with an index-guided construction.

About the same time it was realized that the thin optical cavity of the DH laser diode acted as a guide for electron waves and formed a potential well containing a series of electron modes. This type of DH laser diode was then called a *quantum well* laser diode [22]. Quantum well energy levels required fewer electron/electron hole combinations for lasing, and the emitted laser wavelength was determined only by the thickness of the active layers.

Figure 6.18(a) shows the general construction of a quantum well laser diode powered by charge injection with a double heterostructure and linear contact anode, and the resultant gain region [23]. Proton bombardment produces two regions of insulating material, forming a lateral conductive zone which defines the optical cavity size. The active gain region consists of about 50 thin alternating active layers of narrow bandgap and about 50 passive wider bandgap semiconducting materials. The active region bandgap layers can have values of 5–50 nm, with the narrow bandgap layer thickness being somewhat less than the wide bandgap layer thickness. When this DH laser is forward biased, the electrons and electron holes are optically confined in the active layers and lasing occurs between the lowest bound electron states and the bound hole states in the active region. A detailed view of the active bandgap region is shown in Fig. 6.18(b). The cleaved end windows of the active region are coated such that one cavity end mirror has very high reflectance, and the other cavity end mirror has a lower reflectance. Multiple reflections of coherent light in the cavity increases the stimulated emission intensity, and the coherent laser beam exits through the mirror with the lower reflectance. The refracted emitted beam pattern has the shape of an ellipse, typically with about three

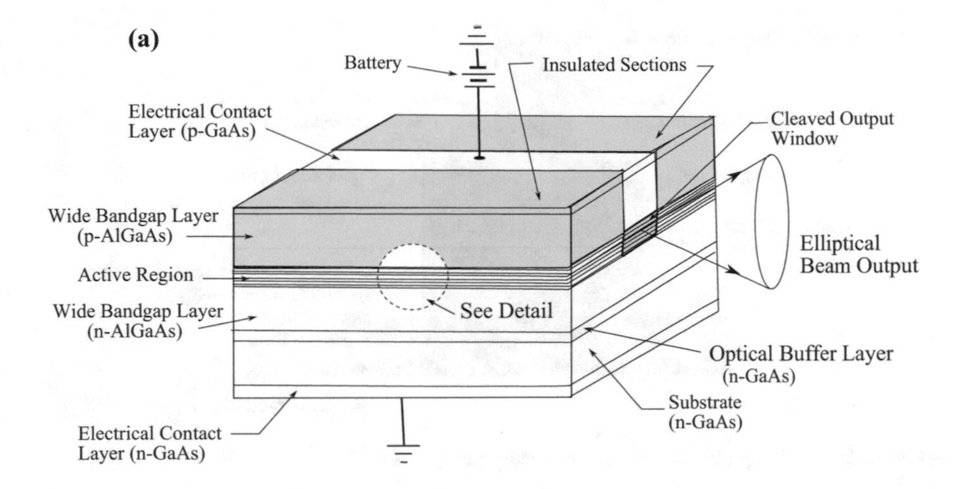

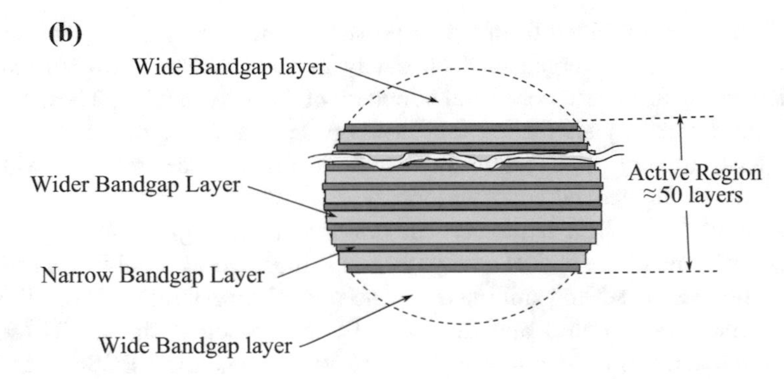

Fig. 6.18 **a** Layer construction of an electrically pumped quantum well heterostructure laser with an active bandgap layer. **b** Detail of the active bandgap layer showing the wide and narrow bandwidth layers

times the angular spread in the vertical direction than in the horizontal direction. This type of laser diode is referred to as an edge-emitting, quantum well injection diode laser. Other specialty laser diodes can be designed with emitted wavelengths over the range 400 nm to 10 μm, for continuous or pulsed operation, and power levels from less that 10 mw to about 5 w continuous wave to about 5-10 kw pulsed mode, with a linewidth of ≈2 nm at the higher powers. Commercial laser diodes are available having picosecond (10^{-12} s) pulse durations.

6.3.2 Vertical Plane Emitting Laser Diodes

In contrast to the edge-emitting laser diodes, the *vertical cavity surface emitting laser* (VCSEL) orients the laser cavity in the direction of current flow. The larger

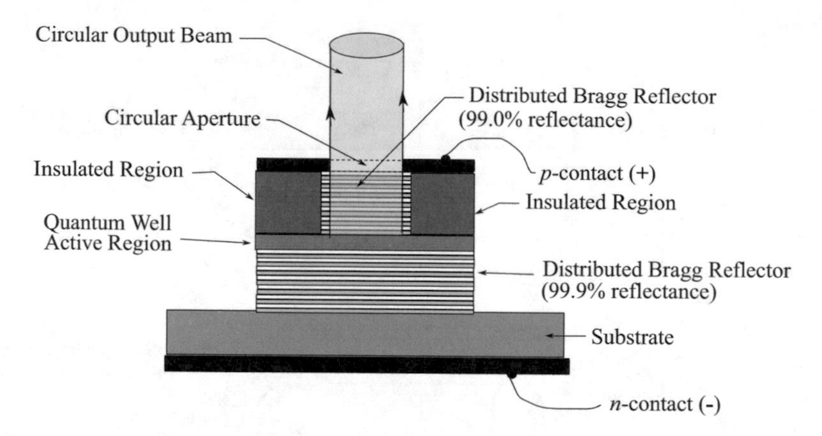

Fig. 6.19 Operating principle of an electrically pumped VCSEL

diameter circular beam emitted from the larger surface area provides higher output power and lower beam divergence. Each cavity mirror consists of a *distributed Bragg reflector* (DBR) at each horizontal boundary of the active region. These DBR cavity mirrors, formed by alternating thin layers of high and low refractive index materials, have a very high reflectance over a narrow spectral bandwidth. Figure 6.19 shows the general construction.

A variation of the VCSEL is the *vertical external cavity surface emitting laser* (VECSEL). Here one of the optical cavity mirrors is mounted external to the active region at a distance of several millimeters. The use of an external mirror allows better transverse mode control and the VECSEL can produce circular TEM_{00} diffraction limited beams of high power. Both the VCSEL and the VECSEL can be designed for electrical or optical pumping. Optical pumping is preferred for the VECSEL, as the active region can be more uniformly pumped to produce a larger diameter and higher power output beam. See Fig. 6.20. VECSEL devices have been reported to produce up to 30 W of output power.

6.3.3 Laser Diode Pumped Solid-State Lasers

With the development of energy efficient laser diodes, most solid-state lasers are optically pumped by laser diodes instead of electrical flashlamps or arc lamps. The laser diode pump can be more easily integrated into the laser package, has a longer lifetime, and is about 20 times more efficient than a flashlamp. The first diode pumped laser was demonstrated by Keyes and Quist in 1964 at the MIT Lincoln Lab. It used a $CaF_2{:}U^{3+}$ rod crystal, and was side pumped by five GaAs laser diodes at $\lambda = 840$ nm to produce laser radiation at $\lambda = 2613$ nm [24]. Common solid-state lasers that are end-pumped using a laser diode are the ruby laser and the Nd:YAG

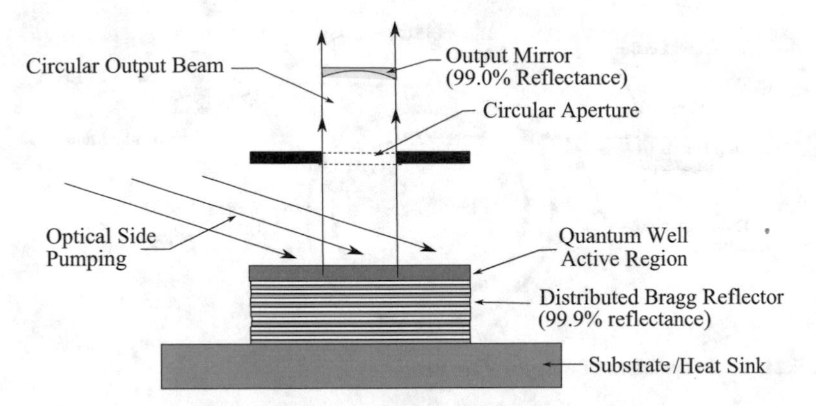

Fig. 6.20 Operating principle of an optically pumped VECSEL

laser and are referred to as *diode pumped solid-state* (DPSS) lasers. Very high-power gas lasers continue to be mostly flashlamp pumped.

For the popular Nd:YAG DPSS laser, a powerful pump GaAlAs laser diode emits a wavelength $\lambda = 808$ nm, which has been tuned to couple the absorbed energy in the crystal

with the energy required to produce the $\lambda = 1064$ nm Nd:YAG crystal transition. Multiple laser diode pumps can be stacked in a two-dimensional array (rack and stack array) to bring the pump energy into the kW range, and many DPSS lasers can either be end pumped or side pumped. DPSS lasers are available with pulsed energy outputs in the 1 J range.

6.4 Some Lasers with Special Properties

6.4.1 Fiber Lasers

The solid-state *fiber laser* was conceptualized around 1960 by Elias Snitzer at American Optical [25]. A fiber laser consists of a length of thin silica glass fiber where the active medium is doped with a rare-earth ion. The small diameter doped fiber core is surrounded by a waveguide cladding material of lower refractive index. The first single core fiber laser was demonstrated in 1961 by Snitzer and Hicks, using the lanthanide ion Nd^{3+} in the doped medium with flashlamp pumping. The output wavelength range was 1030–1100 nm.

Figure 6.21 illustrates the basic operation of a coiled core fiber laser where dielectric input and output mirrors are positioned at the fiber ends. Most modern fiber lasers have a double-core structure with a cladding of low refractive index, where the undoped outer core acts as the waveguide and the smaller diameter doped inner core produces the laser light. End-pumped light from a low-cost laser diode passes through a dielectric bandpass mirror having high transmission at the laser

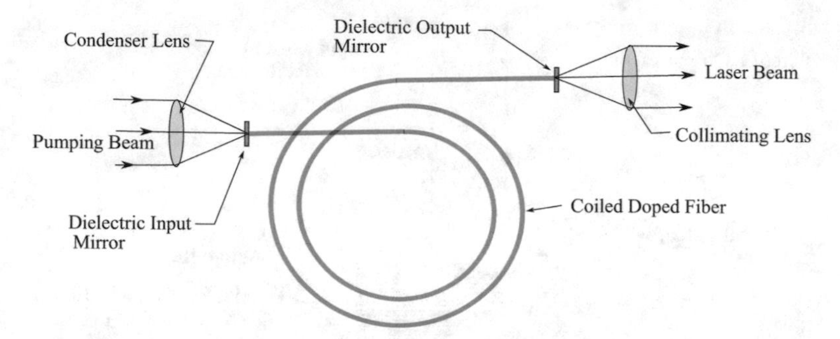

Fig. 6.21 Basic operation of a coiled fiber laser

diode pump wavelength and high reflectance at the fiber core lasing wavelength, and enters the end of the outer pump core. By multiple reflections at the cladding wall, excitation light passes through the inner doped core where lasing transitions occur. During multiple reflections at the cladding interface, a circular cross-section outer core reflects fewer pump photons into the inner core gain region. A rectangular or truncated circular shape of the outer pump core is preferred, since this geometry reflects more light into the inner doped core gain region, resulting in higher lasing efficiency [26]. See Fig. 6.22a and b. The dielectric output mirror has ≈99% reflectance at the fiber core lasing wavelength. Double-core fiber lasers can sustain single transverse mode laser operation.

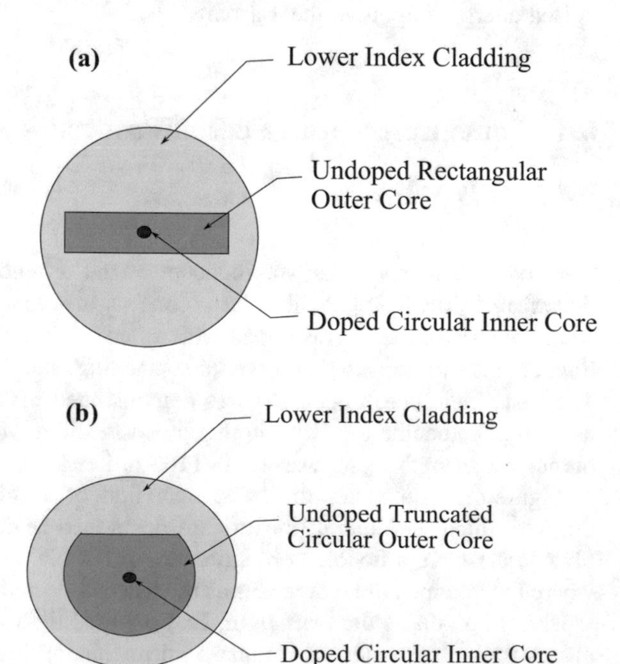

Fig. 6.22 **a** Cross-section of a dual-core fiber with a rectangular shaped outer pump core. **b** Cross-section of a dual-core fiber with a truncated circular shaped outer pump core

For current lanthanide-based lasers, popular doping core materials use the erbium Er^{3+} ion, emitting in the 1540–1560 nm wavelength range, and the ytterbium ion Yb^{3+} with emitting wavelengths in the 980–1070 nm range. These fiber lasers are usually pumped using low-cost laser diodes of appropriate emitting wavelengths. For example, a ytterbium doped fiber laser will emit at $\lambda = 1030$ nm at a diode pump wavelength $\lambda = 940$ nm with a photon conversion efficiency greater than 90% [27].

Output power in the kilowatt range can be obtained using long fibers, and diffraction limited beams can be achieved. Fiber lasers can be Q-switched and either operated continuously or in ultrafast pulse modes. Since the fiber laser is flexible, the output beam direction need not be the same direction as the input pumping beam.

6.4.2 Doped Fiber Amplifiers

A related application of fiber lasers is the *doped fiber amplifier* (DFA). Here the light to be amplified enters one end of the fiber laser and the lasing action of the fiber laser produces significant amplification of the input light. Around 1963 Koester and Snitzer measured the amplification of light from a glass rod laser by a doped fiber laser amplifier [28]. Figure 6.23 illustrates the operating principle of a setup for light amplification of the 1060 nm laser line from a 30.5 cm long neodymium doped core glass rod with a 6.35 mm diameter core and 9.5 mm diameter cladding. The amplifying neodymium doped fiber was coiled as a helix and optically side-pumped in a pulsed mode by a linear flashlamp on the central axis of the coiled helix. Gain was achieved when the doped core underwent stimulated

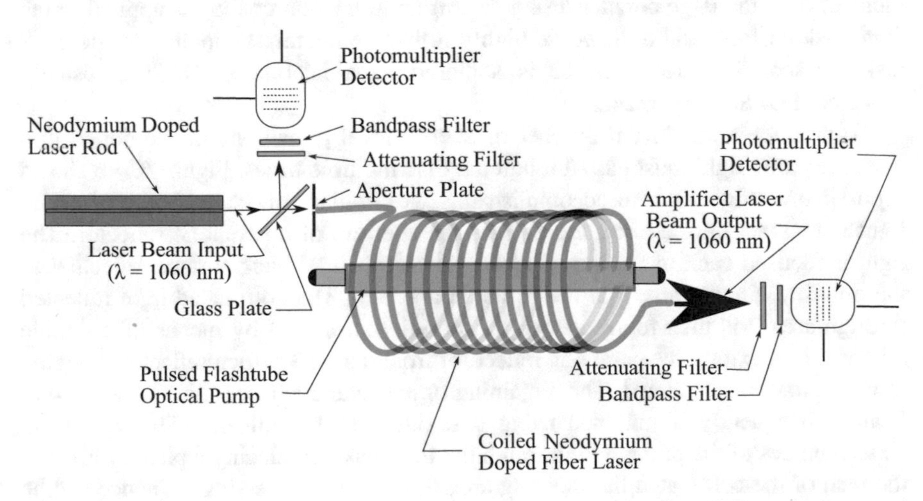

Fig. 6.23 Basic components for the measurement of the gain of a doped fiber amplifier

emissions having the same wavelength as the input wavelength. The gross gain was calculated by measuring the ratio of the output beam intensity from the fiber laser in the pumped and unpumped states. The 1 m long fiber had an active core diameter ≈0.01 mm with a 0.75–1.5 mm diameter cladding. Since resonant cavity lasing in the coiled doped fiber would reduce the gain, fiber laser oscillation was prevented by antireflection coatings on the coiled fiber ends, or by beveling one end of the fiber. A 0.1 mm diameter aperture confined the input beam to the fiber core area, and attenuating filters kept the 1060 nm radiation in a non-saturating range for the fiber laser and photomultiplier detectors. Auxiliary bandpass filters rejected stray light by transmitting only 1060 nm radiation to the photomultiplier detectors. Using this technique, gross gains up to 8.5×10^4 were calculated for the fiber amplifier. Other types of doped fiber amplifiers are currently available using erbium ions as the gain medium (EDFA). These are usually end-pumped by laser diodes using dichroic pump couplers to merge with the light to be amplified. Amplification occurs in the 1550 nm region using a laser diode pump wavelength of 980 nm.

6.4.3 Thin Disc Lasers

The solid-state *thin disk laser* (sometimes called a disk laser or an active mirror laser) uses a thin disk of doped material as the gain medium. The laser beam is emitted normal to the circular disk surface. The first thin disk laser was developed at the University of Stuttgart in the early 1990s using a Nb:YAG disk [29]. It emitted at a laser wavelength $\lambda \approx 1064$ nm using a pump wavelength $\lambda \approx 809$ nm. Another popular disk material having a high gain bandwidth is Yb:YAG, with a disk thickness typically 100–200 μm, and an emission wavelength $\lambda \approx 1030$ nm. A suitable laser diode pump wavelength is $\lambda \approx 940$ nm, and this pump light is incident over the large circular disk area. An antireflection coating is applied to the front disk surface, and a dielectric highly reflective coating is applied to the back disk surface. The back surface is soldered to a heat sink which is usually water-cooled. See Fig. 6.24a.

To produce a significant number of energy level transitions in the doped thin disc, the pump light must pass through the disc multiple times. Figure 6.24b shows a possible configuration to accomplish this. A circular collimated beam of pump light is incident on the circular area **1** at the aperture of a parabolic reflector. The light is focused back to the thin disc where the mirrored back surface reflects it to the opposite circular area **2** at the parabola aperture. The collimated light reflected through area **2** is then retroreflected to the adjacent area **3** by means of a double mirror. Light exiting the parabolic reflector through area **3** is then reflected from the thin disk to opposite area **4**. The remaining opposite and adjacent circular areas **4** to **8** are subsequently illuminated using two other double mirrors. This results in sixteen passes of the pump light through the thin disk. By placing a planar mirror in the path of the collimated light exiting area **8**, the entire reflecting sequence can be reversed, resulting in thirty-two passes of the pump light through the thin disk. In

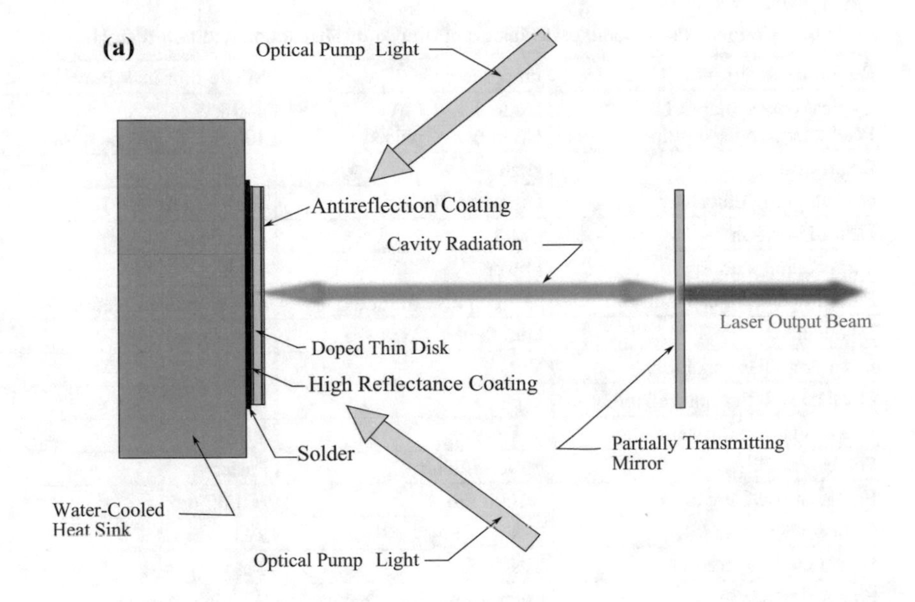

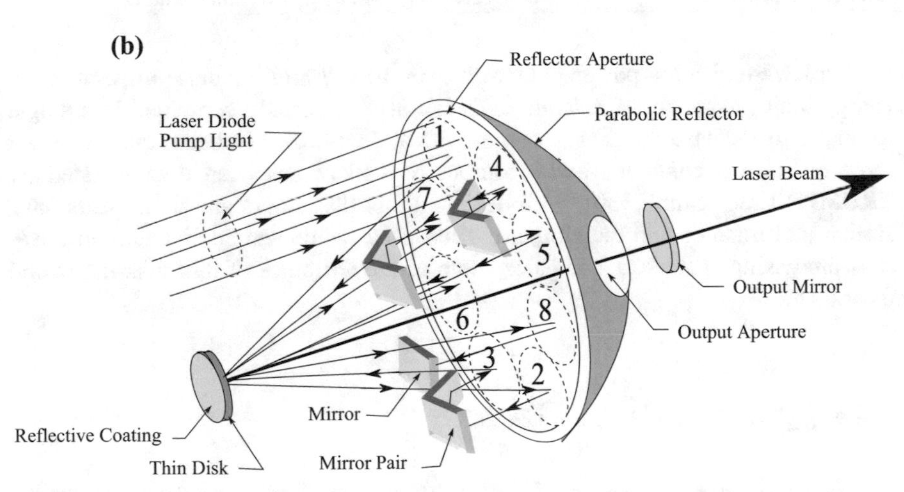

Fig. 6.24 **a** Basic construction of a thin disk laser. **b** Illustration of optical configuration to achieve 32 pump light passes through a thin disk

this manner, over 90% of the directed pump energy can be absorbed by the thin disk. Stimulated emission light from the thin disk passes through a circular opening at the parabola vertex, and a partially transmitting output mirror forms the resonant cavity to produce the laser light.

Thin disk lasers have some advantages over fiber lasers. The thin disk can be efficiently cooled and has a uniform radial temperature distribution from the disk center. Increasing the pumped area on the disk allows higher power levels of pumping light to produce higher laser output. For CW operation of a single disk the

Table 6.1 Comparison of some performance attributes of fiber lasers and thin disk lasers

Performance attribute	Fiber laser	Single thin disk laser
Typical power output CW Peak energy pulsed output	Up to 1 kW CW Up to 5 kW pulsed	4–5 kW 100 mJ @ 300 ns duration
Beam quality	High	Good
Optical pump efficiency	High (60–70%)	High (up to 90%)
Heat dissipation	Good	Excellent
Temperature stability	Lower	High
Pulsed operation capability	ns—ps pulse duration	ns—ps pulse duration
Linewidth	Less than 1 kHz	Less than 1 kHz
Laser amplifier capability	Yes	Yes
Physical stability and reliability	High	High
Wavelength tuning capability	Yes	Yes
Power scalability	More difficult	Easier
Emission wavelength	≈1000 nm	≈1000 nm
Q-switching capability	Yes	Yes
Single mode operation	Yes	Yes
Practical applications	Micromachining, drilling	Welding, cutting

scaling laws indicate a power limit well over 10 kW and for pulsed operation an energy limit greater that 1 J. Thin disks can also be serially combined in a single resonator to increase the energy output with little sacrifice in beam quality. For a three disk laser, continuous wave output of 14 kW has been demonstrated by TRUMPH Laser GmbH [30]. Well-designed thin disk lasers and fiber lasers have similar performance and therefore are often used for related applications in materials processing. Table 6.1 compares some of the attributes of thin disk lasers and fiber lasers.

6.4.4 Quantum Cascade Lasers

The *quantum cascade laser* is related to the quantum well laser diode. However, these heterojunction lasers do not require both p-type and n-type semiconductor materials. For that reason they are referred to as *unipolar*, consisting usually of *n*-type semiconductor materials. The laser emission wavelengths are entirely determined by the width and spacing of quantum wells, having widths from a fraction of a nanometer to several nanometers. When an electrical bias is applied, population inversion occurs for energy levels in the quantum wells. Photon emission from the "active" region results from electron tunneling transitions through "barrier" layers to lower energy subbands, e.g. $E_3 \rightarrow E_2$ and $E_2 \rightarrow E_1$ in sequential quantum wells. A "relaxation" region, consisting of another series of quantum wells, restores the relative energy levels in the next active region. These active/relaxation regions are repeated, usually up to 25 times, and the subsequent energy level transitions produce a "cascading" intensity of laser light emitted from the material. The first quantum cascade laser was demonstrated in 1994 at Bell Labs, emitting at λ = 4.2 μm in the mid infrared [31]. Mid infrared 0.038 GHz (8 μm) to 0.10 GHz (3 μm) continuous wave quantum cascade lasers operating at room temperature can produce output power at watt levels [32].

The conduction band diagram of Fig. 6.25 illustrates the process for three repeating active/relaxation regions and the emitted photons. Figure 6.26 shows a prototypical structure of a quantum cascade laser [33]. The semiconductor layers are precisely deposited by molecular beam epitaxy (MBE), and in this representation some sections designate single layers and other regions. Cladding layers confine the emitted light along the horizontal direction, and the laser light is edge emitted through cleaved end facets having a reflectance ≈0.27.

Quantum cascade lasers have been shown to be the most useful source of terahertz (THz) frequency laser light (often called *T-rays*). Terahertz (one trillion or 10^{12} Hz) radiation lies in the frequency range of 0.3 THz (λ = 1000 μm) to 10 THz (λ = 30 μm). See Fig. 2.1. Both continuous and pulsed quantum cascade lasers have been demonstrated over a frequency range 0.84 THz (λ = 357 nm) to 5.0 THz (λ = 60 μm). Terahertz quantum cascade lasers in this frequency range are generally cryogenically cooled to sustain lasing action. For example, a maximum temperature for continuous operation over this range is 117 K for 130 mW output power, and a maximum temperature for pulsed operation over this range is 169 K for 250 mW output power. Since terahertz radiation can penetrate materials without damage, terahertz quantum cascade lasers find applications in medical and security

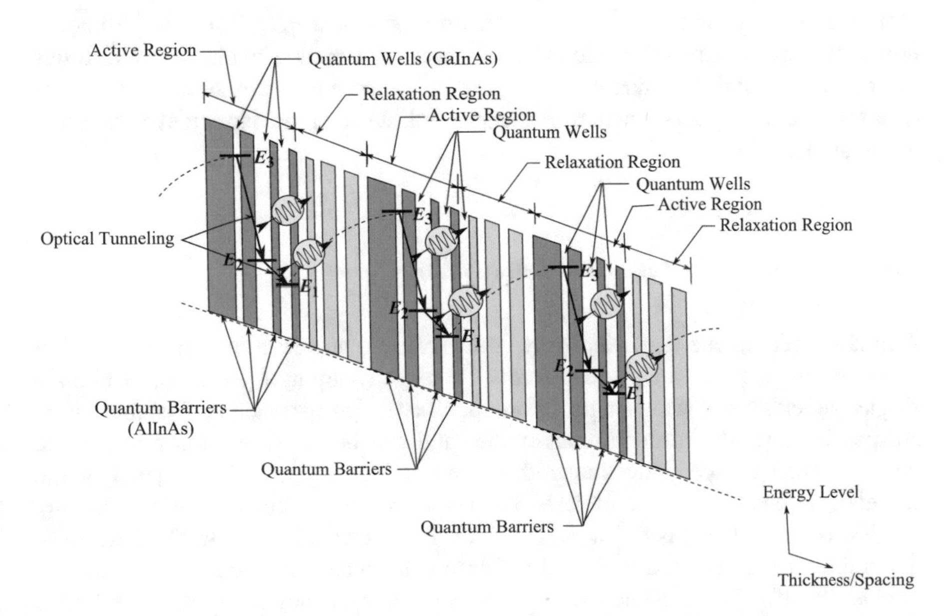

Fig. 6.25 Illustration of quantum cascade laser conduction band diagram for three repeating sections

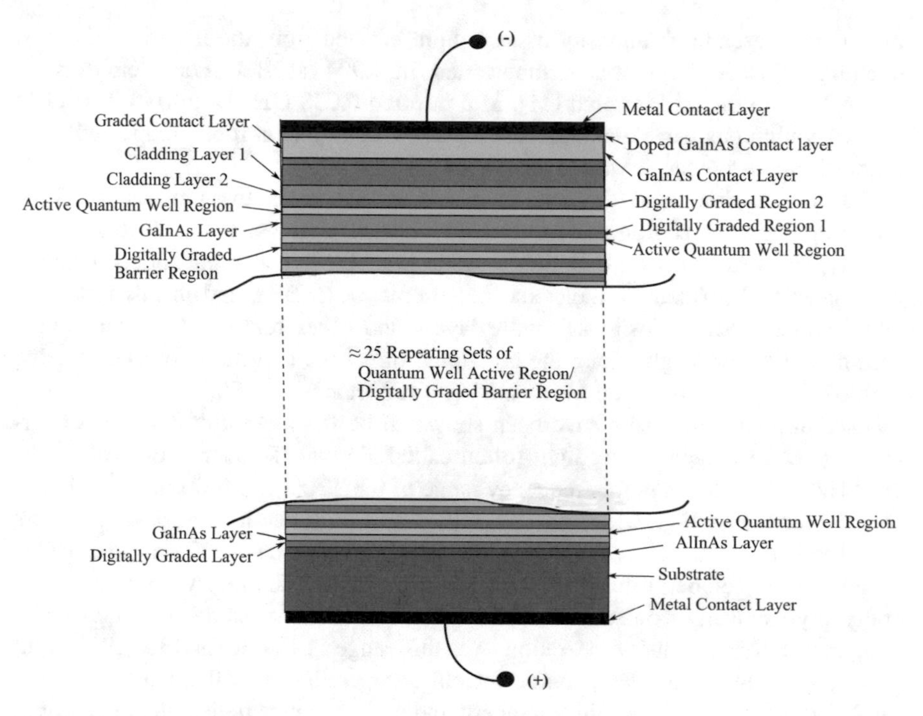

Fig. 6.26 Typical construction of a quantum cascade laser

scanning, cancer detection, mid-IR spectroscopy, astrophysical interferometry, non-destructive testing of semiconductor chips, and hidden explosives and drugs detection. Benjamin Williams has written a comprehensive review on terahertz quantum cascade lasers, including the present stage of development and future expectations [34].

6.4.5 Tunable Lasers and Linewidth

A tunable laser is one that selectively emits different wavelengths smoothly over a range, or emits discrete wavelengths over a range (hopping operation), or emits a single predetermined wavelength using specific design parameters. Tunable lasers are usually operated in continuous mode, and can be classified according to the spectral range covered, the linewidth of the emitted light, and the speed of the wavelength change. A useful laser beam parameter for tunable lasers (or any single-frequency laser) is the laser linewidth. It represents the spectral half-width of the emitted laser line, and is related to beam coherence and phase noise created in the laser itself. When the linewidth is expressed in frequency units [Hz], it is called the *frequency linewidth* Δf. When it is expressed in wavelength units [m], it is called the *spectral linewidth* $\Delta\lambda$. The relationship between the frequency linewidth and the spectral linewidth is related to the uncertainty principle and can be approximated by the following expression:

$$\Delta\lambda \cong \left(\frac{\lambda^2}{c}\right)\Delta f, \tag{6.7}$$

where the speed of light $c = 2.998 \times 10^8$ m/s [35, 36].

Narrow linewidth lasers are loosely grouped as having a frequency linewidth $\Delta f < 1$ GHz, while ultra-narrow linewidth lasers can be further specified for $\Delta f <$ 50 kHz. Ultra-narrow linewidth lasers require single mode operation and special techniques to maintain cavity stability and spectral purity. A 657 nm wavelength diode laser with a frequency linewidth $\Delta f < 1.5$ Hz. has been reported by scientists in Germany [38]. Table 6.2 shows the achievable linewidths of several single wavelength laser types. Table 6.2 shows the frequency and spectral linewidths for several fixed wavelength laser types.

Several types of lasers are continuously tunable over a wavelength range from several nanometers to tens of nanometers. The dye laser was one of the first continuously tunable lasers, and can be tuned using multiple prisms, gratings, or birefringent filters. Pulsed dye lasers can also be tuned, maintaining a narrow linewidth pulse. Theo Hänsch at Stanford University has described a pulsed dye laser, tunable from near UV through visible, and having a spectral linewidth $\Delta\lambda = 4 \times 10^{-4}$ nm [39].

The Ti:sapphire (Ti:Al_2O_3) laser was developed by Peter Moulton in 1982 at the MIT Lincoln Lab, using titanium as the ion dopant, and sapphire as the host crystal. A tunable Ti:sapphire laser was first reported by Rapoport and Khattak in 1986. This laser has a wide gain spectrum and can be tuned in continuous mode over a wide spectral range in the near and mid infrared, $\lambda \approx 650$ nm to $\lambda \approx 1100$ nm, while maintaining a narrow frequency linewidth $\Delta\lambda$. The tuned wavelength is selected by dispersing prisms or gratings.

Table 6.2 A comparison of linewidths for several single wavelength laser types

Laser type	Emission wavelength λ (nm)	Frequency linewidth Δf	Spectral linewidth $\Delta\lambda$ (nm)
Ruby crystal	694.3	330 GHz	0.530
Nd:YAG solid-state	1064	120 GHz	0.453
HeNe gas	632.8	1.5 GHz	2.0×10^{-3}
Optimized solid-state dye [39]	590	350 MHz	4.1×10^{-4}
Er/Yb doped fiber [37]	1555.5	≈1 kHz	$\approx 8.1 \times 10^{-9}$
Single mode locked diode [38]	657	<1.5 Hz	$< 2.2 \times 10^{-12}$

Tuned semiconductor laser diodes are available with center wavelengths from about 400–2800 nm. They can be tuned by several methods. One method uses a frequency-selective grating embedded in the semiconductor structure. For example, a 1040 nm laser diode can be tuned over a 30 nm range with a frequency linewidth $\Delta f \approx 100$ kHz. Laser diodes can also be tuned by varying the refractive index of the laser cavity by precise control of the diode temperature. Other types of lasers that can be tuned are fiber lasers in the near infrared, and quantum cascade lasers in the near and far infrared.

6.4.6 Fast and Ultrafast Pulsed Lasers

Pulsed lasers can be grouped into several categories, depending on the pulse repetition rate and the pulse duration time. If the pulse repetition rate and duty factor are high, the output appears to be almost continuous, and the laser is referred to as "quasi-CW". Lasers having pulse durations in the nanosecond (1 ns = 10^{-9} s) to picosecond (1 ps = 10^{-12} s) range are called *fast lasers*, while lasers having pulse durations in the femtosecond (1 fs = 10^{-15} s) to attosecond range (1 as = 10^{-18} s) are called *ultrafast lasers* or *ultrashort pulse lasers*. Femtosecond ultrafast pulses are in the time frame of molecular dynamics, and can be used to visually "freeze" this motion. Attosecond pulses are in the time domain of the faster motion of electrons, and can visually stop electron motion in time. Pulsed lasers are capable of providing high energy over the pulse duration. The ns and ps pulsed lasers are usually operated with Q-switching. Q-switching can provide high pulse energies, but produces lower pulse repetition rates and longer pulse durations that do not reach the femtosecond level.

For solid-state lasers, femtosecond pulses are usually produced by a technique called *mode-locking* or *phase-locking*. In mode-locking, all possible modes in a laser cavity have the same relative phase. The number of modes is determined by the gain bandwidth of the lasing medium and the length of the resonant cavity (Sect. 6.1.1). Mode-locking is achieved by the intracavity insertion of an optical device that produces lower energy losses, or higher gain in the cavity than for CW operation, producing a very short pulse that moves back-and-forth in the cavity. Short duration pulses require a high gain bandwidth, low group dispersion in the cavity, and a suitable type of mode-locking device. Similar to Q-switching, mode-locking can be active or passive. Active mode-locking uses an electronic controlled modulating device synchronized with the round-trip travel time of light in the cavity, producing a single moving light pulse. Active mode-locking was demonstrated in 1964 for a HeNe laser at Bell Laboratories using an ultrasonic diffraction grating mounted inside the laser cavity [40]. Popular methods of passive mode-locking use a bleachable optical element, such as a saturable Bragg reflector (SBR), or a semiconductor saturable absorber mirror (SESAM). The SESAM preferentially reflects the higher intensity peaks during many light traversals and these peaks are superposed to produce an amplified series of high intensity pulses.

A SESAM device acts as a near-perfect reflector when the peak pulse power exceeds a certain value, reaching saturation [41]. The SESAM has the advantage that the semiconductor material has a broad absorption range, an fast recovery time, and a saturation value that can be controlled. For certain lasing materials, a self-originating optical Kerr lens is used for *Kerr lens mode-locking* (KML), where a Gaussian shaped power pulse produces a transverse refractive index gradient across the gain medium, and the beam can be self-focused to a small internal aperture. The aperture truncates the pulse wings, shortening the pulse duration. The Kerr lens effect acts as a fast saturable absorber. Ultrafast pulses can be produced using crystal lasers, fiber lasers, disk lasers, and semiconductor diode lasers, generating pulses of femtosecond duration.

When the pulses maintain a stable temporal and spectral shape during their propagation, they are called *solitons*. Soliton pulses can be created by minimizing the spectral dispersion in the laser cavity. One method introduces anomalous dispersion in the cavity, which counterbalances the Kerr nonlinear dispersion of the lasing material. The resultant mode-locking with soliton formation produces a *mode-locked soliton laser*. The generated pulses are usually in the femtosecond duration range.

Pulse duration is limited by refractive dispersion in the laser gain medium, producing different velocities for the spectral components of the beam. This normal or positive dispersion, called *group velocity dispersion* or *group delay dispersion* [fs^2], broadens the laser pulse and ultimately limits the spectral linewidth and pulse duration. This positive dispersion can be compensated for by external multiple prisms producing controlled negative dispersion [42]. Also, gratings or special types of negative dispersion high-reflectance mirrors can be used. Mode-locked Ti: sapphire lasers using KML and SESAM elements with dispersion compensation have produced 5–10 fs pulse durations. In 1997 a research team in Switzerland and Germany demonstrated a 200 mW output Kerr lens mode-locked Ti:sapphire laser produced 6.5 fs pulses at a pulse repetition rate *PRR* ≈85 MHz [43]. The 2.3 mm thick, 0.25% doped laser crystal was end-pumped by a 5 W argon ion laser and used a broadband SESAM to initiate self-starting. See Fig. 6.27. A single double-chirped mirror and a variable negative dispersion prism pair supplied negative 314 fs^2 group delay dispersion to compensate for positive group delay dispersion in the laser system. A *chirped mirror* is a highly reflective mirror consisting of thin dielectric layers of high and low refractive index, where the layer depths are varied. It produces negative group delay dispersion. A *double-chirped mirror* also varies the ratio of the layer optical thicknesses, and has a broadband antireflection coating on the front surface [44]. The double-chirped mirror minimizes spectral discontinuities in the group delay dispersion, providing smoother dispersion compensation over a wider wavelength range.

A mode-locked thin disc laser using a SESAM and five dispersion compensating high-reflectance elements having 96 fs laser pulses at a 77.7 MHz repetition rate was described in 2012 by Saraceno [45]. The negative dispersion compensating elements are *Gires-Tournois interferometer* (GTI) mirrors. These GTI mirrors have a partially reflecting front surface and a back surface having a reflectance exceeding

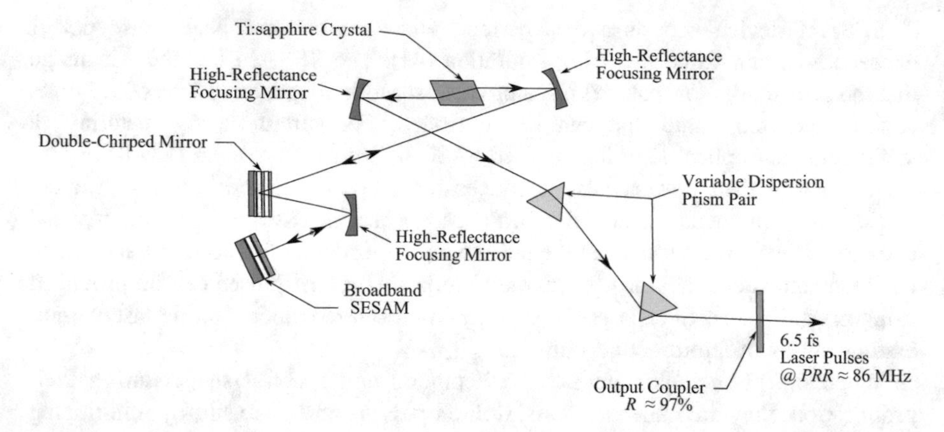

Fig. 6.27 Basic layout for a Ti:sapphire laser system producing 6.5 fs duration pulses. Pump laser is not shown. (Adapted from Ref. [43] with permission from the Optical Society of America)

99.97%, forming a phase changing resonator that introduces negative group delay dispersion. Figure 6.28 illustrates the architecture of this 96 fs pulsed thin disk laser. The 200 μm thick $Yb:LuScO_3$ disk material is mounted on a water-cooled 1.4 mm diamond substrate. A 976 nm wavelength stabilized pump diode with $\Delta\lambda < 0.5$ nm linewidth provides 24 passes through the disk with > 95% absorption efficiency, using the mirror arrangement as shown in Fig. 6.28. The GTI mirrors provide a total of -2800 fs^2 negative group delay dispersion per roundtrip.

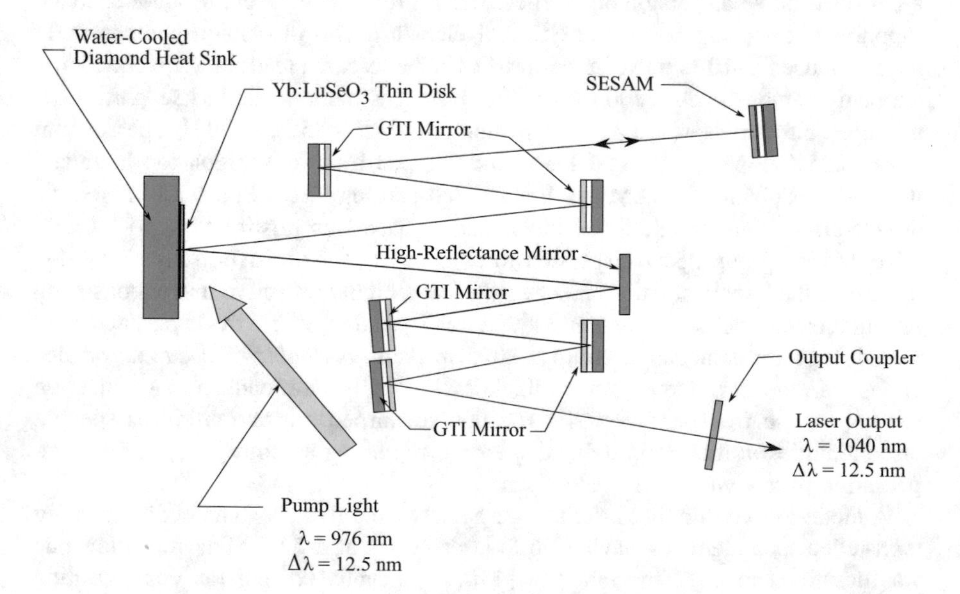

Fig. 6.28 Experimental setup for a multiple-pass thin disk laser to achieve sub-100 fs pulse durations

Laser pulses of attosecond duration represent the current frontier of short pulse generation capability, and are close to the time scale of electron motion (atomic unit of time = 24 as). Attosecond pulses are not created in conventional laser resonant cavities, but by the interaction of ionized electrons and a driving laser at the atomic level. Pulse durations in the attosecond region are shorter than possible with the ≈4 fs optical period of visible light, and occur in the extreme ultraviolet XUV (also called EUV) region, e.g. $\lambda \approx 13$ nm. One technique to produce attosecond pulses consists of bombarding an ionizing neon gas with an intense near infrared (NIR) driving laser. This might typically be a polarized Ti:sapphire laser operating at $\lambda \approx 750$ nm with a pulse duration ≈2.5 fs The oscillating trigger pulses from the driver laser produce acceleration of the neon electrons, and at sufficient field strength electrons are ejected over different trajectories from the atom at high speed, acquiring energy. However, when the field of the driving laser reverses polarity some of the electrons reverse direction and decelerate to recombine with the neon atom, returning to the ground state and emitting an XUV photon. Many photon pulses are emitted from the entire volume of noble gas, and they are coherent with significant additive energy.

The interaction of the trigger waveform and the electron motion is highly nonlinear. Electron ejection is confined to a small region near the maximum energy peak of the trigger pulse, with attosecond pulses emitted every half-cycle of the driving laser light. This results in a train of pulses, with a pair of pulses being emitted every cycle of the driver laser. For rare gas ions, harmonics are produced at odd multiples of the trigger pulse frequency, and these harmonics reach the high XUV and soft X-ray region of attosecond pulses. This technique for generating attosecond pulses is called high-harmonic generation (HHG).

In 2001 cooperating scientists in France and The Netherlands generated the first train of attosecond pulses with HHG, using a 40 fs pulsed 800 nm IR laser focused on a low-pressure argon gas jet. By measuring the relative phase difference of the generated harmonics, they found that the harmonics were locked in phase with a pulse duration ≈250 as [46]. To generate single isolated attosecond laser pulses from noble gases, it is necessary to have an intense phase stabilized or phase-locked trigger laser consisting of only a few wave cycles. Baltuska and cooperating scientists in Austria and Germany in 2003 demonstrated a stable triggering laser consisting of a few wave clusters with the carrier envelope peaks aligned with the pulse power peaks [47]. The laser consisted of a Kerr lens mode-locked Ti:sapphire type with external chirped mirror pulse compression, delivering 5 fs, 0.5 mJ pulses at a carrier wavelength of 750 nm. The negative group dispersion of the chirped mirror was used here to temporally compress a dispersed wave pulse in the trigger laser. Using a few-cycle pulse Ti:sapphire laser with ≈100 eV energy, in 2004 scientists in Germany and Austria generated and measured isolated 250 as, 13 nm XUV pulses from an ionized neon jet [48]. Isolated XUV pulses of ≈80 as duration and ≈0.5 nJ energy were generated in 2008 by a scientific team in Germany and California by controlling the sub-1.5 cycle phase controlled waveform of a linear polarized 3.3 fs, 720 nm NIR driver laser [49]. To obtain these single isolated 80 as pulses, only the driver laser half-cycle having the higher intensity in the cycle was

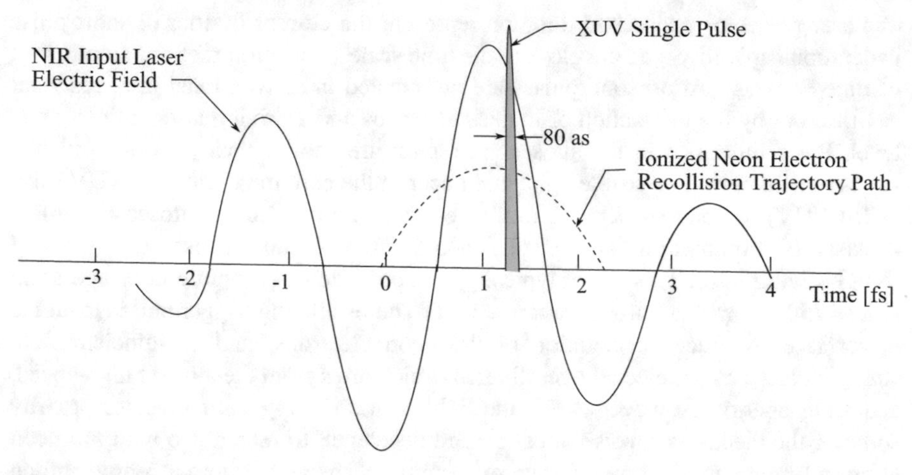

Fig. 6.29 Illustration of the creation of an 80 as XUV single pulse from neon gas ionized by a driving NIR laser

used. This intensity peak was then aligned with the peak of the electron recollision trajectory curve, creating only the highest energy XUV photons. See Fig. 6.29. Light pulses of attosecond duration would allow observations of electron motion processes in the interiors of atoms. Zhao and co-workers at the University of Central Florida have produced an XUV laser pulse duration of 67 as using a double optical grating [50]. In the future it is expected that isolated laser pulses less that the 24 attosecond time scale of electron motion will be generated, and perhaps even in the zeptosecond (1 zs = 10^{-21} s) range. Typical or achieved pulse and energy parameters for several laser types are summarized in Table 6.3.

6.4.7 Surface Plasmons and Nanolasers

Another type of laser where the lasing operation occurs at sub-wavelength levels is the *plasmonic laser* or *nanolaser*. The concept of a nanolaser was proposed in 2003 by David Bergman in Israel and Mark Stockman in the United States. They theorized that a coherent electric field could be created by electron vibrations in a metal/dielectric medium from quasi-static resonances, or eigensolutions of Maxwell's equations [51]. These resonances would have a spatial size smaller than the wavelength of the emitted electromagnetic field. This type of laser is often called a SPASER (surface plasmons amplified by stimulated emission of radiation).

It was realized that the free electron oscillating waves on a surface, so-called *surface plasmons*, could combine with a light wave to form *surface plasmon polaritons* (SSPs). These SSPs could then be guided in metallic nanostructures adjoining a dielectric gain medium and produce a nanolaser. However, the absorption loss in pure metals at optical frequencies could not produce the required

Table 6.3 Typical or achieved pulse and energy parameters for several laser types. Depending on the application, the pulse energy or average pulse power output may be the appropriate measure of performance

Laser type	Wavelength	Pulse repetition rate	Pulse duration	Pulse energy or Av. pulse power
Mode-locked HeNe [40]	632.8 nm	≈1 kHz	300–500 ps	
Eximer	157–351 nm	10 kHz	300 ns	1000 mJ
Mode-locked semiconductor diode		Several GHz		
Mode-locked Erbium fiber	1535 nm	Several GHz	<100 fs	Several mW
Yb:YAG Q-switched thin disk [30]	1030 nm	10 kHz	300 ns	100 mJ
Yb:YAG regenerative amplifier thin disk [30]	1030 nm	Several MHz	25 ns	280 mJ
Nd:YAG Q-switched	1064 nm	5–500 kHz	10–25 ns	Hundreds of mW
Nd:YAG Mode-locked	1064 nm	MHz–100 GHz	6.6–10.3 ps	≈300 mW
Yb:$LuScO_3$ mode-locked thin disc with SESAM [45]	1040 nm	77.5 MHz	96 fs	5.1 W
Ti:sapphire mode-locked with KML and SESAM [43]	750–800 nm	≈85 MHz	5–10 fs	≈200 mW
Ti:sapphire trigger and electron recollision [48]	13 nm (XUV)	–	250 as	–
Ti:sapphire trigger and electron recollision [49]	13 nm (XUV)	–	80 as	–
Double optical grating [50]	XUV super-continuum	–	67 as	–

oscillations and feedback for lasing operation. The first experimental evidence of laser type emission using SSPs was observed in 2008 using optically pumped molecules at a polymer film/silver film interface. In 2009 a research team demonstrated a true nanolaser using 44 mm diameter gold core nanoparticles surrounded by a dye doped silica shell [52]. The oscillations of the surface plasmons produced a nanolaser having a visible wavelength λ = 531 nm.

Shortly thereafter, a nanolaser was demonstrated at the University of California, Berkeley, by producing stimulated amplification of SSPs in a waveguide of nanometer thickness [53]. The waveguide was a 5 nm thick MgF_2 insulating film between a 129 nm diameter cadmium sulfide nanowire and a silver substrate. See Fig. 6.30. The high-gain semiconducting CdS nanowire was pumped with a focused beam from a pulsed Ti:sapphire laser at λ = 405 nm having sufficient power for laser oscillation of the CdS emission line at λ = 489 nm. The device required cryogenic cooling at a temperature <10 K. The resulting confined optical

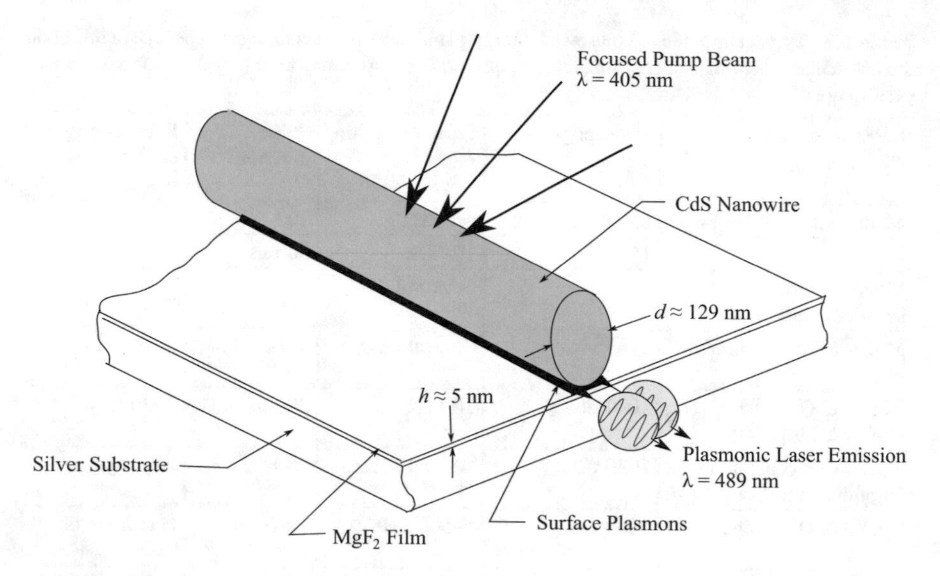

Fig. 6.30 A nanolaser emitting light at $\lambda = 489$ nm, using a hybrid plasmonic waveguide consisting of a thin insulator between a CdS nanowire and a silver substrate

modes in this *hybrid plasmonic waveguide* produce the emitted visible coherent light of this nanolaser.

By increasing the cavity feedback and decreasing metal absorptive losses, a nanolaser operating at room-temperature was demonstrated in 2011, also at the University of California, Berkeley [54]. The construction is illustrated in Fig. 6.31. A CdS square having ≈ 1 μm sides and 45 nm thickness sits atop a silver substrate layer, with a 5 nm MgF_2 interface. The interface forms a cavity between the CdS and Ag layers. Laser light at $\lambda = 405$ nm from a pulsed Ti:sapphire laser was focused to an area larger than the CdS nanosquare. The generated surface plasmons with TM oscillations were confined to the cavity by total internal reflection. This sub-diffraction mode confinement provided strong feedback for single-mode laser emission of resonant wavelengths of 495.5 and 508.4 nm visible light.

The first continuous wave (CW) visible light semiconductor nanolaser was reported in 2012 by a research team in Taiwan, the United States, and China [55]. As shown in Fig. 6.32, a plasmonic cavity was formed between an atomically smooth epitaxially deposited 28 nm thick silver film and a single nanorod in contact with the Ag film. The hexagonal cross-section nanorod consisted of a 480 nm long shell of GaN, with a 170 nm long core of InGaN as the green-emitting gain medium. A 5 nm thick SiO_2 dielectric layer over the Ag film preserved the Ag film smoothness and formed an atomically smooth plasmonic nanocavity between a flat side of the nanorod and the silver film. The nanorod was optically pumped by a CW laser diode at $\lambda = 405$ nm, and the nanolaser was cryogenically cooled to a temperature of 78 K. The generated coherent SPPs produced continuous wave bimodal laser radiation which was emitted from the edges of the cavity at wavelengths

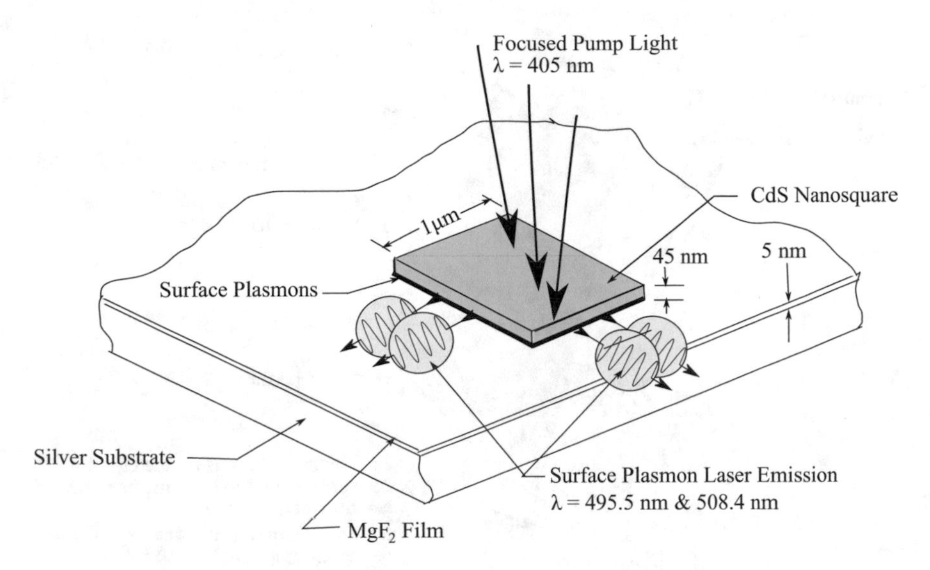

Fig. 6.31 Illustration of a room-temperature nanolaser consisting of a thin CdS nanosquare, a thin MgF_2 interface, and an Ag substrate

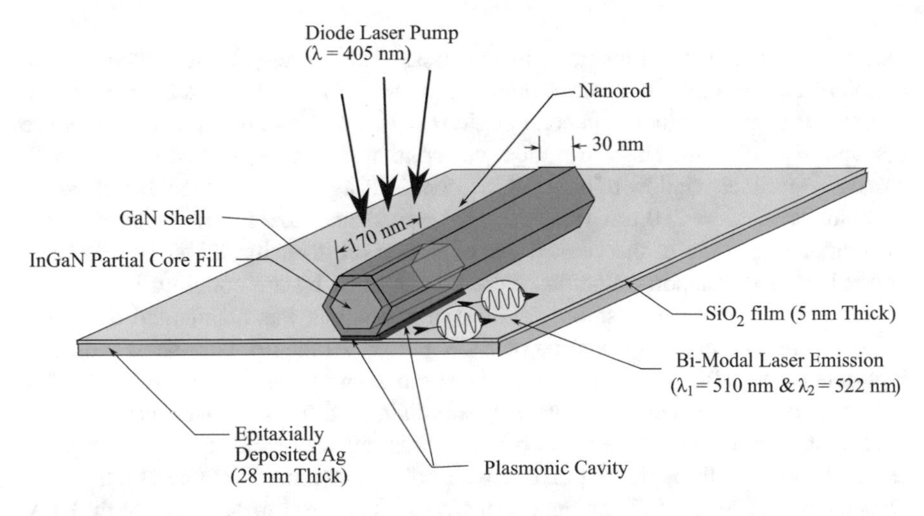

Fig. 6.32 Operating principle of a continuous wave nanolaser using a nano-rod and epitaxially grown silver film

$\lambda = 510$ nm and $\lambda = 522$ nm. The radiation produced by this SPASER was polarized in the direction of the nanorod axis.

Plasmonic laser devices also have the capability of producing attosecond duration pulses in the XUV region. This has been accomplished by In-Yong Park and team members in South Korea, Germany, and the United States, by constructing a tapered silver waveguide for surface plasmon polaritons [56]. The conical

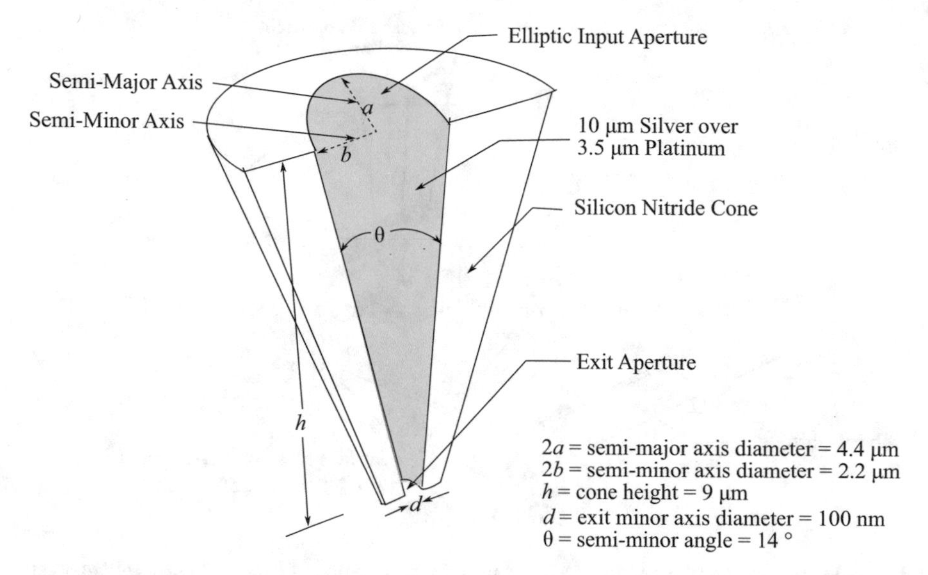

Fig. 6.33 Construction of a nano-conical waveguide for harmonic generation of ultrashort XUV light pulses

waveguide geometry is illustrated in Fig. 6.33. The waveguide described had an elliptical cross-section with an elliptic ratio $r = b/a = 1.1\ \mu\text{m}/2.2\ \mu\text{m} = 0.5$, a height h = 9 μm, a minor-axis cone angle θ = 14°, and a minor-axis exit aperture diameter d = 100 nm. The waveguide was contained in a neon-filled gas cell with an Al_2O_3 entrance window to transmit incident NIR light from a 75 MHz pulsed Ti: sapphire laser at λ = 780 nm This NIR light was focused to a 5 μm diameter spot at the entrance aperture of the conical waveguide at an intensity $\approx 5 \times 10^{11}$ W/cm^2. Guided surface plasmon polaritons were then formed by the coupling of the NIR energy with the surface plasmons. A flow of xenon gas was maintained down the guide for high harmonic generation of XUV pulses. At the tip, ultra-short pulses of high-intensity XUV radiation in the 10–70 nm range were emitted at a 75 MHz pulse repetition rate. The peak intensity was increased fourfold over the incoming NIR light. The longer wavelength NIR light was reflected backward, and the pulse exited the gas cell through a 1 mm diameter hole. The emitted light had high spatial dispersion, but high spatiotemporal coherence. This nanometer-scale high XUV intensity at the tip of the cone is very useful for applications such as near-field spectroscopy.

In 2012 scientists in France demonstrated the generation of attosecond XUV pulse trains using surface plasmons [57]. The apparatus is illustrated in Fig. 6.34. High-intensity 5 fs duration NIR light ($\lambda \approx$ 800 nm) pulses of a few cycles length at a 1 kHz repetition rate was focused to a 1.7 μm diameter spot at the surface of a spinning optically polished silica glass disc. The peak intensity I at the focused spot was $\approx 10^{18}$ W/cm^2, and the entire apparatus was enclosed in a vacuum chamber. A plasma layer is formed at the surface of the glass, forming a plasma mirror for

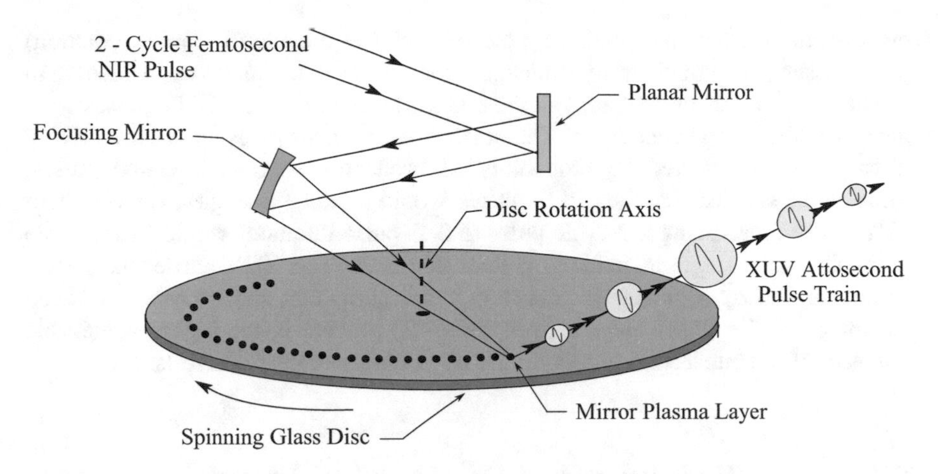

Fig. 6.34 Operating principle of a laser producing XUV attosecond pulse trains

incident wavelengths greater than a few tens of nanometers. The mirror was rotated such that clean polished surface was available for each pulse of the driving NIR laser. The normal component of the oscillating electromagnetic field of the focused light causes bunches of plasma electrons to be periodically pulled in and out of the plasma-vacuum interface. This movement of the plasma surface electrons and the sub-cycle movement of the electrons in the dense plasma produce the emission of a train of coherent attosecond duration XUV pulses. Known as *coherent wake emission* (CWE), it consists of the XUV attosecond pulses superposed on the reflected NIR laser light. These XUV pulses are then spectrally separated.

Another method to generate attosecond high-intensity pulses from a plasma layer is the *relativistic oscillating mirror* (ROM) process, where the laser pulse contains only a few cycles and the focused peak intensity at the plasma layer is in the 10^{19}–10^{21} W/cm^2 range. When the laser field interacts with the electrons in the plasma mirror layer, the electrons oscillate through a stationary point at velocity $v = 0$ to a maximum velocity v approaching the speed of light c. The interaction of the spectral components of the oscillating laser field with the oscillating plasma electrons generates phase-locked harmonics with attosecond pulse durations in the reflected beam. These high-frequency harmonics are generated in the short time interval when the relativistic electron velocity v is closest to the speed of light c. The attosecond pulse duration τ is related to the pulse duration t of the driving laser by $\tau \approx t/\gamma^3$, where γ is the relativistic Lorentz factor $\gamma = \frac{1}{\sqrt{1-v^2/c^2}}$. For example, if the ratio of the maximum electron velocity v to the velocity of light c is 0.995, and the driving laser has a pulse duration $t = 8$ fs, the calculated pulse duration of the reflected XUV pulse would be $\tau \approx 8$ as.

The ROM method of generating isolated attosecond XUV light pulses in the 17 nm spectral range was demonstrated in 2012 by team members located in Germany, Russia, Greece, and the United Kingdom [58]. They used a spinning polished fused silica disc, with a p-polarized 16 TW 3-cycle (8 fs pulse duration) pulsed laser (815 nm central wavelength) focused at a 45° angle of incidence, to produce ≈60 mJ of pulse energy at the disc surface. Attosecond XUV pulses were generated in the reflected beam. The smaller the number of cycles in the driving laser pulse, the greater the probability of producing single attosecond pulses. Simulations showed that these 8 fs pulses would produce a single as pulse about 17% of the time. Using a 2-cycle pulse (5.3 fs pulse duration) would increase the probability of a single as pulse to greater that 50%. The XUV attosecond pulses were analyzed by a grazing-incidence extreme ultraviolet spectrometer. A stated advantage of the ROM process over the CWE process is the higher achievable intensity of the attosecond pulses by the use of high-intensity drive lasers.

6.4.8 Ultra-High Power and Energy-Pulsed Lasers

The highest power lasers, those with peak output powers exceeding ≈100 TW (1 TW = 10^{12} watts) and reaching to petawatt levels (1 PW = 10^{15} watts), have a short pulse duration τ on the order of several femtoseconds. From Eq. (6.6) it is seen that for a laser having moderate pulse energy E, the peak pulse power P^{max} can be brought to a very high level for ultra-short femtosecond duration pulses τ where $P^{max} = E/\tau$. For example, the peak power of a laser with 500 mJ output and 5 fs pulse duration is 100 TW. To produce pulses at the terawatt and petawatt levels at short pulse durations requires energy amplification of the output of a seed laser having a short pulse duration but low pulse energy. This is accomplished by using a regenerative amplifier—a switchable laser using several optical active and passive components. The gain material in the amplifier is usually Ti:sapphire and a series of amplifiers is often used for additional pulse energy amplification. However, the high peak powers generated in the amplifier(s) would damage the amplifier optical components by self-focusing, causing distortion of the pulse waveform. To circumvent this effect, a clever process called *chirped pulse amplification* (CPA) is used. In this process, a seed laser pulse produces pulses having moderate pulse energy and short pulse durations. The seed laser must have a wide gain bandwidth. The pulses are then stretched in time by a dispersive optical device that introduces spectral time delays. Reflective diffraction gratings are often used, since they can produce a 10^3–10^4 pulse stretch ratio. The stretched pulses are then sent through the regenerative laser amplifier, where the peak power is kept at a reasonable level due to the longer pulse duration. The pulses from the amplifier are then recompressed by optical components of opposite dispersion, often a pair of reflective diffraction gratings. The resultant pulses can have very high peak power.

Chirped pulse amplification in the optical domain was first demonstrated by Strickland and Mourou at the University of Rochester in 1985 [59]. A mode-locked Nd:YAG laser produced 150 ps width pulses at a pulse repetition rate of 85 MHz with 5 W average power (61 nJ pulse energy at 410 W peak power). The seed pulses were stretched to a duration ≈300 ps without gratings by the use of a 1.4 km long coiled single-mode dispersive fiber. The resulting rectangular-shaped pulses were then amplified by a Nd:glass regenerative amplifier at $\lambda = 1.062$ μm, and then recompressed by a double-grating arrangement. The final compressed pulse had a 1 mJ output with a 2 ps pulse width. This corresponds to a peak output power of about 500 GW.

Most modern ultra-high power CPA type lasers are based on Ti:sapphire, since this material has excellent stability and a wide bandwidth range from 600 nm to 1100 nm. A single mode-locked Ti:sapphire seed laser can produce <10 fs pulses with an average power at the 200 mW level. Figure 6.35 illustrates a typical arrangement for a CPA based high peak power laser. In 1999 scientists at Lawrence Livermore Laboratory developed a Ti:sapphire-Nd:glass laser system using CPA that produced 1.5 PW pulses at 440 ps duration, with 660 J of pulse energy [60]. To handle the high generated power, large metallic diffraction gratings 94 cm in diameter were utilized during the compression process. Two CPA systems can be combined in series to form a doubled chirp amplification system, often with an intermediate nonlinear filter to improve the temporal contrast of the pulse.

A related CPA technique for generation of high power short duration pulses is *optical parametric chirped pulse amplification* (OPCPA). In this method the regenerative laser amplifier is replaced by an *optical parametric amplifier*, where a nonlinear optical crystal is pumped by a laser emitting energetic nanosecond pulses. The energy from this pulse is then efficiently stretched to a femtosecond pulse by a nonlinear frequency difference parametric process without the gain limitations of a

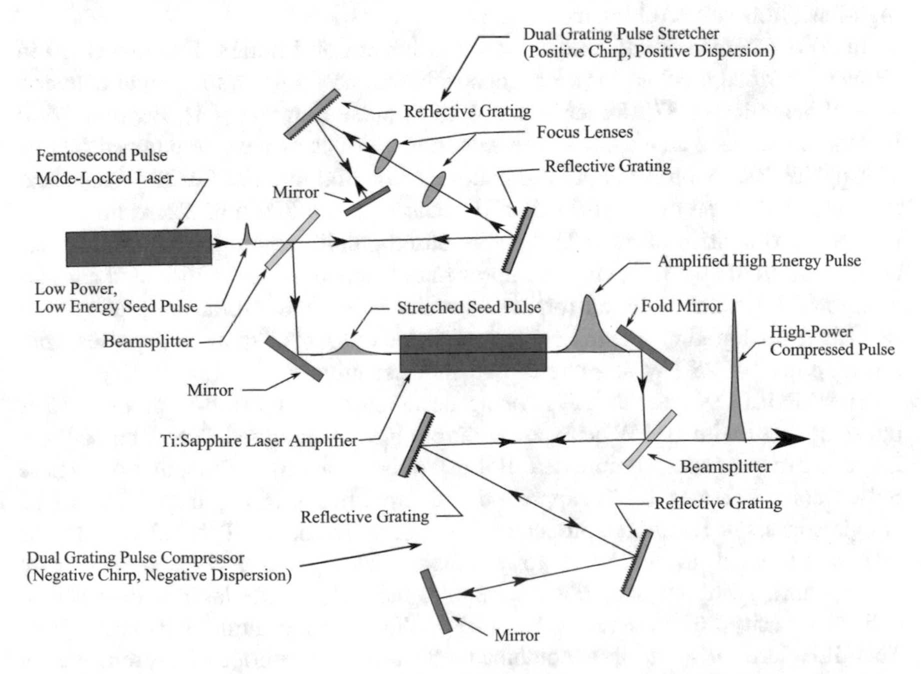

Fig. 6.35 Typical arrangement for a CPA based high peak power short-pulse laser

Ti:sapphire crystal-based amplifier. In addition, OPCPA can be serially combined with CPA for increased energy and peak power outputs.

The SCARLET laser at Ohio State University combined OPCTA with several Ti: sapphire multipass amplifiers to produce 500 TW 30 fs pulses with 15 J energy [61]. The repetition rate was reported to be one pulse or shot per minute. The seed pulse was stretched to 800 ns by diffraction gratings, and amplified by a doubled CPA system and several Ti:sapphire multi-pass amplifiers. The compressor consisted of a pair of 36 cm and 56 cm long diffraction gratings. The final output had an intensity of about 10^{22} W/cm^2, for use in the relativistic region as a laser fusion source.

Many lasers are being developed to exceed the petawatt level. The Texas Petawatt Laser (TPL) developed at the University of Texas at Austin produced 186 J in a 167 fs pulse, yielding a peak power of 1.1 PW [62]. Output from a mode-locked 2 nJ, 100 fs oscillator was stretched by a double grating system and fed into broadband OPCTA stages yielding $\approx$1 J output ($\approx 10^{10}$ gain). This amplified output was then sent through several mixed Nd:glass amplifiers to yield $\approx$250 J output ($\approx$400 gain), and compressed by a double grating system to achieve $\approx$1.1 PW, 190 J, 170 fs final output. A hybrid system demonstrated in 2011 at the Beijing National Laboratory produced 28 fs output at 1.16 PW by the use of doubled CPA and OPCTA [63].

The BELLA (Berkeley Laser Lab Accelerator) petawatt laser at the University of California Berkeley Lab is being built to provide a 40 J, 40 fs pulse, to yield a peak power of about 1 PW. The pulse rate is reported to be a rapid one shot per second (1 Hz). It uses CPA and large diameter Ti:sapphire crystal amplifiers pumped by high-repetition Nd:YAG lasers.

In 2007 Chambaret and associates at the Institut de Lumiére Extreme (ILE) in France described a layout for a Ti:sapphire-based laser system that could deliver a 25 PW beamline at 375 J energy and a 15 fs pulse duration [64]. Figure 6.36(a) illustrates the basic architecture. The front end consists of a diode pumped 500 μJ Ti:sapphire laser with 30 fs pulse duration at λ = 800 nm. An OPCPA amplification device then produces 100 mJ, 10 fs pulses at λ = 800 nm. These pulses are then fed into a chain of three Ti:sapphire-based amplifiers producing a final pulse energy output of 600 J. The final compressor is detailed in Fig. 6.36(b) and consists of spectral filters, focusing mirrors, two smaller monolithic diffraction gratings and two larger tiled grating assemblies. The predicted output is 376 J, 15 fs pulses, with a peak power of 25 PW at a rate of one shot per minute.

The APOLLON laser at ILE is being designed as a first step to produce 15 fs pulses at 150 J with 10 PW peak power output [65]. A modified Ti:sapphire laser at λ = 800 nm produces compressed 100 mJ pulses having $\approx$6 fs duration. These pulses enter a series of Ti:sapphire-based amplifier stages pumped by 0.8 kJ Nd:glass lasers at 1 shot/mn, producing 300 J of pulse energy. This pulsed output is then compressed to a 150 J, 15 fs pulse, yielding 10 PW peak power at λ = 800 nm. At about one shot per minute, the APOLLON laser is expected to deliver intensity on the order of 10^{24}–10^{25} W/cm^2 in the ultra-relativistic region. This ILE laser is part of a combined effort of the European Extreme Light Infrastructure (ELI) to develop high-power lasers.

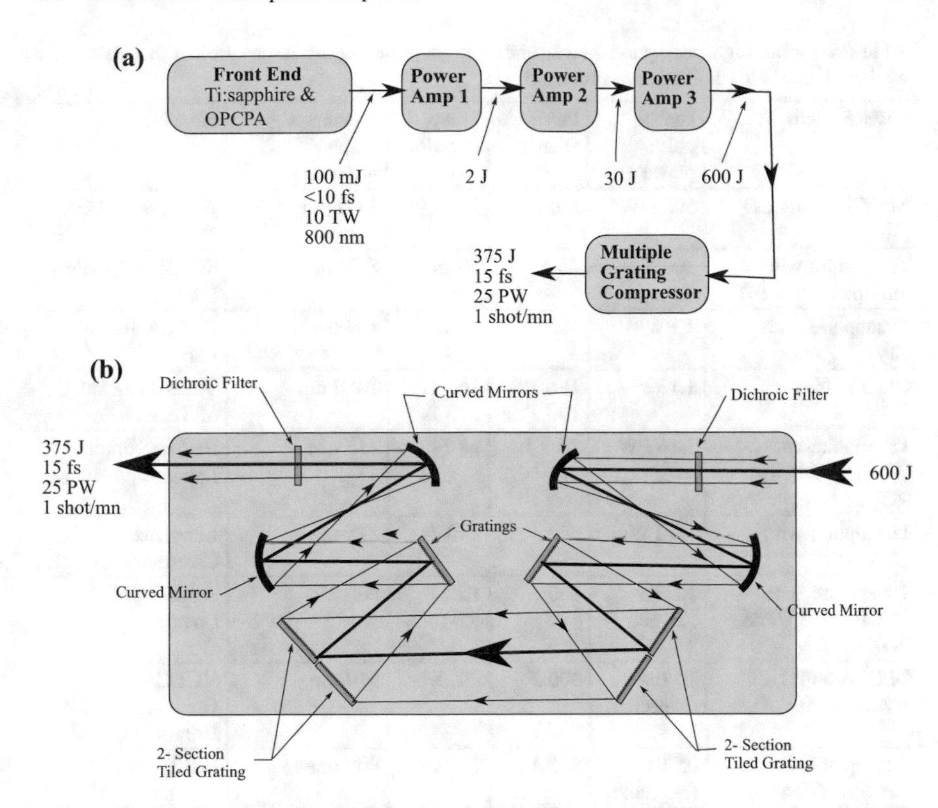

Fig. 6.36 **a** Basic layout of proposed ILE 25 PW laser. **b** Final compressor stage of proposed ILE 25 PW laser

The VULCAN laser project at the Central Laser Facility (CLF) in Great Britain is being developed to deliver 10 PW power at 300 J at 30 fs pulse duration [66]. The system uses OPCPA joule-level amplification using non-linear frequency doubling lithium triborate (LBO) crystals. It is expected to deliver an intensity of 10^{23} W/cm^2.

A very high energy source is being developed at the National Ignition Facility (NIF) at Lawrence Livermore National Laboratory. The driving force behind this laser is the fusion ignition of the hydrogen isotopes deuterium and tritium. In 2012 this system produced 1.85 MJ of energy of several nanoseconds duration at 500 TW peak power, focused on a 2 mm diameter target. This was achieved by the integration of 192 laser beams, grouped into two 96-beam bays. Each 96-beam bay consists of two 48-beam clusters—each cluster having six 8-beam bundles, where each bundle consists of two 4-beam quads [67]. To generate a beam, a single pulse is split into parallel beams which are amplified by a series of Nd:glass amplifiers. To obtain the required UV radiation at 351 nm, the third harmonic of the amplified near-infrared pulse is used. The UV beams are auto-aligned and can be controlled separately or as a sub-group. They are focused and managed by an immense set of

Table 6.4 Some high peak pulse power and/or high pulse energy laser systems that have been developed, are being built, or are proposed

Laser System	Peak pulse power	Pulse energy	Pulse duration	Dominant output wavelength	Notes
Nd:YAG/Nd:glass [59]	500 GW	1 mJ	2 ps	1.06 μm	First use of CPA
Ti:sapphire with multipass CPA [61]	500 TW	15 J	30 fs	815 nm	SCARLET Ohio State U
Ti:sapphire with CPA	1 PW	42.4 J	40 fs	≈800 nm	BELLA Berkeley Lab
OPCPA [62]	1.12 PW	186 J 190 J	167 fs 170 fs	1060 nm	Texas Petawatt U. of Texas
Ti:sapphire with CPA and OPCPA [63]	1.16 PW	32.3 J	28 fs	815 nm	Beijing National Lab
Ti:sapphire with CPA [60]	1.5 PW	660 J	440 fs	815 nm	Lawrence Livermore Lab
Ti:sapphire with OPCPA and CPA [65]	10 PW (expected)	150 J	15 fs	800 nm	APOLLON ILE France
OPCPA amp LBO crystals [66]	10 PW (expected)	300 J	30 fs	910 nm	VULCAN CLF Great Great Britain
Ti:sapphire with OPCPA [64]	25 PW (expected)	375 J	15 fs	800 nm	ILE France ELI Europe
192 integrated beams [67]	500 TW	1.85 MJ	3–10 ns	UV 351 nm	NIF Lawrence Livermore Lab

optical glass components, which must remain damage-proof at energy density levels up to 10 J/cm^2. The laser developed at the National Ignition Facility is the most powerful operating laser system in the world. Table 6.4 lists the relevant parameters for several high power/energy laser systems.

6.5 Some Additional Laser Types

6.5.1 Free-Electron Lasers

The *free-electron laser* (FEL) uses an oscillating electron beam to produce laser light. It was first demonstrated in 1976 at the High Energy Physics Laboratory at Stanford University [68]. The process is related to synchrotron radiation as described in Sect. 5.2.7, and the basic method of operation is illustrated in Fig. 6.37. Electrons are accelerated to relativistic speeds close to the speed of light by a LINAC linear accelerator. These electrons then enter an *undulator* or *wiggler*, consisting of a series

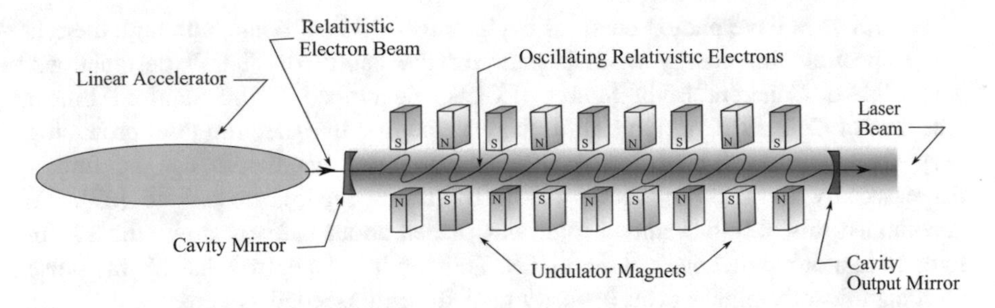

Fig. 6.37 Operating principle of a free-electron laser. All components are in vacuum

of alternating polarity field magnets. These electrons oscillate due to the changing Lorentz force, and these electrons become the lasing medium. Usually the undulator is placed between cavity mirrors for multipass beam amplification. The relativistic oscillating electrons radiate coherent synchrotron radiation, where the emission wavelength depends on the electron energy, the magnetic undulator period, and the strength of the magnets. The original 1976 FEL used a 24 meV electron beam and a 5 m long undulator. Free-electron lasers are capable of high power emission over a wide spectral range, from microwave to X-ray. A lack of suitable mirrors in the X-ray region requires single-pass operation without mirrors using a much longer undulator length. In this case, laser emission occurs through another process called *self-amplified stimulated radiation* (SASE), where the interaction between the spontaneous emission of bunched electrons and their oscillations produces a coherent beam of pulsed femtosecond high intensity light.

The Office of Naval Research has initiated development of a free-electron laser with 100 kW average power output. As a ship-based military weapons system, this FEL could seriously damage airborne targets using a precision beam traveling at the speed of light. From a single FEL, about 10 directed beam lines could be produced and directed to multiple targets. The system is expected to eventually reach gigawatt peak power levels.

6.5.2 Pulsed X-Ray Lasers

The European *X-ray free electron lase*r (XFEL) under construction in Germany is a FEL-SASE type laser which uses a 2.3 km long superconducting linear accelerator at −271 °C to achieve ≈17.5 GeV, followed by an undulator over 100 m long, consisting of alternating polarity quadrupole magnets. The total XFEL length will be about 3.4 km, operating in underground tunnels in a vacuum. With three to five possible undulator beamlines, separate output wavelengths can be in the 0.5–6 nm range, having a pulse duration of several femtoseconds, and an output about one million times the intensity of other types of X-ray sources. A future target is to produce *hard X-ray* output at $\lambda \approx 0.085$ nm with pulse durations less than 100 fs.

Hard X-rays have photon energies in the ≈10–100 keV range, although there is some spectral and energy overlap between the "soft" or "hard" designations. The LINAC Coherent Light Source (LCLS), developed at the Stanford Linear Accelerator Center (SLAC) in California, was the first free-electron laser producing very short pulses of hard X-rays. The X-ray pulse bandwidth can be further decreased by a factor of about 40–50 by a self-seeding technique [69]. To accomplish this, a thin diamond filter was placed about halfway down the 130 m long undulator magnet bank of the LCLS. This produced a narrow hard X-ray pulse with significantly higher peak intensity than for an unseeded system.

A tabletop technique to generate coherent soft and hard X-ray radiation using HHG and coherent phase matching has been described by Popmintchev and joint-researchers at JILA (called the Joint Institute for Laboratory Physics before 1995), University of Colorado, and NIST (National Institute of Standards and Technology), Boulder, Colorado [70, 71]. A mid-infrared laser (λ = 1300 nm) source generating femtosecond pulses were focused into an optimized waveguide containing a nonlinear HHG medium. See Fig. 6.38. The waveguide contained a mixture of molecular and atomic noble gases, such as argon or neon, pressurized up to 10 atmospheres. The noble gases were moderately ionized, and HHG upconversion produced fields of X-ray radiation. These HHG fields were coherently added by velocity phase matching with the driving laser to produce intense soft and hard X-ray laser beams at a designated wavelength. A related technique using a higher pressure helium gas medium has generated about 5000 harmonic pulses covering the UV to X-ray range, with the X-rays having energy exceeding 1.6 keV [72].

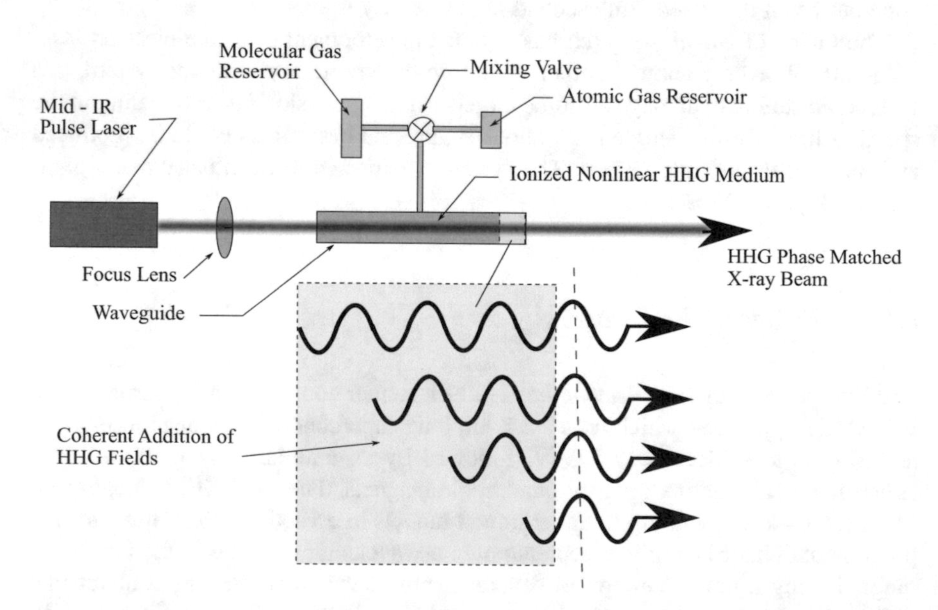

Fig. 6.38 Illustration of an X-ray laser using a mid infrared pulse laser and HHG phase matched X-ray output

Using a powerful LCLS free-electron X-ray laser as a pump source, experimenters at SLAC demonstrated X-ray emission from electron transitions in an ionized atomic gas to produce the first *atomic X-ray laser* [73]. A capsule of neon gas was bombarded by LCLS pulses, producing an atomic population inversion, with many of the inner shell electrons being ejected from the neon atoms. The filling of the resultant electron holes by other electrons produced femtosecond duration coherent hard X-rays of wavelength $\lambda \approx 1.46$ nm.

6.5.3 Visible Color Laser

A laser that can produce pure red, green, or blue colors was developed at Brown University, using an optically pumped quantum dot nanocrystal film in a VSCEL configuration [74]. The quantum dots were deposited on a glass plate from a colloidal solution, and consisted of s CdSe core with a ZnCd alloy shell. The same materials are used for each color, with an inner core diameter of 4.2 nm for red light emission, 3.2 nm for green light, and 2.5 nm for blue light. The resulting *colloidal quantum dot vertical cavity emitting laser* (CQD-VCSEL) was optically pumped by a diode laser at 400 nm. When the pump threshold energy exceeded a certain value, the emitted light changed from photoluminescence to amplified spontaneous emission, and spatially coherent laser light was emitted from the CQD-VCSEL at a specified narrow bandwidth visible color.

6.5.4 White Light, Supercontinuum, and Multi-frequency Lasers

The term "white light laser" refers to a laser that exhibits spatial, spectral, and temporal coherence over the visible and near infrared spectrum. When the emitted light is from the infrared to ultraviolet spectral regions, it is generally called a *supercontinuum white light laser* or *supercontinuum laser*.

One type of supercontinuum laser uses a glass fiber with an integrated photonic crystal. The photonic crystal fiber was first described, fabricated and analyzed in 1996 in Great Britain [75]. It consisted of a solid central silica core surrounded by a silica photonic crystal lattice of hexagonal air holes extending along the length of the fiber. Figure 6.39 shows the cross-section of the fiber and the photonic crystal cladding. The center solid glass hexagonal core acted as a lattice defect, with the transmitted light confined to this core. The hexagons had a center-to-center distance ≈ 2 μm, with a band gap at $\lambda \approx 1.5$ μm. They demonstrated efficient single-mode core transmission over a wide spectral range using focused laser wavelengths of 457.9, 632.8, 850, and 1550 nm through a meter long fiber. However, they did not describe the generation of any visible continuum from a single wavelength input. In

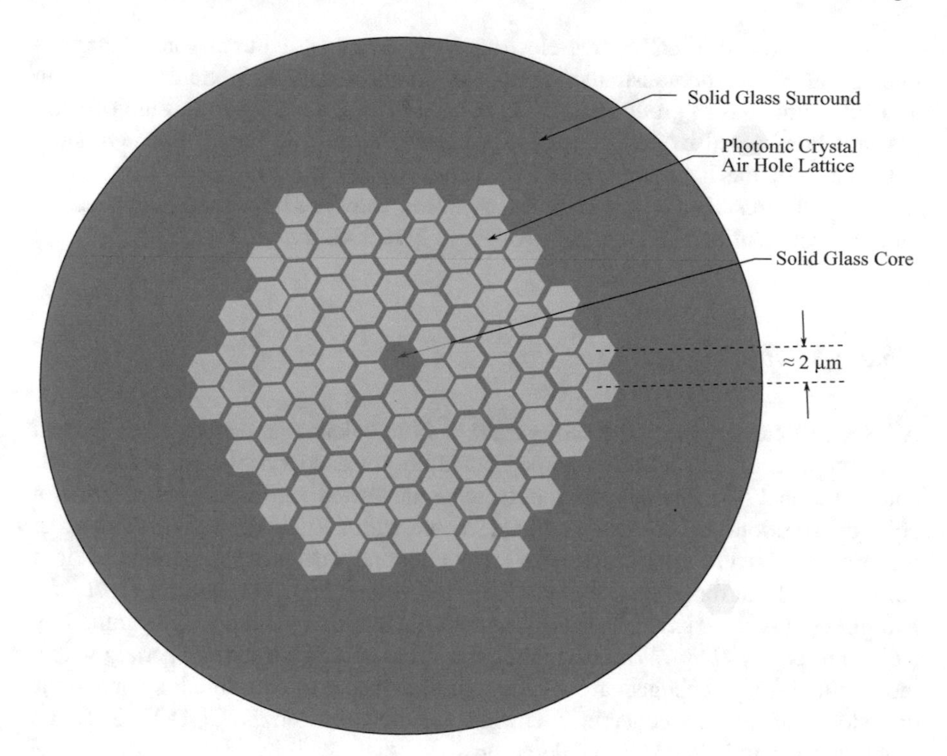

Fig. 6.39 Cross-section of an optical fiber with photonic crystal cladding to produce supercontinuum laser output

2000 Ranka and co-workers at Bell Laboratories generated the first infrared-visible-ultraviolet continuum using a photonic crystal fiber [76]. A 100 fs pulsed 770 nm mode-locked Ti:sapphire laser was coupled to a 75 cm long photonic crystal fiber. The resulting supercontinuum was in the 400–1500 nm range with a 550 THz frequency bandwidth ($\Delta\lambda \approx 1174$ nm).

The production of a supercontinuum from narrow band input depends on complex nonlinear optical properties of the photonic crystal. The photonic crystal structure in the fiber forms photonic band gaps producing high-order dispersion. Although the physics of supercontinuum generation from a photonic crystal fiber is not completely understood, other contributing factors are Raman scattering and four-wave mixing. A review article by John Dudley and associates describe the theory, operating parameters, and generation of supercontinuum light using photonic crystal fiber lasers [77].

Another method to generate a supercontinuum laser was demonstrated in 2000 at the University of Bath in Great Britain.using a tapered fiber [78]. Figure 6.40 illustrates the tapered fiber, which was heat-drawn from a standard telecommunications fiber. The untapered input and output ends transitioned to a a central 1.8 μm diameter, 90 mm long waist over 35 mm long transition regions. An unamplified

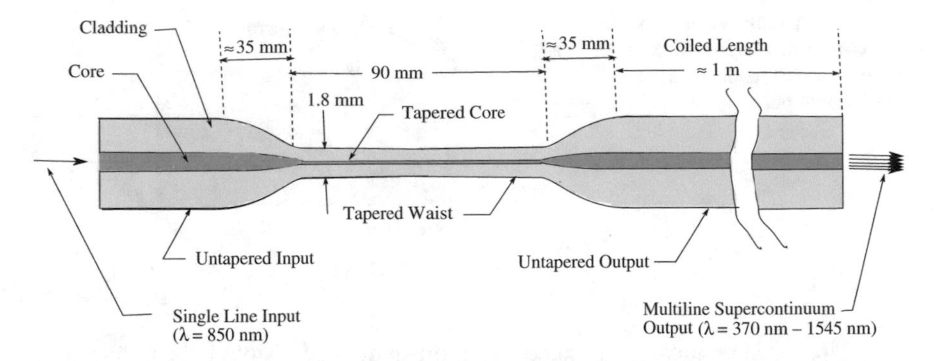

Fig. 6.40 A tapered silica fiber to produce supercontinuum laser output

76 MHz Ti:sapphire laser at $\lambda = 850$ nm with 200–500 fs pulse duration was coupled into the shorter input fiber end. Nonlinear processes in the waist region generated significant spectral broadening to produce a multiline supercontinuum output beam over the wavelenght range $\lambda = 370$–1545 nm. 80 MHz pulsed output energy was 3.9 nJ at 300 mW average power.

In 2006 researchers in Stuttgart, Germany extended the ultraviolet range of a tapered fiber supercontinuum laser [79]. The waist length was 90 mm and the waist diameter was varied at 2.0, 2.7, and 4.3 μm. For the 2.7 μm diameter waist, the supercontinuum wavelength range was 400–1650 nm, with 600 mW average power input and 290 mW average power output. The thicker 4.3 μm waist diameter produced 400 mW average power output with the supercontinuum spectrum shifted to the red, and the 2.0 μm waist diameter produced about 320 mW average power with a substantial shift of the supercontinuum spectrum into the ultraviolet. Femtosecond pulses were produced at a 20 MHz repetition rate with energy close to 39 nJ.

An ion laser using a mix of krypton and argon can function as a white light laser. Krypton contains 16 possible laser emission lines between 337.5 and 799.3 nm, and argon has 15 emission lines between 334.5 and 528.7 nm. When used in multiline mode, the output closely resembles white light. A pulsed ion laser containing xenon to produce white light has also been demonstrated.

Another white light laser concept investigated at the University of Salford in Great Britain uses stimulated Raman scattering in a cell with cavity feedback to produce a broad band multiline output. The stimulated Raman scattering produces up and down shifting of frequency lines which are multiples of the vibrational or rotational energy levels of the material in the cell. As shown in Fig. 6.41, a modified Raman cell has four corner mirrors with reflectance ≈0.93 to produce cavity feedback. For an input frequency bandwidth Δf_{in} of two incident laser beams, self-synchronous up and down frequency shifts within the Raman cavity produce a significant increase in the number of lines and bandwidth Δf_{out} of the output beam. The multiline output of over 100 frequencies, extends from the infrared to ultraviolet regions, and the output would appear as a white light laser source.

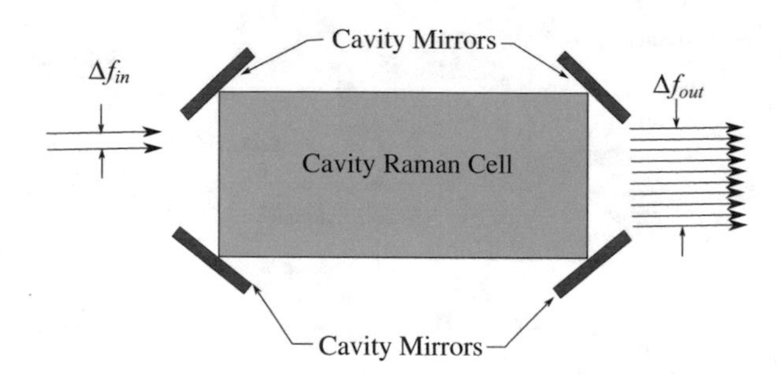

Fig. 6.41 A Raman cell with four corner mirrors to produce cavity feedback and broad multiline output

A type of laser mixing can generate multi-frequency visible light by electron-hole recollision. This was demonstrated by cooperating scientists at the University of Santa Barbara, California, and the Chinese University in Hong Kong [80]. First, a Coulomb bound electron-hole pair quasiparticle, called an *exciton*, was created by bombarding a gallium arsenide semiconductor nanostructure with a weak NIR free-electron laser. Subsequent bombardment by a high energy THz frequency laser caused the electron to recollide and recombine with the hole at a higher energy than its energy level in the exciton. Upon recombination, the excess electron energy produced new frequencies of emitted light. Up to 11 new frequencies were simultaneously emitted, appearing as white light.

White light and supercontinuum lasers are used in diverse applications such as illumination, laser light shows, spectroscopy, microscopy, and control of sub-cycle pulses for visible light femtosecond pulse generation. Supercontinuum lasers are available commercially that produce a light spectrum that extends from 400–2400 nm wavelength. White light illumination of objects could also be achieved by simultaneous illumination by a set of four separated narrow bandwidth diode lasers, each having a blue, green, yellow, or red color. An experiment to determine the effectiveness of this approach is described in Appendix 6A [81].

6.5.5 Random Illumination Laser

When lasers are used for general illumination of objects, or as light sources for projectors, their high spatial coherence can produce a visible speckle on the viewed objects or projected image. Speckle occurs when coherent photons incident on a scattering surface introduces random phase delays. Researchers at Yale University developed a dye *random laser* with reduced spatial coherence to avoid this undesirable effect [82]. The laser consisted of a colloidal solution of 240 nm diameter polystyrene spheres with Rhodamine 640 as the lasing dye in diethyl glycol solution. A series of irregular non-homogeneous spatial modes were produced in the laser, where these multiple modes had uncorrelated phases.

A relevant parameter for random lasers is *photon degeneracy* δ, which describes the number of photons per coherence volume. Non-lasing light sources, e.g. incandescent lamps, have low photon degeneracy and low spatial coherence.

Conventional lasers have a high photon degeneracy and high spatial coherence, while the random laser maintains high photon degeneracy, but has a low spatial coherence. For comparison, an incandescent source at 4000 K has $\delta \approx 10^{-3}$, an efficient light-emitting diode has $\delta \approx 10^{-2}$, a pulsed Ti:sapphire laser has $\delta \approx 10^{6}$, and a 1 mW HeNe CW laser has $\delta \approx 10^{9}$. The MHz repetition rate random dye laser of the Yale researchers had $\delta \approx 10^{3}$. With a moderately low spatial coherence, this random laser produced no speckle when used as a laser illuminator.

6.5.6 *Dark Pulse Lasers*

Bright soliton pulses can be propagated from lasers with peak intensity that exceeds that of the continuous wave mode intensity baseline. When soliton pulses are produced having peak intensities much less than the continuous wave intensity, they are called *dark soliton pulses* or *dark solitons*. See Fig. 6.42a and b. For an optical fiber exhibiting nonlinear properties such as group velocity dispersion, traveling soliton pulses can be described by the nonlinear Schrödinger equation, after the Austrian physicist Erwin Schrödinger (1887–1961). Dark soliton pulses arise from certain solutions of this nonlinear Schrödinger equation. Propagation of dark solitons in an optical fiber requires positive group velocity dispersion and soliton power laser input, where the input pulses have the shape of the fundamental dark soliton. Propagated dark soliton pulses through optical fibers were first observed in 1988 by Weiner and co-workers at Bellcore [83]. A mode-locked pulsed dye laser and 8.6 kHz repetition rate amplifier supplied 75 fs pulses at $\lambda = 620$ nm. These pulses were shaped by lenses, gratings, and a spatial mask before input to the optical fiber. At proper input power, 185 fs dark soliton pulses at $\lambda = 620$ nm were propagated through a 1.4 m long polarization preserving single-mode optical fiber with negligible change in shape.

Dark pulse solitons can also be generated directly from a doped fiber laser, where the fiber dispersion properties are closely controlled. In 2012 researchers in China generated dark pulses from a 14.5 m long erbium doped fiber ring laser having net anomalous group velocity dispersion [84]. A 4 m long section of erbium doped fiber was coupled to standard single-mode fiber, along with two fiber-based polarization controllers. The resulting net cavity dispersion was ≈ -0.106 ps^2 at $\lambda = 1560$ nm. The laser output was changed from bright soliton pulse emission to CW emission, and then to dark soliton pulse emission by increasing the pump power and adjusting the polarization controllers. It was determined that the dark pulse solitons were shaped in the cavity, and were of the type resulting from solutions of the nonlinear Schrödinger equation.

A train of dark pulses has also been generated from a mode-locked quantum dot laser at JILA and NIST in Colorado [85]. The pulses in a mode-locked laser cavity are described by solutions of the Haus equation, named after the Slovene-American physicist Hermann Haus (1925–2003). Using negative solutions to the Haus equation, a pulse train of dark pulses can be generated. The researchers constructed

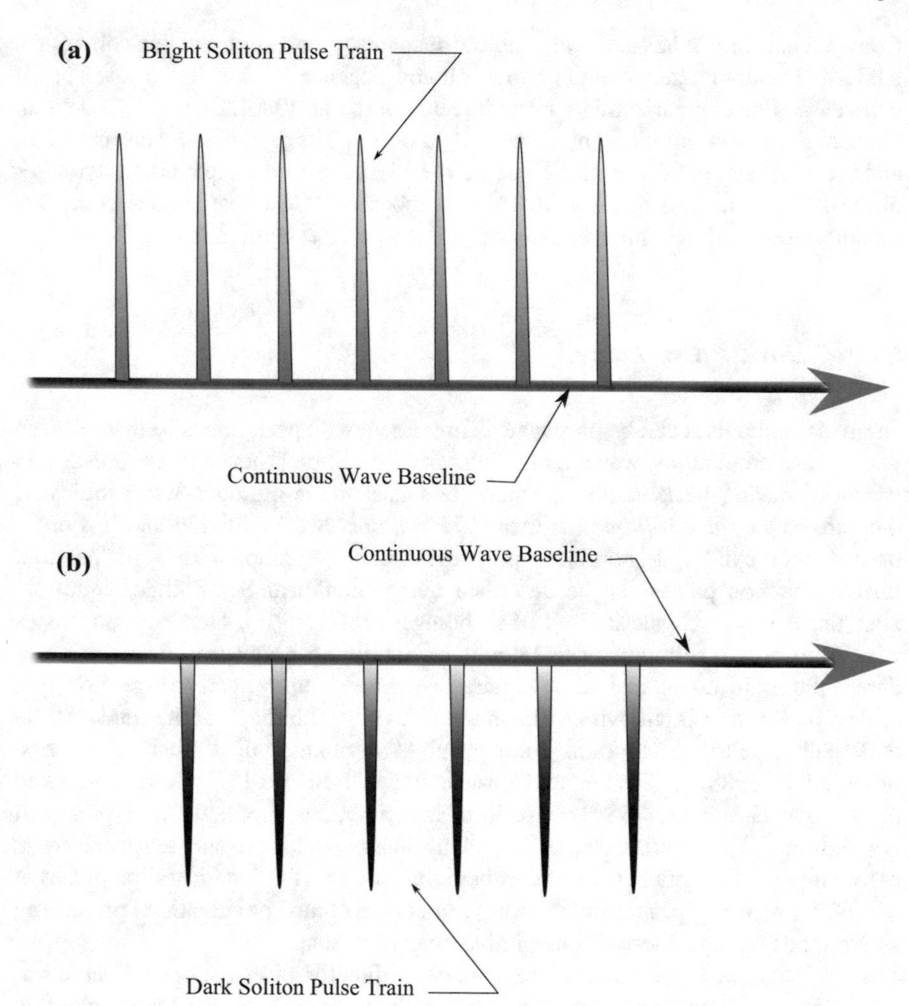

Fig. 6.42 **a** Illustration of a train of bright soliton laser pulses. **b** Illustration of a train of dark soliton laser pulses

an indium gallium arsenide quantum dot with a 5 mm long semiconductor waveguide. A train of dark pulses of ≈90 ps duration was generated and recorded at about 30% of the CW intensity.

6.5.7 Coupled Lasers

It is possible to couple two lasers such that both maintain coherent output. One method is pump-coupling, where a fraction of the output of each laser is used to

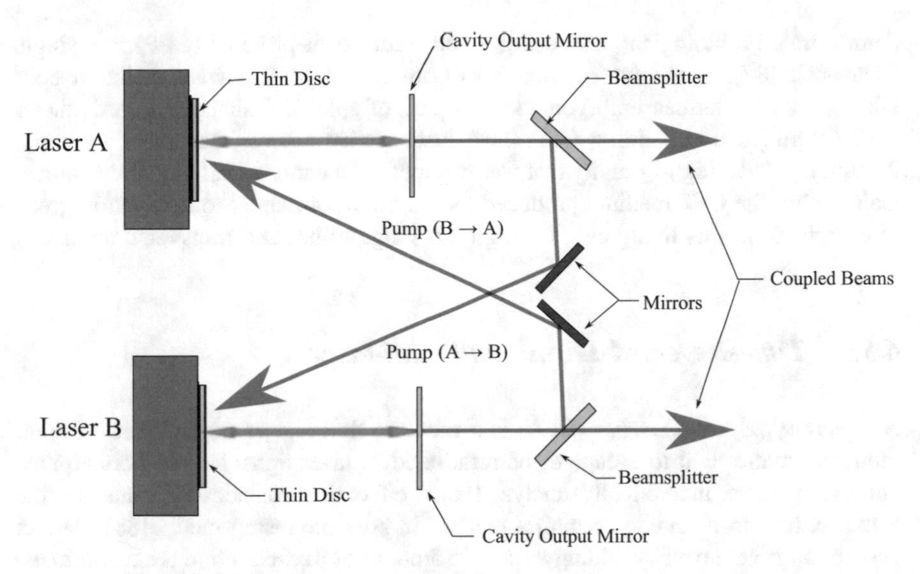

Fig. 6.43 A possible arrangement for a pair of pump-coupled thin disc lasers

pump the other laser. A possible method is shown in Fig. 6.43 for coherent coupling of two thin disk lasers. The intensity of each pumping beam determines whether the coupling for each laser is weak or strong.

A pair of spatially separated edge-emitting single-mode semiconductor lasers can be mutually and bidirectionally coupled in series to produce a single coupled beam. One method is by coherent optical coupling, in which the TE mode of each laser cavity is injected into the other laser cavity. The equations describing the resultant nonlinear lasing processes can be quite complex. Investigators at the Vienna University of Technology, Princeton University, and Yale University have theoretically analyzed two coherently coupled semiconductor lasers where the cavities are independently pumped [86]. The non-Hermitian operators describing laser operation can produce singularities or *exceptional points* which control the laser emission above the lasing threshold. With one laser pumped for normal bright operation, gradually increasing the pump power to the second laser can induce an exceptional point affecting the interaction between the coupled lasers. At the critical pump power which induces the exceptional point, both lasers are switched off and a total blackout occurs.

6.5.8 Biological Laser

There is an ongoing search for new types of materials that support lasing. One of the more unique types is based on a biological living cell for laser operation. Gather and Yun at the Wellman Center for Photomedicine at Harvard University have demonstrated a biological laser using green fluorescent proteins (GFP) in a single living cell [87]. A cavity resonator was constructed from a water solution containing a single human embryonic kidney cell of spherical shape with a diameter ≈15–20 μm, and containing GFP. Two laser cavity mirrors, separated by about 20 μm, formed a high-Q cavity that was pumped by a nanosecond blue light source. Gain within the GFP medium produced the emission of nanosecond duration green laser light from this living cell, having both longitudinal and transverse modes.

6.5.9 Time-Reversed Lasing and Anti-lasers

A *coherent perfect absorber* (CPA) is a medium that allows no incident coherent monochromatic light to escape or be reradiated as laser light. Researchers at Yale University have theoretically analyzed this effect by considering states of the S-matrix (electromagnetic scattering matrix) in possible lasing media [88]. Perfect absorption is achieved by adding a specific amount of dissipation to the medium, so that "time-reversed lasing" occurs near the lasing threshold. From the interaction of wave interference and optical absorption in this loss medium, certain interference patterns are produced that completely capture the incident radiation. Such a CPA device is often referred to as an *anti-laser*. A suitable material is silicon, which does not easily support lasing, and can completely retain the incident laser radiation as heat or an electron-hole pair. A silicon slab CPA was analyzed using coherent monochromatic illumination in the 500 nm to 900 nm range.

Based on this analysis, the first operational anti-laser was constructed in 2011 at Yale University [89]. Two collimated counter-propagating laser beams from a tunable Ti:sapphire laser (800–1000 nm range) were incident on opposite faces of a 110 μm thick silicon resonant cavity wafer, which functioned as a two-channel low-Q Fabry-Perot etalon. Using the time-reversal symmetry property of the system, the amplitudes and relative phases of the coherent incident beams were adjusted until destructive interference occurred between the scattered beams within the cavity wafer. At a precise cavity threshold, radiation could not escape from the cavity wafer, the CPA condition occurred, and all radiation was absorbed within the bands of the silicon material.

6.5.10 Superradiant Laser

The so-called *superradiant laser* is actually quite dim compared to conventional lasers, but has some unique physical and optical properties. A cooled mass of atoms is contained between two cavity mirrors. These atoms are pumped to a stable excited state, and when pumped by a separate laser, self-synchronized interband transitions emit a directional coherent wave of photons. The photons exit the mirror so quickly that on the average less than one photon is contained in the cavity at any instant of time. Thus superradiant lasers do not depend on a large number of photons in the cavity, as do conventional lasers. This laser radiation is insensitive to any thermal vibration of the cavity mirrors, resulting in an output with high stability and ultra-narrow spectral linewidth on the order of one MHz.

A prototype superradiant laser was built in 2012 at NIST/JILA in Colorado [90]. A group of over a million rubidium-57 atoms was cryogenically cooled to $\approx 2 \times 10^{-4}$ K and contained as a condensate in a mirrored cavity about 2 cm long. Using low power tuning lasers, photons were emitted at a high rate by continuous switching of the atoms between two energy levels. Collective synchronization of the atomic dipoles in the condensate produced a common coherence of all emitted photons. At any instant of time there was an average of 0.2 photons in the cavity. The spectral linewidth was measured to be about ten thousand times less than achievable with a conventional laser. Although this prototype did not exhibit the full stability potential of a superradiant laser, future designs could be useful for advanced atomic clocks to perform more accurate special and general relativity time dilation measurements.

References

1. F.L. Pedrotti, L.S. Pedrotti, *Introduction to Optics*, 2nd edn. (Prentice Hall, New Jersey, 1993)
2. T.H. Maiman, Simulated optical radiation in ruby. Nature **187**, 493–494 (1960)
3. T.H. Maiman, Ruby laser systems, U.S. Patent No. 3,353,115, November 1967
4. F.J. McClung, R.W. Hellwarth, Giant optical pulsations from ruby. J. Appl. Phys. **33**(3), 828–829 (1962). doi:10.1063/1.17777174
5. L.F. Johnson et al., Continuous operation of a solid-state optical maser. Phys. Rev. Lett. **126**, 1406–1409 (1962). doi:10.1103/PhysRev.126.1406
6. J.E. Geusic, H.M. Marcos, L.G. Van Uitert, Laser oscillations in nd-doped yttrium aluminum, yttrium gallium and gadolinium garnets. Appl. Physics Lett. **4**(10), 182–184 (1964). doi:10.1063/1.1753928
7. A.L. Schawlow, C.H. Townes, Masers and maser communications system, U.S Patent No. 2,929,922, March 1960
8. G. Gould, Optically pumped laser amplifiers, U.S Patent No. 4,053,845, October 1977
9. A. Javan, W.R. Bennett, D.R. Herriott, Population inversion and continuous optical maser oscillation in a gas discharge containing a He-Ne mixture. Phys. Rev. Lett. **63**, 106–110 (1961). doi:10.1103/PhysRevLett.6.106
10. A.D. White, J.D. Rigden, Continuous gas maser operation in the visible. Proc. IRE **50**, 1697 (1962)
11. C.K.N. Patel., "Continuous wave laser action in vibrational-rotational transition in CO_2", *Phys. Rev.*, **136**(5A), 1187–1193 (1964)
12. W.B. Bridges, Laser oscillation in singly ionized argon in the visible spectrum. Appl. Phys. Lett. **4**, 128–130 (1964)
13. L.A. Newman, Waveguide laser having a capacitively coupled discharge, U.S Patent No. 4,381,564, April 1983
14. M. Boiteux, O. de Witte, A transverse flow repetitive dye laser. Appl. Opt. **9**(2), 514–515 (1970). doi:10.1364/AO.9.000514
15. H.W. Furumoto, H.L. Ceccon, Ultra-long flashlamp-excited pulse dye laser for therapy and method therefore, U.S Patent No. 5,624,435, April 1997

16. T. Ilorine, Dye laser, U.S Patent No. 6,101,207, August 2000
17. F.P. Schäfer, H. Müller, Tunable ring dye laser. Optics Comm. **2**(8), 407–409 (1970). doi:10.1016/0030-4018(71)90256-2
18. B. Fan, S. M Faris, Solid-state dye laser, U.S Patent No. 6,141,367, October 2000
19. Y. Sun, X. Fan, Distinguishing DNA by analog-to-digital conversion by using optofluidic lasers. Angew. Chem. Int. **51**, 1236–1239 (2012). doi:10.1002/anie.201107381
20. R.N. Hall et al., Coherent light emission from GaAs junctions. Phys. Rev. Lett. **9**(9), 366–368 (1962). doi:10.1103/PhysRevLett.9.366
21. H. Kroemer, A proposed class of heterojunction injection lasers. Proc. IEEE **51**, 1782–1783 (1963). doi:10.1109/PROC.1963.2706
22. R. Dingle, W. Wiegmann, C.H. Henry, Quantum states of confined carriers in very thin $Al_xGa_{1-x}As$-GaAs-$Al_xGa_{1-x}As$ heterostructures. Phys. Rev. Lett. **33**, 827–830 (1974)
23. R. Dingle, C.H. Henry, Quantum effects in heterostructure lasers, U.S. Patent No. 3,982,207, September 1976. doi:10.1103/PhysRevLett.33.827
24. R.J. Keyes, T.M. Quist, Injection luminescent pumping of CaF_2: U^{3+} with GaAs diode lasers. Appl. Phys. Lett. **4**, 50–52 (1964). doi:10.1063/1.1753958
25. E. Snitzer, Optical maser action of Nd^{+3} in a barium crown glass. Phys. Rev. Lett. **7**, 444–446 (1961). doi:10.1103/PhysRevLett.7.444
26. H. Zellmer et al., Double-core light-conducting fiber, process for producing the same, double-core fiber laser, and double-core fiber amplifier, U.S. Patent No. 5,864,645, January 1999
27. J. Hecht, Fiber lasers: the state of the art. Laser Focus World **48**(4), 57–60 (2012)
28. C.J. Koester, E. Snitzer, Amplification in a fiber laser. Appl. Opt. **3**(10), 1182–1186 (1964). doi:10.1364/AO.3.001182
29. A. Giesen et al., Scalable concept for diode-pumped high-power solid-state lasers. Appl. Phys. B **58**(5), 365–372 (1994). doi:10.1007/BF01081875
30. A. Killi et al., Current status and development trends of disk laser technology. SPIE Proc. **6871**, 68710L (2008). doi:10.1117/12.761664
31. J. Faist et al., Quantum cascade laser. Science **264**(5158), 553–556 (1994). doi:10.1126/science.264.5158.553
32. A. Lyakh et al., 1.6 W high wall plug efficiency, continuous-wave room temperature quantum laser emitting at 4.6 μm, Appl. Phys. Lett. **92**(11), 111110-1-3 (2008). doi:10.1063/1.2899630
33. F. Capasso et al., Unipolar semiconductor laser, U.S. Patent No. 5,457,709, October 1995
34. B.S. Williams, Terahertz quantum-cascade lasers. Nat. Photonics **1**, 517–525 (2007). doi:10.1038/nphoton.2007.166
35. F.J. Duarte, *Tunable Laser Optics* (Elsevier Science, San Diego, 2003), pp. 49–54
36. F.J. Duarte, Multiple-prism grating solid-state dye laser oscillator: optimized architecture. Appl. Opt. **38**(30), 6347–6349 (1999). doi:10.1364/AO.38.006347
37. J. Geng, C. Spiegelberg, S. Jiang, Narrow linewidth fiber laser for 100-km optical frequency Domain Reflectometry. IEEE Photonics Technol. Lett. **17**(9), 1827–1829 (2005). doi:10.1109/LPT.2005.853258
38. H. Stoehr et al., Diode laser with 1 Hz linewidth, Optics Lett. **31**(6), 736–738 (2006). doi:10.1364/OL.31.00073641
39. T.W. Hänsch, Repetitively pulsed tunable dye laser for high resolution spectroscopy. Appl. Opt. **11**, 895–898 (1972). doi:10.1364/AO.11.000895
40. L.E. Hargrove, R.L. Fork, M.A. Pollack, Locking of He–Ne laser modes induced by synchronous intracavity modulation, Appl. Phys. Lett. **5**(1), 4–5 (1964). doi:10.1063/1.1754025
41. U. Keller et al., Semiconductor saturable absorber mirrors (SESAMs) for femtosecond to nanosecond pulse generation in solid-state lasers. IEEE J. Sel. Top. Quantum Electron. **2**(3), 435–453 (1996). doi:10.1109/2944.571743
42. F.J. Duarte, *"The Physics of Multiple-Prism Optics", Chapter 4 in Tunable Laser Optics* (Elsevier Science, San Diego, CA, 2003)

43. I.D. Jung et al., Self-starting 6.5 fs pulses from a Ti: sapphire laser. Opt. Lett. **22**(13), 1009–1011 (1997). doi:10.1364/OL.22.001009
44. F.X. Kärtner et al., Design and fabrication of double-chirped mirrors. Opt. Lett. **22**(11), 831–833 (1997). doi:10.1364/OL.22.000831
45. C.J. Saraceno et al.,Sub-100 femtosecond pulses from a SESAM modelocked thin disk laser. Appl. Phys. B **106**, 559–562 (2012). doi:10.1007/s00340-012-4900-5
46. P.M. Paul et al., Observation of a train of attosecond pulses from high harmonic generation. Science **292**(5522), 1689–1692 (2001). doi:10.1126/science.1059413
47. A. Baltuska et al., Attosecond control of electronic processes by intense light fields. Nature **421**, 611–615 (2003). doi:10.1038/nature01414
48. R. Kienberger et al., Atomic transient recorder. Nature **427**, 817–821 (2004). doi:10.1038/nature02277
49. E. Goulielmakis et al., Single-cycle nonlinear optics. Science **320**(5883), 1614–1617 (2008). doi:10.1126/science.1157846
50. K. Zhao et al., Tailoring a 67 attosecond pulse through advantageous phase-mismatch". Opt. Lett. **37**(18), 3891–3893 (2012). doi:10.1364/OL.37.003891
51. D.J. Bergman, M.I. Stockman, Can we make a nanoscopic laser? Laser Phys. **14**(3), 409–411 (2004)
52. M.A. Noginov et al., Demonstration of a spaser-based nanolaser. Nature **460**, 1110–1112 (2009). doi:10.1038/nature08318
53. R.E. Oulton et al., Plasmon lasers at deep subwavelength scale. Nature **461**(1), 629–632 (2009). doi:10.1038/nature08364
54. R.-M. Ma et al., Room-temperature sub-diffraction-limited plasmon laser by total internal reflection. Nat. Mater. **10**, 110–113 (2011). doi:10.1038/nmat2919
55. Y.-Jung Lu et al., Plasmonic nanolaser using epitaxially grown silver film. Science **337**(6093), 450–453 (2012). doi:10.1128/science.1223504
56. I.-Y. Park et al., Plasmonic generation of ultrashort extreme-ultraviolet light pulses. Nat. Photonics **5**, 677–681 (2011). doi:10.1038/nphoton.2011.258
57. A. Borot et al., Attosecond control of collective electron motion in plasmas. Nat. Phys. **8**, 416–421 (2012). doi:10.1038/nphys2269
58. P. Heissler et al., Few-cycle driven relativistically oscillating plasma mirrors—a source of intense, isolated attosecond pulses. Phys. Rev. Lett. **108**(23), 23500-3-7 (2012). doi:10.1103/PhysRevLett.108.235003
59. D. Strickland, G. Mourou, Compression of amplified chirped optical pulses. Opt. Commun. **56**(3), 219–221 (1985). doi:10.1016/0030-4018(85)90120-8
60. M. Perry et al., Petawatt laser pulses. Opt. Lett. **24**(3), 160–162 (1999). doi:10.1364/OL.24.000160
61. G. Overton, OSU's SCARLET laser aims for 500 TW/15 J/30 fs pulses. Laser Focus World **48**(6), 15–16 (2012)
62. E.W. Gaul et al., Demonstration of a 1.1 petawatt laser based on a hybrid optical parametric chirped pulse amplification/mixed Nd: glass amplifier. Appl. Opt. **49**(9), 1676–1681 (2010). doi:10.1364/AO.49.001676
63. Z. Wang et al., High-contrast 1.16 PW Ti: sapphire laser system combined with a doubled chirped-pulse amplification scheme and a femtosecond optical-parametric amplifier. Opt. Lett. **36**(16), 3194–3196 (2011). doi:10.1364/OL.36.003194
64. J.P Chambaret et al., "ILE 25 PW single laser beamline: the French step for the European Extreme Light Infrastructure", Lasers and Electro-Optics, CLEO 2007, *Photonic Applications Systems Technologies Conference*, Paper JWC4, Baltimore, May 6–11, 2007. doi:10.1109/CLEO.2007.4453722
65. J.P Chambaret et al., "The extreme light infrastructure project ELI and its prototype APOLLON/ILE: 'the associated laser bottlenecks'", *Frontiers in Optics—High Peak Power Laser Technology,* Paper FM12, San Jose, Oct 11–15, 2009. doi:10.1364/FIO.2009.FM12
66. C. Hernandez-Gomez et al., The Vulcan 10 PW project. J. Phys: Conf. Ser. **244**, 032006 (2010). doi:10.1088/1742-6596/244/3/032006

67. J. Hecht, NIF is up and running at last. Laser Focus World **45**(11), 33–39 (2009)
68. D.A.G. Deacon et al., First operation of a free-electron laser. Phys. Rev. Lett. **38**, 892–894 (1977). doi:10.1103/PhysRevLett.38.892
69. J. Amann et al., Demonstration of self-seeding in a hard-X-ray free-electron laser. Nat. Photonics **6**, 693–698 (2012). doi:10.1038/nphoton.2012.180
70. T. Popmintchev et al., Phase matching of high harmonic generation in the soft and hard X-ray regions of the spectrum. Proc. Natl. Acad. Sci. U.S.A. **106**(26), 10516–10521 (2009). doi:10.1073/pnas.093748106
71. T. Popmintchev et al. Phase-matched generation of coherent soft and hard X-rays using IR lasers, U.S. Patent Application Publication 2011/0007772, 13 Jan 2011
72. T. Popmintchev et al., Bright coherent ultrahigh harmonics in the keV X-ray region from mid-infrared femtosecond lasers. Science **8**(336), 1287–1291 (2012). doi:10.1126/science1218497
73. N. Rohringer et al., Atomic inner-shell X-ray laser at 1.46 nanometres pumped by an X-ray free-electron laser. Nature **481**, 488–491 (2012). doi:10.1038/nature10721
74. C. Dang et al., Red, green and blue lasing enabled by single-exiton gain in colloidal quantum dot films. Nat. Nanotechnol. **7**, 335–339 (2012). doi:10.1038/nnano.2012.61
75. J.C. Knight et al., All-silica single-mode optical fiber with photonic crystal cladding. Opt. Lett. **21**(19), 1547–1549 (1996). doi:10.1364/OL.21.001547
76. J.K. Ranka, R.S. Windeler, A.J. Stentz, Visible continuum generation in air-silica microstructure optical fibers with anomalous dispersion at 800 nm. Opt. Lett. **25**, 25–27 (2000). doi:10.1364/OL.25.000025
77. J.M. Dudley, G. Genty, S. Coen, Supercontinuum generation in photonic crystal fiber. Rev. Mod. Phys. **78**, 1135–1184 (2006). doi:10.1103/RevModPhys.78.1135
78. T.A. Birks, W.J. Wadsworth, P.St. J. Russell, Supercontinuum generation in tapered fibers. *Opt. Lett.* **25** (19), 1415–1417 (2000). doi:10.1384/OL.25.001415
79. F. Hoos, S. Pricking, H. Giessen, Compact 20 MHz solid-state femtosecond whitelight-laser. Opt. Express **14**(22), 10913–10920 (2006). doi:10.1364/OE.14.010913
80. B. Zaks, R.B. Liu, M.S. Sherwin, Experimental observation of electron-hole recollisions. Nature **483**, 580–583 (2012). doi:10.1038/nature10864
81. A. Neumann et al., Four-color laser white illumination demonstrating high color rendering quality. Opt. Express **19**(S4), A982–A990 (2011). doi:10.1364/OE.19.00A982
82. B. Redding, M.A. Choma, H. Cao, Speckle-free laser imaging using random laser illumination. Nat. Photonics **6**, 355–359 (2012). doi:10.1038/nphoton.2012.90
83. A.M. Weiner et al., Experimental observation of the fundamental dark soliton in optical fibers. Phys. Rev. Lett. **61**(21), 2445–2448 (1988). doi:10.1103/PhysRevLett.61.2445
84. H.P. Li et al., Dark pulse generation in a dispersion-managed fiber laser. Laser Phys. **22**(1), 261–264 (2012). doi:10.1134/S1054660X12010094
85. M. Feng et al., Dark pulse quantum dot diode laser. Opt. Express **18**(13), 13385–13395 (2010). doi:10.1364/OE.18.013385
86. M. Liertzer et al., Pump-induced exceptional points in lasers. Phys. Rev. Lett. **108**, 17390-1-5 (2012). doi:10.1103/PhysRevLett.108.173901
87. M.C. Gaither, S.H. Yun, Single cell biological lasers. Nat. Photonics **5**, 406–410 (2011). doi:10.1038/nphoton.2011.99
88. Y.D. Chong et al., Coherent perfect absorbers: time-reversed lasers. Phys. Rev. Lett. **105**(5), 053901–053904 (2010). doi:10.1103/PhysRevLett.105053901
89. W. Wan et al., Time-reversed lasing and interferometric control of absorption. Science **331** (6019), 889–892 (2011). doi:10.1126/science.1200735
90. J.G. Bohnet et al., A steady-state superradiant laser with fewer than one intracavity photon. Nature **484**, 78–81 (2012). doi:10.1038/nature10920

Chapter 7
Variation and Control of Light Propagation Properties

7.1 Slowing the Speed of Light

7.1.1 Refractive Index

One of the most well-known methods of reducing the speed or velocity of light is the propagation of light through a medium having a refractive index greater than unity. The speed of light is reduced while traveling through the medium. The refractive index n is defined as the ratio of the speed of light v in the medium to the cosmic speed of light c in vacuum (2.998×10^8 m/s).

$$n = \frac{c}{v} \tag{7.1}$$

Since the speed of light in the medium $v = c/n$, for common crown glass ($n = 1.52$), the internal light speed is 1.97×10^8 m/s. For diamond ($n = 2.42$), the internal light speed is 1.24×10^8 m/s. Germanium has one of the highest refractive indices of natural materials ($n = 4.01$), and transmits in the infrared region. It produces an internal light speed of 7.48×10^7 m/s, a reduction to $\approx 25\%$ of the cosmic speed c. The refractive index of a material depends on the wavelength of the transmitted light, resulting in dispersion of non-monochromatic light.

Around 1880 Lorentz developed a theory of frequency dispersion in dielectric media which described this reduction in light velocity [1]. This approach used the effect of an electromagnetic wave on the dielectric polarization of the medium.

7.1.2 Phase Velocity and Group Velocity

The distinction between *phase velocity* v_p (sometimes called *wave velocity*) and *group velocity* v_g in optics was first formalized by Lord Rayleigh in 1881. The

D.F. Vanderwerf, *The Story of Light Science*,
DOI 10.1007/978-3-319-64316-8_7

velocity used in Eq. (7.1) is the phase velocity. The phase velocity represents the speed of any point on a sinusoidal wave of infinite length. It is usually given by the expression

$$v_p = (f)(\lambda) = (\omega/2\pi)(2\pi/k) = \omega/k, \tag{7.2}$$

where

λ = wavelength [m],
f = frequency [s^{-1}],
ω = angular frequency [rad/s],
k = propagation angular wavenumber = $2\pi/\lambda$ [rad/m].

Consider two sinusoidal waves of close wavelength propagating through a dispersive medium at slightly different speeds, v_{p1} and v_{p2}, as shown in Fig. 7.1a. Then the superposition of these waves forms a modulated envelope, as shown in Fig. 7.1b. The peak of the modulated envelope corresponds to the coincidence peaks of each sinusoidal wave. Both waves move at their own phase velocity until another coincidence peak is formed, corresponding to a new peak position of the modulated envelope. The envelope moves at a *group velocity* v_g, which is different than the phase velocity v_p of either sinusoidal wave. It is defined as

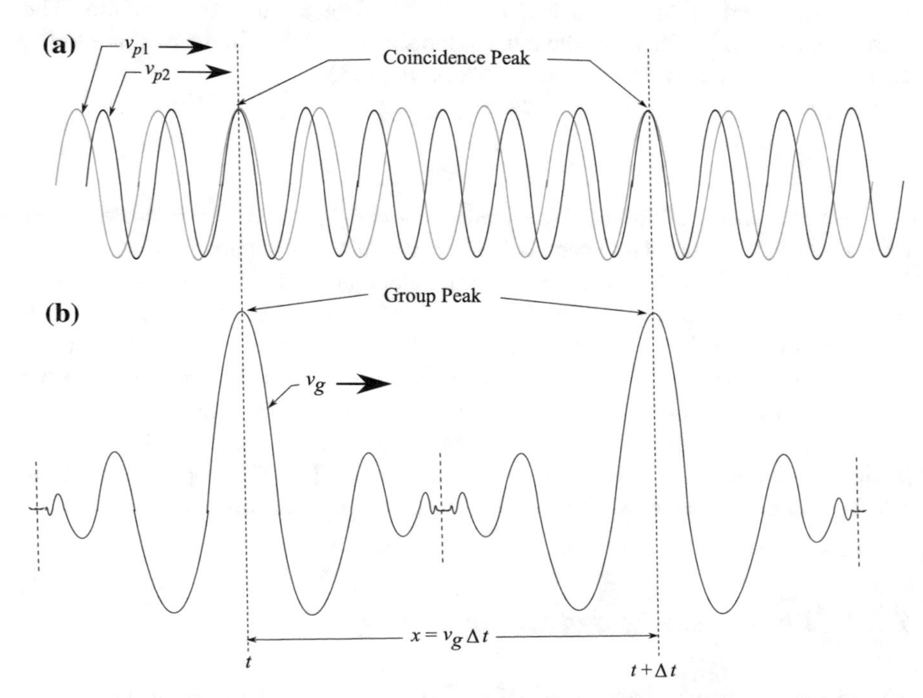

Fig. 7.1 **a** Two sinusoidal waves of close wavelength propagating through a dispersive medium at slightly different speeds, v_{p1} and v_{p2}. **b** Superposition of two sinusoidal waves to form a group wave traveling at a group velocity v_g

$$v_g = \frac{d\omega}{dk}. \tag{7.3}$$

The phase and group velocities can then be related by

$$v_g = v_p\left(1 + \frac{\lambda\, dn}{n\, d\lambda}\right), \tag{7.4}$$

where $\frac{dn}{d\lambda}$ represents the change in the material refractive index with wavelength. Most materials have normal dispersion, where $\frac{dn}{d\lambda} > 0$, and the group velocity exceeds the phase velocity. However, some materials can be engineered to have absorption bands over a specific spectral region where $\frac{dn}{d\lambda} < 0$ where the phase velocity exceeds the group velocity. This is called *anomalous dispersion.* Of course, where there is no dispersion, $\frac{dn}{d\lambda} = 0$, and there is no distinction between phase and group velocities.

Two useful quantities are the *group delay* $T_g = \frac{d\varphi}{d\omega}$ [s], the *group velocity dispersion* $GVD = \frac{\lambda^3}{2\pi c^2}\left(\frac{d^2n}{d\lambda^2}\right) = \frac{d}{d\omega}\left(\frac{1}{v_g}\right)$ [s^2/m], and the *group delay dispersion* $GDD = \frac{d}{d\omega}\left(\frac{1}{v_g}\right)L$ [s^2], where L is the length of the dispersive medium.

7.1.3 Some Slow Light Methodologies

In Eq. (7.1) the refractive index n refers to the phase velocity. The *phase refractive index n* of natural materials seldom exceeds a value of about four, e.g. $n = 3.96$ for silicon and $n = 4.01$ for germanium. A *group refractive index* n_g can be defined as

$$n_g = n + \omega\frac{dn}{d\omega}. \tag{7.5}$$

By control of the value of the dispersive component $\frac{dn}{d\omega}$, very high values of n_g can be attained with magnitudes up to the order of 10^8. These extremely high values of n_g can produce significant reduction of the group velocity $v_g = c/n_g$ of light in certain media. This phenomenon, where $v_g << c$, is usually called *slow light.* Several methods have been demonstrated to produce ultra-slow group velocities using material dispersion. In contrast, the phase velocity v_p cannot be significantly reduced in dispersive media.

One technique for increasing the group refractive index and dispersion in a transparent material is *stimulated Brillouin scattering* (SBS), which can produce a significant reduction in light speed in a crystal or optical fiber. As shown in Fig. 7.2a, a pump laser beam of frequency ω_{pump} enters an optically transparent material. This pump beam interacts with a counter-propagating signal laser of frequency ω_{signal} (sometimes called the *Stokes photon*), where ω_{signal} is slightly less than ω_{pump}. From the interaction of these waves, a process called *electrostriction* produces a variation in the density or refractive index of the medium. Electrostriction forms a moving acoustic interference wave (sometimes called an *acoustic phonon*) that functions as a Bragg grating. Light is backscattered from this acoustic grating as a Stokes wave with a frequency $\omega_{Stokes} = \omega_{pump} - \omega_{signal}$. This process produces Brillouin scattering [2]. For the slowing of light, the scattered wave propagates in the opposite direction of the pump laser. As the input pump laser power increases, the nonlinear interaction between the acoustic wave and the scattered Stokes wave produces a *Stokes resonance* or *Stokes gain* and SBS occurs. See Fig. 7.2b.

Near the Stokes resonance position, there is a rapid change in the phase and group refractive indices. For significant light speed reduction, the much greater

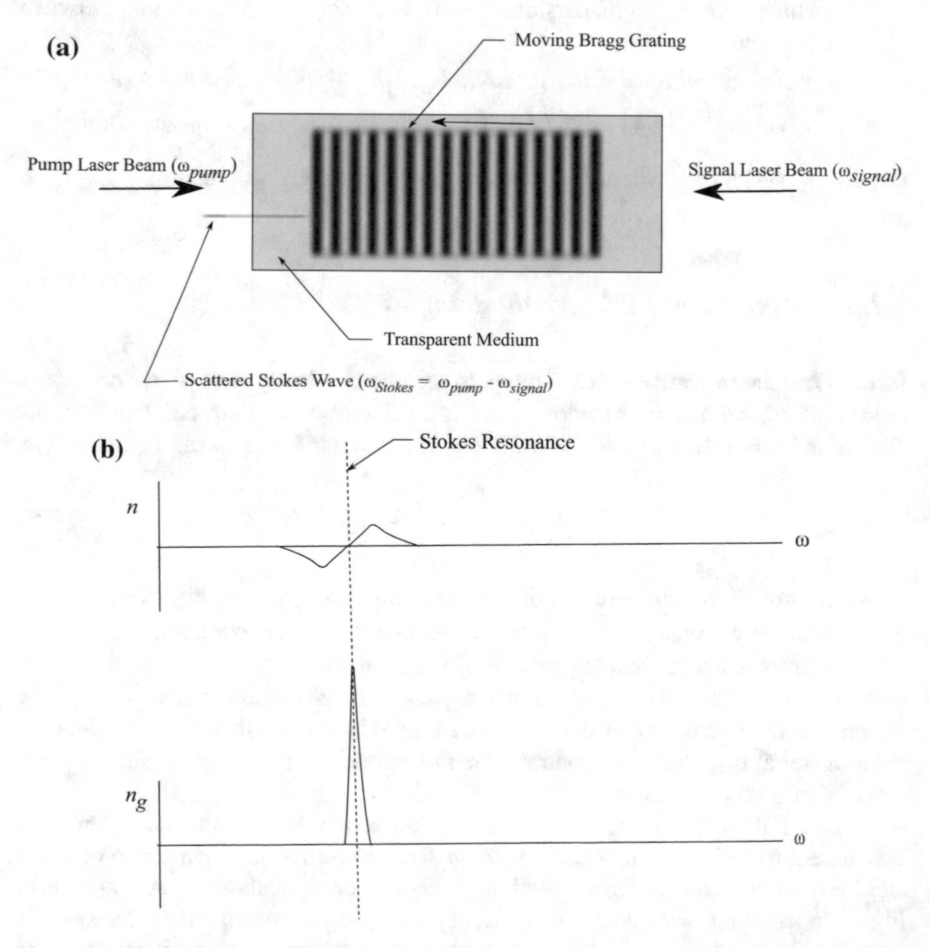

Fig. 7.2 **a** Illustration of the formation of a moving acoustic interference wave in a transparent material. **b** Variations of the phase and group refractive indices near the Stokes resonance position

increase in the group refractive index is the important factor. SBS can be produced in gases, liquids, or solids, and has found practical use in producing time delays in optical communication fibers [3–5].

Another technique for reducing the group velocity uses *electromagnetic induced transparency* (EIT). By applying an electromagnetic field to certain metal vapors (e.g. rubidium) having three energy levels, two narrow resonances (*hyperfine levels*) with a few gigahertz separation can occur at each side of a central absorption line tuned to group frequency ω_g. The value of ω_g is determined by a weak probe laser. By tuning a strong coupling or control laser to one of these side resonances, a narrow high-transmission region is formed at the center of the absorption line. The rapid change in refractive index along the steep slopes of this transparent region produces a slow group velocity v_g. Figure 7.3a illustrates the transitions for a three-level gas, and Fig. 7.3b illustrates the side resonances and narrow region of low absorption or high transparency for EIT. This is sometimes called *double atomic resonance*. In 1992 researchers at Stanford University used a 10 cm long Pb vapor cell with 7×10^{15} atoms/cm^3 to obtain a value $v_g \approx c/250$ at the 283 nm resonance transition [6]. This produced a group velocity $v_g \approx 1.2 \times 10^6$ m/s.

A third technique for ultraslow light propagation in a medium uses *coherent population oscillations* (CPO). Two independently modulated laser beams of unequal intensity at a common single wavelength are passed through a material. The interaction of the pump laser with the probe laser produces a periodic modulation of the pumped ground state population. This creates a spectral transparency or hole in the material at the beat frequency between the two beams. This narrow spectral hole has steep side slopes, producing a rapid change in the group refractive index and a significant slowing of the group velocity. CPO was realized at the University of Rochester in 2003 using a single laser and a ruby crystal at room temperature [7]. Figure 7.4 illustrates the basic experimental setup. An argon-ion laser at a single line wavelength of 514.5 nm was modulated to produce stronger continuous millisecond pump pulses and weaker sinusoidal modulated probe pulses. These beams were focused into a 7.25 cm long ruby rod. The beat frequency modulation between these two pulses created a narrow ($\approx$36 Hz) spectral hole centered at the laser wavelength $\lambda = 514.5$ nm. The time difference between the normal and slowed beams was measured using a beam splitter, a pair of detectors, and an oscilloscope. This delay was attributed to a group velocity v_g of 57.7 m/s in the ruby crystal spectral hole, which corresponds to $v_g \approx c/(5.2 \times 10^6)$ m/s.

7.1.4 Thermal deBroglie Wavelength and Bose-Einstein Condensates

Several techniques to produce ultraslow light propagation require the use of a Bose-Einstein condensate. In 1924 Satyendra Nath Bose derived Planck's law of blackbody radiation by considering the quantum mechanical properties of groups of

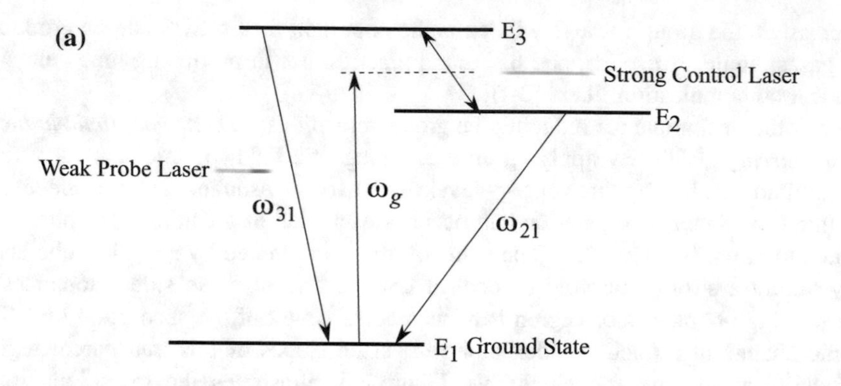

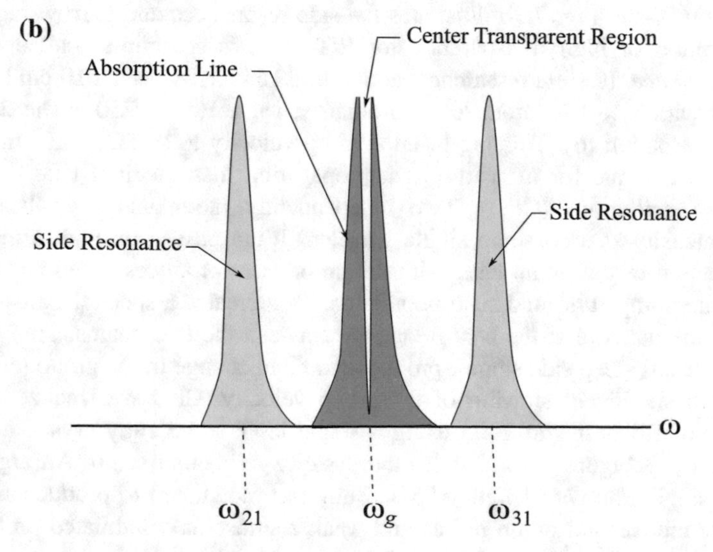

Fig. 7.3 **a** Illustration of the transitions between three energy levels to produce an electromagnetically induced transparency at frequency ω_0. **b** Illustration of the side resonances and the central absorption line, with the formation of the EIT region

photon particles [8]. These photon particles belong to a class of elementary particles called *bosons*, having a zero or integral spin number s. Photons have a spin number $s = 1$. Bosons are described by *Bose-Einstein statistics*, where the particles can occupy the same quantum or energy state. This is contrasted to another class of particles called *fermions*, which have half-integral spin numbers and are described by *Fermi-Dirac statistics*, where any two particles cannot occupy the same quantum or energy state. An example of a fermion is the electron, having a spin number $s = ½$.

A useful quantity for determining the quantum mechanical description of an atomic gas is the *thermal deBroglie wavelength* λ^T_{dBrog}, given by

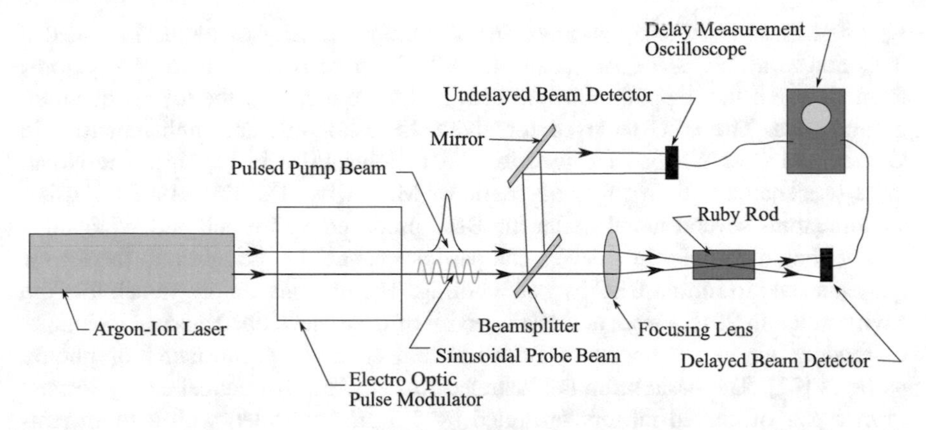

Fig. 7.4 Experimental setup for ultraslow light propagation using a single laser and a ruby crystal (This figure adapted with permission from Ref. [7]. Copyrighted by the American Physical Society.)

$$\lambda_{dBrog}^{T} = \frac{h}{\sqrt{2\pi mkT}}, \tag{7.5}$$

where h is Planck's constant, m is the mass of the atomic particle, k is Boltzmann's constant, and T is the temperature of the gas in degrees Kelvin. If the average distance between atoms is d, the classical theory of the gas applies when $d \gg \lambda_{dBrog}^{T}$, and the quantum mechanical theory prevails when $d \approx \lambda_{dBrog}^{T}$. The quantum state can be achieved by super-cooling of the gas particles. When certain monatomic boson gases are cooled close to absolute zero (0 K or −273.15 C), a new state of matter is created which is known as the (BEC). This state of matter was predicted by Einstein in 1924 in a supportive paper to Bose [9, 10]. In a *Bose-Einstein condensate* a significant fraction of the atoms would condense into the lowest quantum state and form a single "super-atom".

In 1995 the first observable BEC was produced in the laboratory at JILA-NIST and the University of Colorado, using a super-cooled dilute vapor of alkali atoms [11]. In order to achieve a BEC the dimensionless phase space density ρ_{ps} of the atoms must satisfy the following relationship:

$$\rho_{ps} = N(\lambda_{dBrog})^3 > 2.612, \tag{7.6}$$

where N is the number density of atoms [atoms/cm^3].

Rubidium atoms were cooled to a few millionths of a degree above absolute zero by placing a sample of atoms in a magnetic trap with subsequent evaporative cooling. A magnetic trap formed by the intersection of six laser beams confined the gas sample to a compact, high-density volume. A radio frequency magnetic field of frequency f_{evap} was injected into the sample while elastic scattering provided efficient evaporative cooling. At a temperature of about 170 nK and a value

$f_{evap} \approx 4.23$ MHz, a sharp increase in the number density of atoms indicated a transition to a BEC state. At $f_{evap} \approx 4.1$ MHz, a pure BEC of about 2000 atoms formed, where the trapping potential confined these atoms to the lowest quantum ground state. The BEC survived for about 15 s. For this accomplishment, Eric Cornell and Carl Wieman received the 2001 Nobel Prize in Physics. The Nobel Prize was shared with Wolfgang Ketterle of MIT, who also formed a BEC using sodium atoms several months after the BEC produced by Cornell and Wieman.

For photon particles in a cavity, the particles cannot be brought into the lowest quantum state to form a BEC by pure cooling. The photons simply vanish into the cavity walls. In 2010 a team at the University of Bonn developed a novel technique to produce a room temperature Bose-Einstein condensate consisting of photon particles [12]. The basic setup is illustrated in Fig. 7.5a. An optical cavity formed from a pair of curved mirrors separated by 1.5 μm was filled with a room temperature dye solution. Photons were then injected into the cavity from a variable intensity pump laser. The microcavity had a cutoff wavelength of 585 nm and supported only seven half-wavelengths. The supported photon energy range was determined by resonance with the dye molecules and the cavity spacing, where the curved mirrors formed a trapping potential for the photon particles. Since the photon is a massless boson that does not interact with other photons, the required thermal equilibrium of the photons for a BEC was achieved by scattering with the dye molecules during the time when the photons were trapped within the microcavity. The number of photons in the cavity was then increased by increasing the pump laser intensity to a level where a sharp peak wavelength of 585 nm leaked through the output coupler mirror. This was recorded by a spectrophotometer. This sharp peak indicated that the cavity photons had condensed into the lowest quantum state, and a photon BEC had formed at room temperature. Figure 7.5b illustrates the formation of an observable Bose-Einstein photon condensate.

7.1.5 Extreme Reduction of Light Speed in a Super-Cooled Atomic Gas

The propagation of light pulses through an opaque Bose-Einstein condensate can produce an extreme reduction in light speed. In 1999 Lene Hau and colleagues at the Rowland Institute of Science, Cambridge, Massachusetts explored pulse propagation speeds in a super-cooled BEC of sodium atoms [13]. The experimental setup is illustrated in Fig. 7.6. A sodium atomic gas was confined in a magneto-optical trap by a group of ring magnets. These atoms were initially cooled to a temperature of $\approx$1mK, yielding a number density $N \approx 6 \times 10^{11}$ atoms/cm^3. The atoms were then polarization gradient cooled to $\approx$1mK, and are brought to a single atomic state by optical pumping. Next the cloud was cooled by evaporation for 38 s to a transition temperature of $T_c = 435$ nK, where Bose-Einstein

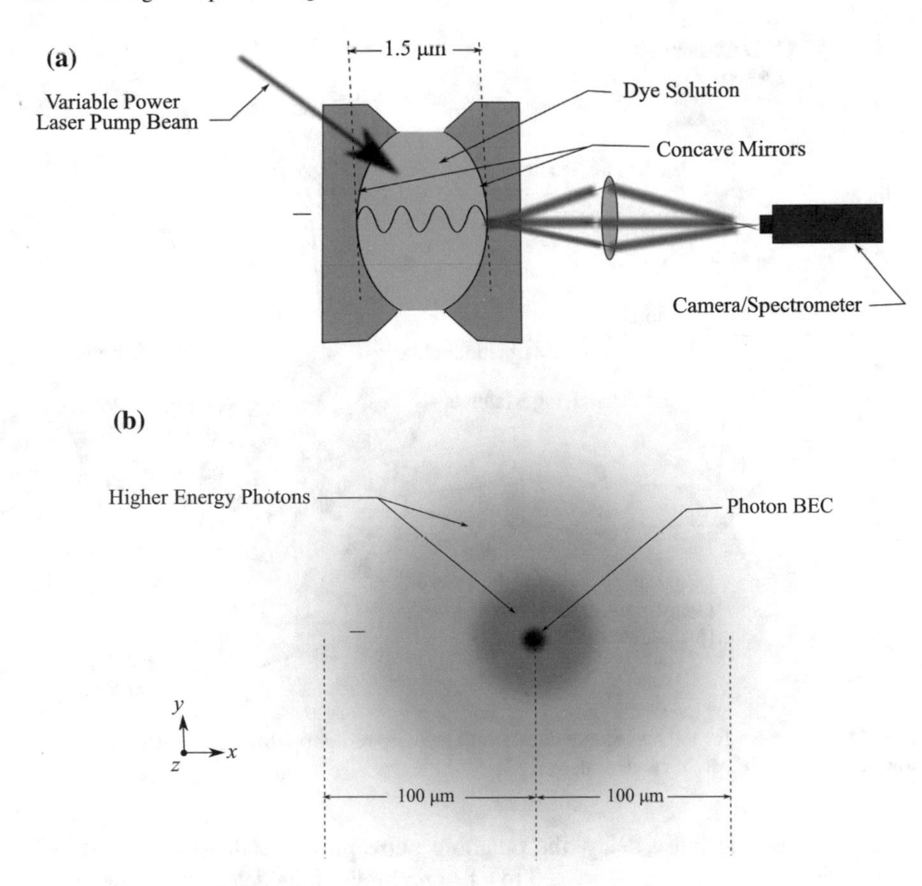

Fig. 7.5 **a** Illustration of apparatus producing the first room temperature Bose-Einstein condensate of photon particles. **b** Illustration of a formed Bose-Einstein condensate of photon particles

condensation began with a number density $N \approx 5 \times 10^{13}$ atoms/cm^3. The cloud had a prolate ellipsoid shape elongated along the z-axis.

A pulsed probe beam of frequency ω_p was tuned to an absorption resonance line in the sodium atomic gas. A strong coupling laser beam was also tuned to this absorption line, where quantum interference produced a narrow dip in the absorption line. This yielded a very high transparency peak (EIT) with a very steep dispersion curve. The group velocity v_g was related to the probe frequency and refractive index using the following equation:

$$v_g = \frac{c}{n + \omega_p \frac{dn}{d\omega_p}}, \tag{7.7}$$

where c is the speed of light and n is the refractive index at the probe frequency ω_p.

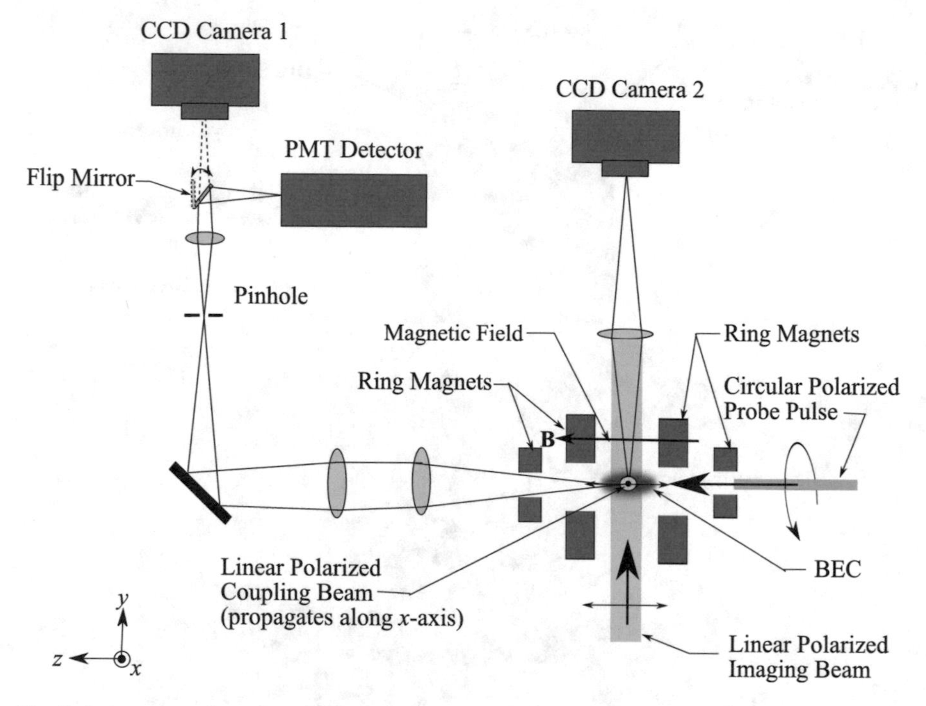

Fig. 7.6 Apparatus for the measurement of group velocity of light pulses propagated through an ultra-cooled Bose-Einstein condensate

To measure the pulse delay, the outgoing pulse probe beam was focused to a 15 μm diameter pinhole and directed to a fast multiplier tube detector. By the use of a flip mirror, the pinhole was imaged in the *x*-*y* plane of the BEC and formed an image of a 15 μm diameter central section of the BEC at CCD camera **1**. CCD camera **2** captured an image of the BEC in the *x*-*z* plane. Pulse group velocities were determined by measuring a series of pulse delay values for a BEC further cooled to temperatures between 50 nK and 2.5 μK. For each pair of pulse delay and temperature values, a corresponding value of v_g was calculated. At a condensate temperature of ≈50 nK, more than 90% of the atoms were in a BEC state, and a value of $v_g \approx 17$ m/s was obtained for pulse durations of several microseconds.

7.1.6 Ultraslow Light Propagation in Higher Temperature Atomic Gases

Around the same time as the ultra-cold BEC group velocity measurements of Hau, an experiment was performed to observe ultraslow light in an optically opaque hot atomic gas [14]. The experimenters used a 2.5 cm long cell filled with isotropic ^{87}Rb atoms and a buffer Ne gas at 30 Torr, with a ^{87}Rb density $\approx 2 \times 10^{12}$ atoms/cm^3.

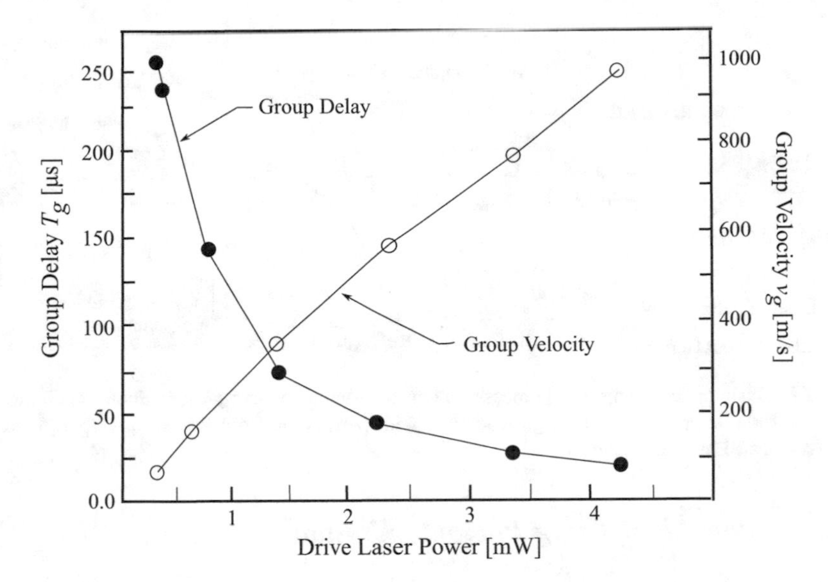

Fig. 7.7 Group delays and average group velocities for various drive laser powers in the experiment of Kash (This figure adapted with permission from Ref. [14]. Copyrighted by the American Physical Society.)

A 2 mm diameter drive laser and a weaker probe laser were used to achieve EIT in this highly nonlinear phase coherent group of atoms (sometimes called a *phaseonium*). Group delays T_g for the D_1 resonance line ($\lambda = 765$ nm) of the ^{87}Rb atoms at ≈360 K (87 C) were determined by measurements of the time retardation of the amplitude modulation through the cell. From these group delays, average values of v_g as low as 90 m/s were inferred. Figure 7.7 shows the observed group delays and average group velocities for various drive laser powers.

A method to achieve slow light propagation in a room temperature atomic gas uses the technique of resonant light propagation, which produces an effect similar to EIT. In 1999 a research team from Berkeley, California, and Gatchina, Russia propagated light pulses through a cell filled with room temperature ^{85}Rb gas [15]. The cell wall was coated with an anti-relaxation wall coating (e.g. paraffin) to maintain a long-life coherence state. The long coherence state produced very narrow line resonances between three energy levels in the gas, yielding very large group time delays T_g and very slow group velocities v_g. A weak magnetic field controlled the group velocity. Figure 7.8 illustrates the basic optics in the experiment. A continuous wave diode laser of 795 nm wavelength produced a D_1 transition line in the gas. Using a Faraday rotator energized by a current pulse, a linear polarized pulse probe beam and an orthogonal linear polarized control beam interacted within the gas, and the transmitted probe intensity was measured with a photodetector. Pulse delay measurements were obtained using the photodetector signal timing referenced to the pulse generation timing. This experiment yielded a group time delay $T_g \approx 13$ ms and a group velocity $v_g \approx 8$ m/s.

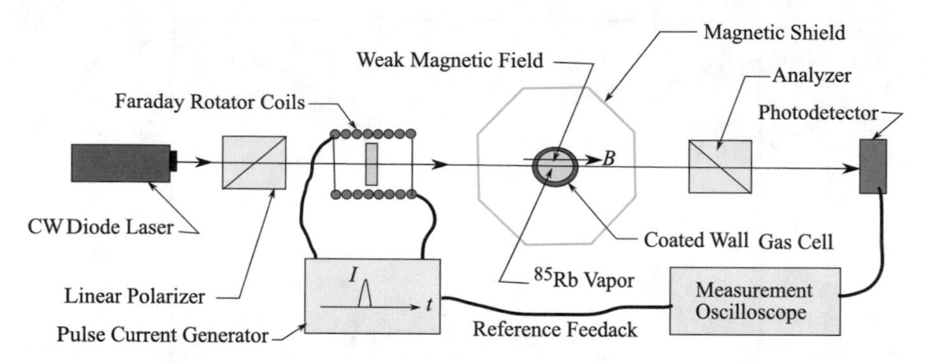

Fig. 7.8 Basic optical setup for the measurement of group delay and group velocity in an atomic vapor at room temperature (This figure adapted with permission from Ref. [15]. Copyrighted by the American Physical Society.)

7.1.7 Slow Light Using Photonic Crystals

The use of a photonic crystal was described in Sect. 5.2.6 to control the direction of emitted light from a liquid crystal diode. Around the year 2001 it was discovered that photonic crystals could also produce very high group refractive indices, yielding a significant reduction in light speed. Rather than using material resonances to produce very high group refractive indices, slow light can be achieved using a photonic crystal waveguide (PCW). The slow light is enhanced when a predetermined pattern of holes is missing from the crystal. When the pattern is linear, it is called a *line defect* and a photonic wave guide is created, having a photonic band gap. Slow light occurs near the edge of this photonic band gap. Figure 7.9 illustrates a single line defect in a silicon/air/silicon photonic crystal slab with thickness ≈0.2 μm, as fabricated by Notomi and co-workers in 2000 [16].

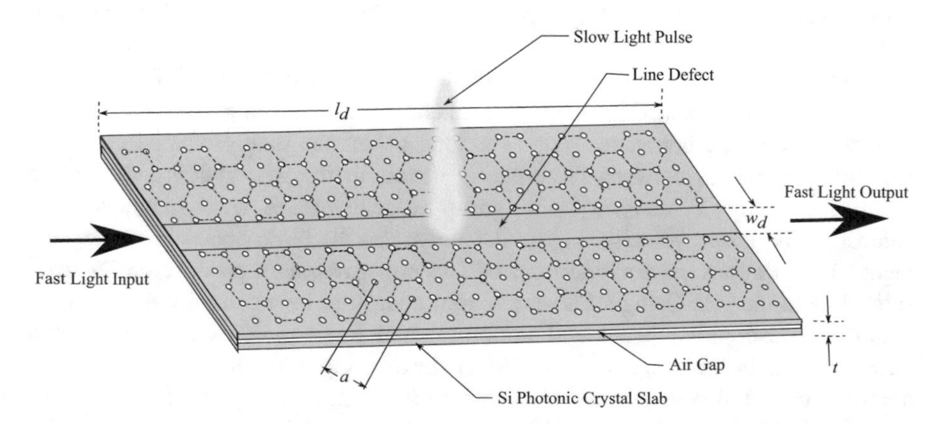

Fig. 7.9 Illustration of a photonic crystal waveguide with a linear defect to produce slow light (hole diameters and spacings are not to scale)

The holes were arranged in a repeating hexagonal pattern. The hexagons had a lattice constant $a = 0.39$, the holes had a diameter of $0.55a$, and the line defect had a varying width w_d with a length $l_d = 172$ μm. The line defect formed a waveguide with a defined cutoff. In the waveguide, it was found that the dispersion of the group refractive index n_g was very large, resulting in a deduced group velocity v_g up to 90 times slower than in air for a wavelength near 1520 nm.

For slow light photonic crystals to be useful, the slow light must have a wide frequency bandwidth that is free of distortion. However, high-order dispersion in the waveguide that accompanies slow light generation contributes to increased distortion of the signal pulse. To generate distortion-free wide-band dispersion of slow light, several methods are employed. The first is *dispersion compensated slow light*, and is obtained by balancing positive or negative group dispersion in the input section of the guide with opposite dispersion in the output section of the guide. This can be accomplished by offsetting holes in the front and back sections of the line defect by opposing half-periods, and then coupling the sections using a center patterned chirped region. A second technique is *zero-dispersion slow light*, where

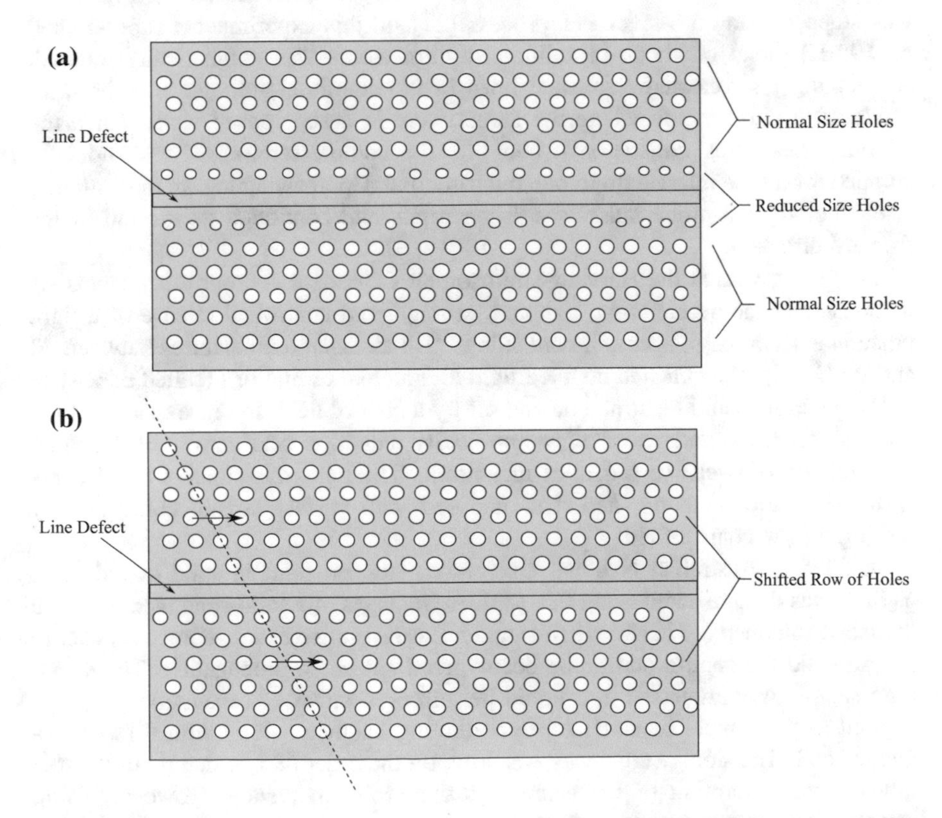

Fig. 7.10 **a** A method to produce zero-dispersion slow light in a photonic crystal using a variation of hole size. **b** Another method to produce zero-dispersion slow light in a photonic crystal using shifted rows of holes

the hole pattern geometry adjacent to the sides of the line defect is altered. As shown in Fig. 7.10a, the holes on either side of the line defect have a smaller diameter than the normal hole pattern of the photonic crystal [17]. Figure 7.10b illustrates another zero-dispersion technique where a row of holes on both sides of the line defect waveguide has a period shift relative to the normal hole pattern [18]. Two excellent review articles on slow light in photonic crystals have been written [19, 20].

7.2 Stopped Light and Storage of Light Pulses

In the extreme limit of slow light, light pulses have been completely stopped and stored for later use. The light pulses are stored as stable coupled atomic excitations in a medium, which exactly replicate the properties of the incident light pulse, and can subsequently be reconverted into a light pulse traveling at normal speed. In 2001 at the Roland Institute for Science, Cambridge, Massachusetts, a light pulse was stopped using a 3-level EIT process [21]. In this experiment a super-cooled ($\approx$0.9 μK) cloud of about 11 million sodium atoms was magnetically trapped. A coupling laser created resonance lines in the atomic cloud, and a probe laser produced a pulse region where the atoms were in a superposed state. When the coupling laser was quickly switched off, the compressed pulse was suddenly stopped for a time interval up to one microsecond. Upon switching on the coupling laser, the stopped probe pulse was regenerated, and continued its motion in the forward direction.

That same year at the Harvard-Smithsonian Center for Astrophysics, scientists demonstrated the deceleration, trapping, storage, and controlled release of a light pulse in a 4 cm long vapor cell containing ^{87}Rb atoms at temperatures between 70 and 90 C [22]. They created coupled light and atomic excitations (called dark-state polaritons) that could be turned on and off by a control field. In this experiment the input signal pulse underwent a group velocity reduction near zero, with a spatial compression of over five orders of magnitude. The light was stored up to 0.5 ms with the control field on, and then released with perfect atomic coherence by turning off the control field.

In 2005 at Australian National University, stopped light in solid doped Pr^{3+}: Y_2SiO_5 was demonstrated using EIT [23]. Advantages of a solid-state medium are a thousand-fold increase in storage time and a less restrictive beam geometry. Also, in a rigid solid the control and probe beams need not be co-propagating. The 4 mm long sample was immersed in a liquid helium cryostat. A DC magnetic field was applied to the sample by three superconducting magnets, along with a radio frequency coil. The storage efficiency was low, on the order of 1%, mainly due to the low optical absorption of the sample at the probe frequency. However, using counter-propagating control and probe beams, the light pulse was stopped for several seconds, which is several orders of magnitude greater than reported for vapor media.

A unique process developed in Germany has demonstrated the storage of a light pulse up to a minute in a rare-earth ion-doped crystal using EIT and an evolutionary genetic algorithm [24]. An external three-dimensional DC magnetic field on a Pr^{3+}: Y2SiO5 crystal caused the hyperfine levels to split and shift into 36 components. Two separate probe and control laser pulses drove the system into an atomic coherence of hyperfine states. The probe pulse was stored by EIT. To retrieve the stored probe pulse, a control read pulse formed beats with various atomic coherences to generate an output beam having identical properties as the probe beam. The output signal was read by a detector and profiled by a CCD camera. The atomic coherences produced up to 1296 lines in the absorption spectrum. To determine the optimum conditions to produce EIT storage duration approaching the regime of the population lifetime $T_1 \approx 100$ s, an evolutionary approach was applied using a genetic algorithm. To obtain an optimized preparation pulse, the three-dimensional magnetic field was optimized by an evolutionary algorithm to obtain long coherence times through a self-learning loop. The loop went through several hundred generations, until the *gene sequences*, represented by pulse shapes, converged toward an optimum preparation pulse sequence. Using this technique, light storage times up to one minute were observed.

7.3 Time-Reversed Light Using a Photonic Crystal

In 2005 a time-reversed light pulse was produced at Stanford University using a two-dimensional photonic crystal with a modulated refractive index [25]. The coherence and information encoded properties of the pulse are completely retained during the time-reversal process. The principle of the device is illustrated in Fig. 7.11. The substrate has a fixed refractive index $n = 1.5$. There is a multiplicity of dielectric rods of $n = 3.5$ with a radius $r = 0.2a$, where a is the lattice constant. In addition, there are several larger holes with a radius $r = 0.5a$ having a slightly tunable index ($\delta n/n < 10^{-4}$) around $n = 3.5$. The top and bottom halves of the array form subsystems **A** and **B**, each of which functions as a coupled resonator optical waveguide. The evanescent coupling rates between nearest neighbors in each subsystem are given by α_A and α_B. When $\alpha_A = -\alpha_B$, the two subsystems have opposite dispersion. Several air holes of radius $r = 0.2a$ in subsystem **B** yield $\alpha_B = -\alpha_A = 1.89 \times 10^{-3}(2\pi c/a)$.

In operation, an input pulse with wave vector $\boldsymbol{k}$ enters subsystem **A**, which is tuned to the resonance frequency ω_k. It is then detuned by refractive index modulation, while the pulse in system **B** is tuned into resonance by an amount $-\omega_k$. The spectrum of this pulse in subsystem **B** then evolves into the same wave vector state, but with an opposite detuning. The resulting time-reversed spectrally inverted wave moves backward through subsystem **B**, and exits the photonic crystal parallel to the input pulse. However, this time-reversal of light does not imply that the input pulse and the exit pulse occur at the same moment in time.

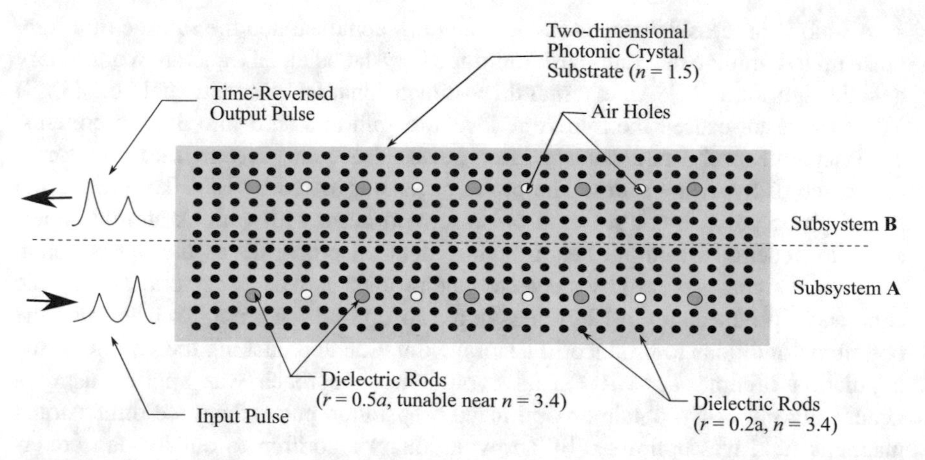

Fig. 7.11 Time-reversed light pulse using a two-dimensional photonic crystal with a modulated refractive index (This figure adapted with permission from Ref. [25]. Copyrighted by the American Physical Society.)

7.4 Exceeding the Cosmic Speed of Light in a Medium

7.4.1 The Speed of Light and Causality

According to Einstein's 1905 special theory of relativity, the speed of light in vacuum (free-space or cosmic light speed c) is the maximum speed achievable by any object.

Any light pulse exceeding the cosmic light speed is called *superluminal,* and particles that could surpass the cosmic speed are often called *tachyons*. If a light pulse can exceed the speed of light c, then the question arises whether transmitted information or a signal from an observer could influence an event that could never be observed. This is seemingly a violation of the widely-accepted law of causality, since the effect (event) would precede the cause (signal), and the event lies in the unknown zone or non-causal region of the light cone, as described in Sect. 3.6.1. For an electromagnetic wave propagating with a phase velocity $v_p > c$, no information can be transmitted since there is no variation in the wave pattern. In certain anomalous dispersive media, a Gaussian pulse envelope can travel at a group velocity $v_g > c$, where the group velocity occurs at the pulse peak However, the velocity of the half-maximum point of the leading edge of the pulse, or *signal velocity*, determines the arrival of information. Since the signal velocity cannot exceed the cosmic speed of light c, a signal in any dispersive material cannot travel faster than c. As the pulse propagates, it undergoes continuous reshaping such that the pulse peak can never overtake the leading edge of the pulse, even for $v_g > c$. Thus the principle of causality is preserved [26].

7.4.2 *Superluminal Light in Dispersive Media*

Figure 7.12 illustrates a cell of length L composed of a dispersive medium having a group refractive index n_g. The positive time delay ΔT of the light pulse traveling through the cell compared to the pulse traveling through free-space is given by $\Delta T = (n_g - 1)(L/c)$. From this equation, it can be inferred that if one had a material with a negative group reflective index, then ΔT would be negative and the pulse would move faster than the value of c. This superluminal light would appear to leave the cell before it entered.

In 2000 at the NEC Research Institute, New Jersey, superluminal light pulses were produced using gain-assisted anomalous dispersion in atomic cesium vapor at 30 C [27]. The glass cell containing the vapor was coated with paraffin and had a length $L = 6$ cm. Two continuous wave strong pump beams with energy and frequencies $E_1(\omega_1)$ and $E_2(\omega_2)$ entered the three energy-level cesium vapor. A tunable CW probe laser beam E_p created two closely-spaced maximum gain points where the probe beam frequency was resonant with transitions caused by the pump beam frequencies ω_1 and ω_2. A low-absorption or "transparent" region then formed between the gain peaks, where the anomalous dispersion took place. The top curve of Fig. 7.13 shows the resulting gain curve with a peak frequency separation of 2.7 MHz. Using a radio-frequency interferometric technique, the group refractive index n_g was measured as a function of the probe frequency. The lower curve of Fig. 7.13 shows the negative change in n_g, occurring over a probe frequency range of 1.9 MHz. This negative group refractive index in the anomalous dispersion region generated about a 62 ns pulse advancement shift, which yielded an effective value of $n_g \approx -310$, with no observable pulse distortion. Thus the superluminal

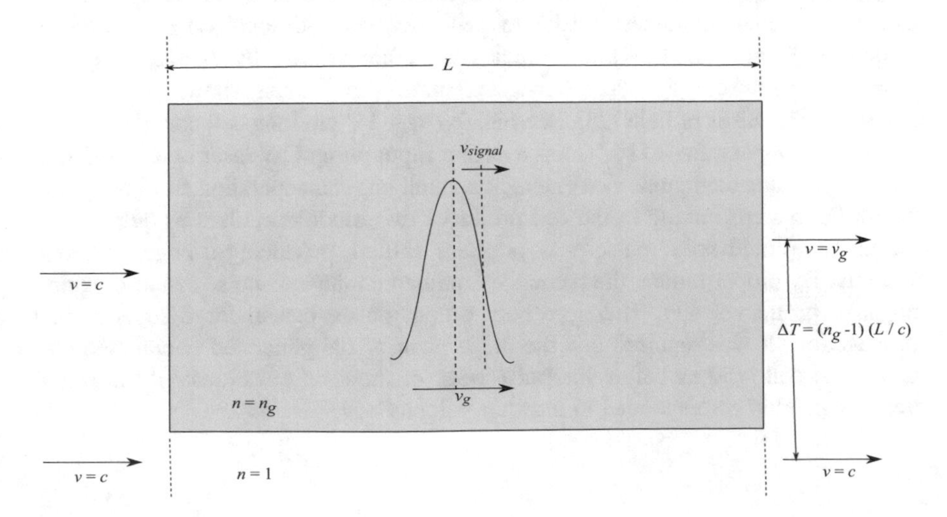

Fig. 7.12 A cell of length L composed of a dispersive medium having a group refractive index n_g

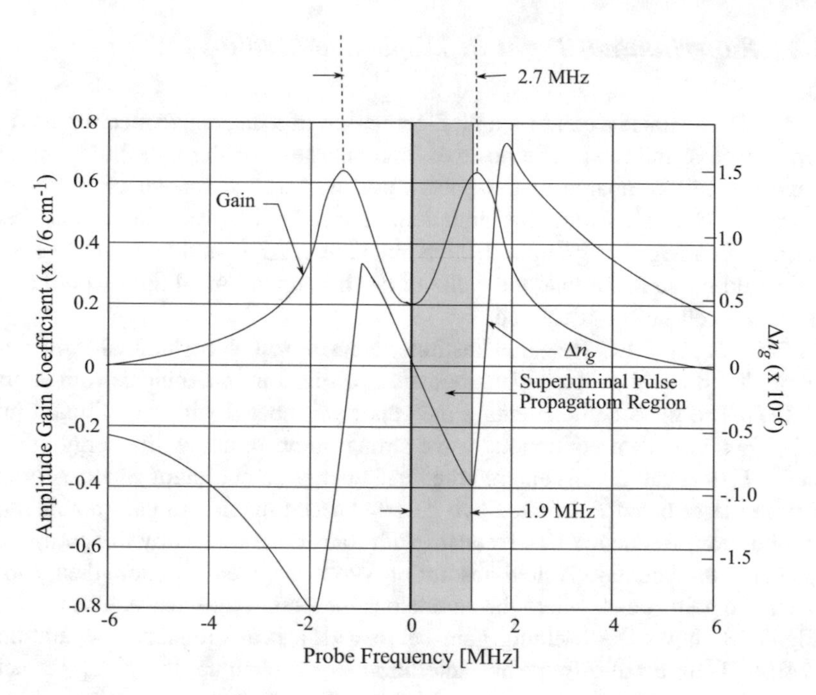

Fig. 7.13 Curves showing superluminal light pulses using gain-assisted anomalous dispersion in atomic cesium vapor at 30 C. *Top curve* resulting gain curve with a peak frequency separation of 2.7 MHz. *Lower curve* negative change in n_g occurring over a probe frequency range of 1.9 MHz

pulse appeared to leave the cell before it entered, and could be explained entirely by the wave nature of light.

In the process of *four-wave mixing*, two input light beams of frequency f_1 and f_2 interact in certain nonlinear media to generate two additional output beams of frequency f_3 and f_4. The four output beams are related by $f_3 = 2f_1 - f_2$ and $f_4 = 2f_2 - f_1$, where $f_2 > f_1$. Four-wave mixing has been shown to produce superluminal pulses of light [28]. See Fig. 7.14. A 1.7 cm long cell containing ^{85}Rb vapor at a temperature ≈116 °C has a strong input pump CW laser at λ = 795 nm. Another weaker seed pulse also enters at a small angular separation from the pump beam. Four-wave mixing in the cell produces two modified pulses as output. The output "amplified seed" pulse peak is phase shifted, producing a negative group velocity. By proper tuning, the second "generated conjugate" pulse can also exhibit negative group velocity. Therefore both output pulses can undergo superluminal propagation. It was reported that the pulse peak of the generated conjugate pulse exited the cell ≈50 ns before the pulse peak of the seed pulse entered the cell at velocity c. This corresponded to a group velocity $v_g = -c/880$.

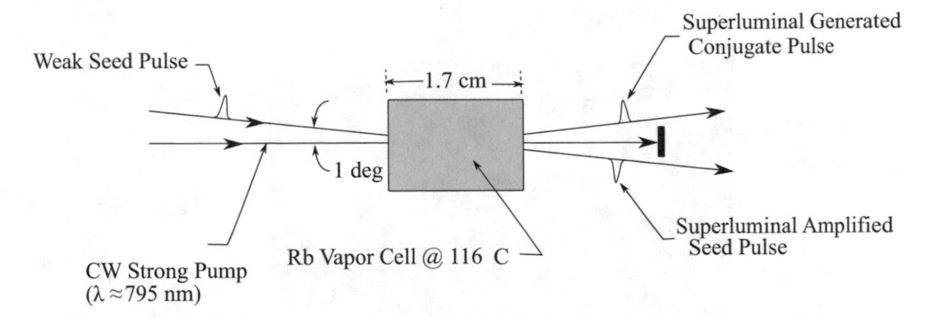

Fig. 7.14 Generation of superluminal light pulses from a Rb cell using four-wave mixing

7.5 Light Path Manipulation

7.5.1 *Light Propagation and Negative Refraction*

In addition to the negative refractive index realized by anomalous dispersion in a vapor, a negative refractive index can be achieved using a conductive material where both the permittivity ε and permeability μ have negative values over a defined frequency range $\Delta\omega$. This was theoretically predicted by Veselago in 1968 for a homogeneous material [29]. For simultaneous negative values of ε and μ, the material must exhibit a frequency dependence of ε and μ, such that $n(\omega) = -\sqrt{\varepsilon(\omega)\mu(\omega)} < 0$, where both $\varepsilon(\omega)$ and $\mu(\omega)$ are complex quantities. When the real parts of $\varepsilon(\omega)$ and $\mu(\omega)$ are negative, then the refractive index is also negative. Figure 7.15 illustrates how a planar block of material having $n = -1.5$ can function as a focusing lens for a point source, where Snell's law is applied at both planar surfaces. By adjusting the negative index slab thickness, the focus can be shifted along the optical axis. Shown for comparison, a slab of glass with $n = 1.5$ produces light divergence upon exit. More generally, diverging lenses of positive index material become converging using negative index material, and converging lenses using $n > 0$ material become divergent when $n < 0$.

It was shown by Pendry in 2000 that a converging planar negative index slab lens could produce a more perfect focus than a conventional diffraction-limited positive lens with curved surfaces [30]. For visible light, simulations showed that the focus could be reduced to a few nanometers wide. Many other electromagnetic properties are reversed using negative index media, such as wave phase velocity, Doppler shift, and the direction of emitted Cherenkov radiation.

A nano-structured *metamaterial* is capable of achieving simultaneous negative values of $\varepsilon(\omega)$ and $\mu(\omega)$, where the material would be transparent, as is the case where $\varepsilon(\omega)$ and $\mu(\omega)$ are both positive. The metamaterial is usually composed of conducting repeating metallic cells with a period less than the light wavelength of interest. Under this restriction, the metamaterial can be considered to be fairly homogeneous at this wavelength. The first successful negative index metamaterial

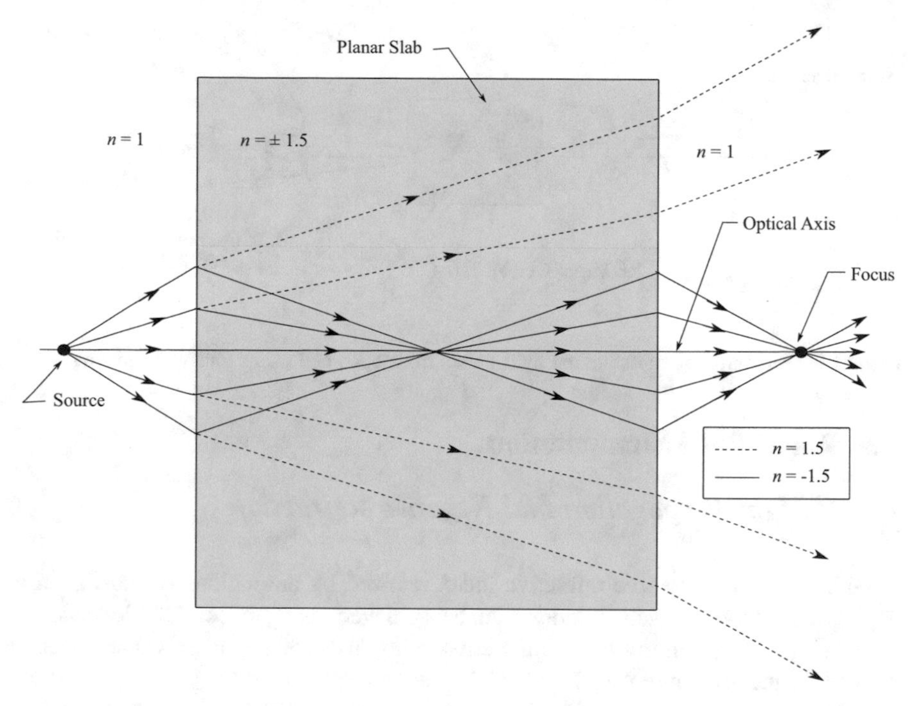

Fig. 7.15 Light ray paths for a planar slab of material with $n = -1.5$, which functions as a convergent lens. Ray paths for $n = 1.5$ are shown for comparison

was fabricated in 2000 in the microwave region using a periodic array of split-ring resonators and thin wires [31]. Figure 7.16 illustrates the operation of a unit cell in the array. The split-ring resonator (SRR) is composed of a non-magnetic conductor, such as copper. A thin wire is oriented across from the split-ring. An electromagnetic wave is incident as shown, where the magnetic vector is perpendicular to the split ring and the electric vector is parallel to the wire. Near the resonance frequency, the magnetic permeability $\mu(\omega)$ becomes negative at the high frequency side of the resonance. Also, the thin wire behaves as a plasma with a frequency ω_p, which exhibits negative $\varepsilon(\omega)$ when the plasma frequency is less than the frequency ω of the incident electromagnetic wave. The combination of these two elements produces a cell having simultaneous negative values of permittivity and permeability, and hence exhibits a negative refractive index.

The first experimental verification of a negative refractive index was performed at the University of California, San Diego using GHz microwave radiation having a center wavelength of 3 cm [32]. A periodic structured metamaterial was constructed using interlocked circuit board material, where split-rings and thin wire patterns were mask- etched on opposite sides of the circuit boards. Figure 7.17 shows a section of the metamaterial, where each unit cell had a side dimension of 5 mm. A prism cut from the metamaterial was secured by upper and lower circular aluminum plates of 15 cm radius and 1.2 cm separation. For a measurement reference,

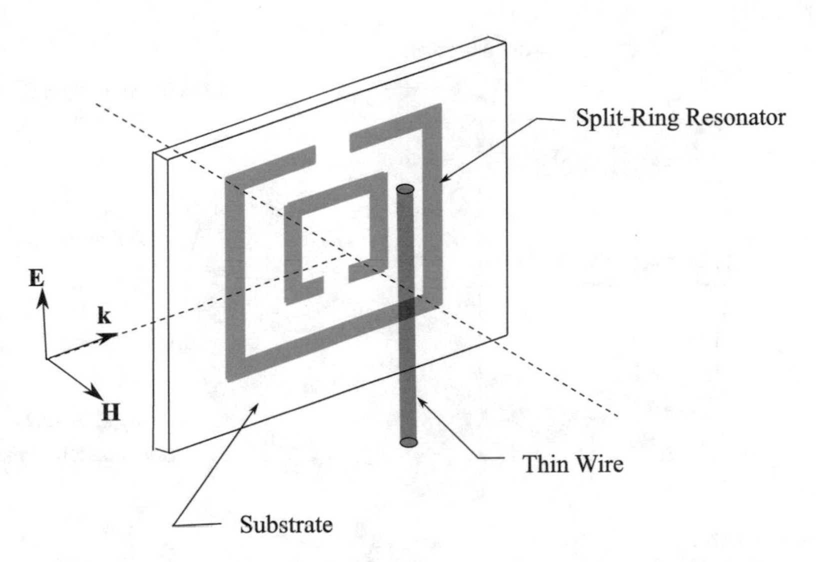

Fig. 7.16 Components of a unit cell of a metamaterial having negative refractive index

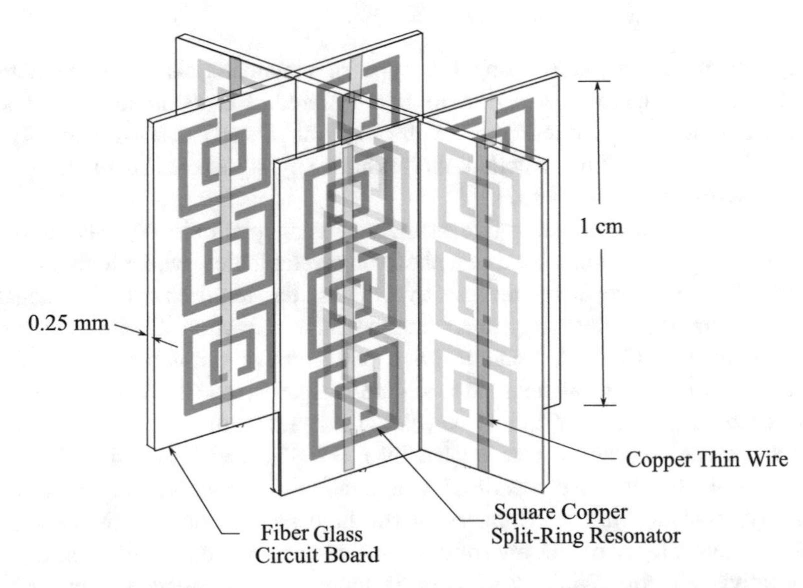

Fig. 7.17 Construction of an array of mounted split-ring resonators and thin wires to form a metamaterial having negative refractive index

a similar prism of positive refractive index Teflon material ($n \approx 1.4$) was formed. A detector measured the power spectrum for each prism as a function of rotation angle. Figure 7.18 illustrates the experiment. The expected ray deviation for the negative index metamaterial was clearly demonstrated, and indicated a value of $n \approx -2.7$.

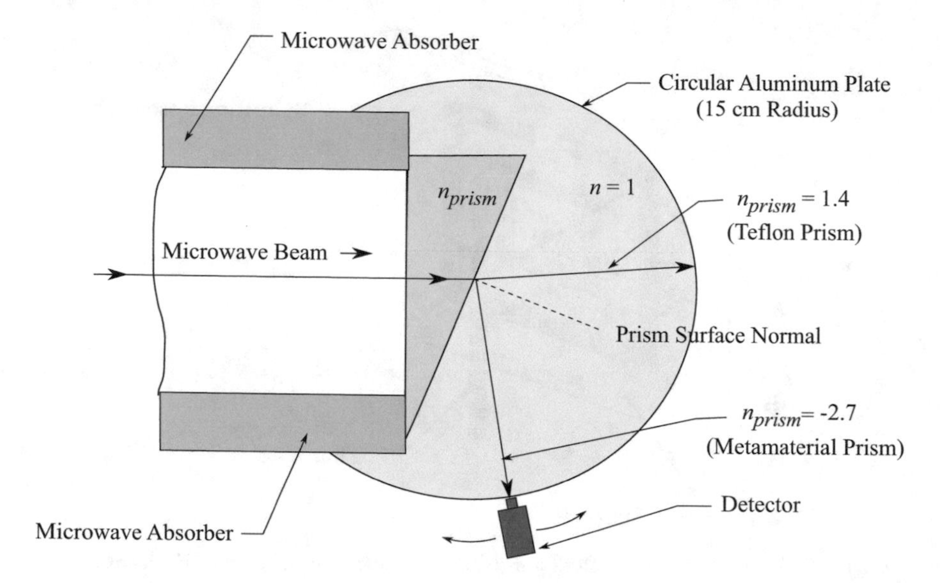

Fig. 7.18 Measurement apparatus to verify negative refractive index

Negative refractive index using this type of metamaterials has been demonstrated from the microwave region to near-infrared. Extension into the visible region is a topic of current research—with only limited success. This is mainly due to the difficulty in finding materials that exhibit any significant magnetic permeability at visible frequencies [33].

Light can also undergo negative refraction using photonic crystals. This can occur in the lowest photonic band of a photonic crystal lattice, where both the group velocity and refractive index are positive. Thus, the requirement of a negative refractive index is not necessary to achieve negative refraction. Co-investigators at MIT and Imperial College, London, have performed a simulation analysis of a square array of circular air holes in a silicon dielectric ($\varepsilon = 12.0$) with TE mode incident radiation at a near-optical wavelength of 1.55 μm [34]. The hole radius r and the lattice constant a were related by $r = 0.35a$, as illustrated in Fig. 7.19. Two photonic bands were identified where all-angle negative refraction could occur. By requiring mirror symmetry of the hole pattern, and a thickness of an integer or half-integer of the transmitted wavelength, reflections were minimized and negative refraction could occur over an incident angle range of about $\pm 40°$.

By the construction of a two-dimensional photonic crystal array of aluminum oxide rods in air, negative refraction can be achieved, where the positive permittivity ε is periodically modulated and the permeability μ is unity [35]. The incident electromagnetic waves had the electric vector parallel to the rods. For a horn antenna with an incident microwave frequency of 13.7 GHz, measurements indicated a negative refractive index of -1.94 and a transmission efficiency $\approx 63\%$ for incident angles >20°.

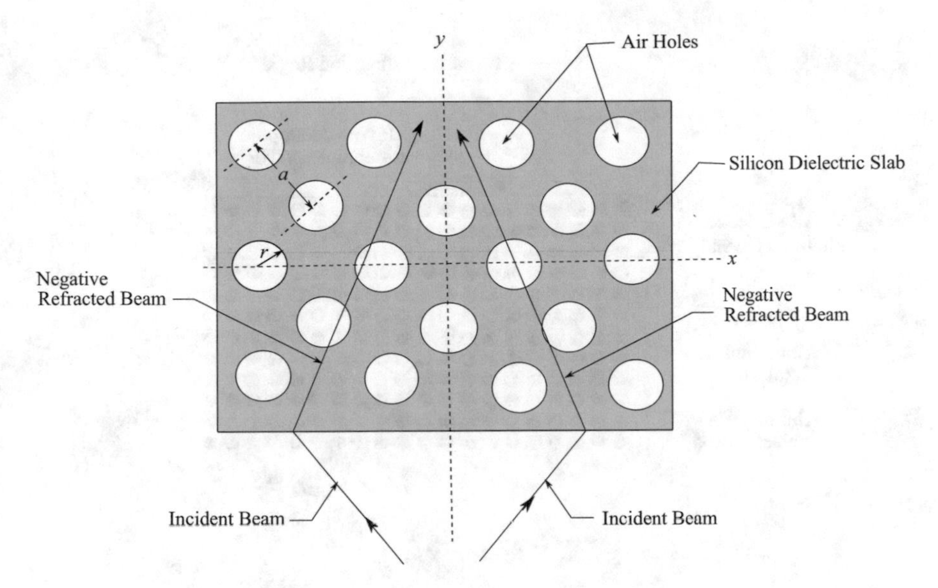

Fig. 7.19 Illustration of a photonic crystal producing negative refraction. The air hole pattern is symmetric about the *x* and *y* axes

Another type of photonic crystal producing negative refraction was modeled using an array of alternating metallic and dielectric rods [36]. The aluminum metal rods had a radius $r = 0.14a$ (1.5 mm), and the aluminum dioxide (alumina) dielectric rods had a radius $r = 0.136a$ (1.55 mm), where a is the lattice constant. A microwave electromagnetic field was applied at varying angles to the edge of the photonic crystal, with the electric vector parallel to the rods. Figure 7.20 shows the two-dimensional photonic crystal structure and measurement technique to observe negative refraction. A monopole antenna receiver sliding along the exit edge of the photonic crystal measured the transmitted electric field intensity as a function of incident angle. It was found that for frequencies between 9.25 and 10.3 GHz, negative refraction occurred between incident angles of 15° and 45°. At a frequency $\omega = 9.7$ GHz and an incident angle of 45°, the measured angle of refraction was $n = -0.96$.

A negative refracting photonic crystal can function as a lens near optical frequencies. Fabrication of a photonic crystal using air holes etched in a silicon layer on an insulating substrate was accomplished by developers in Yokohama and Tokyo [37]. Total internal reflection in the silicon layer then confined the light to the plane of propagation, and a focus was obtained at near-infrared wavelengths. They also cascaded two photonic crystals with opposing aberrations to reduce the focal spot size.

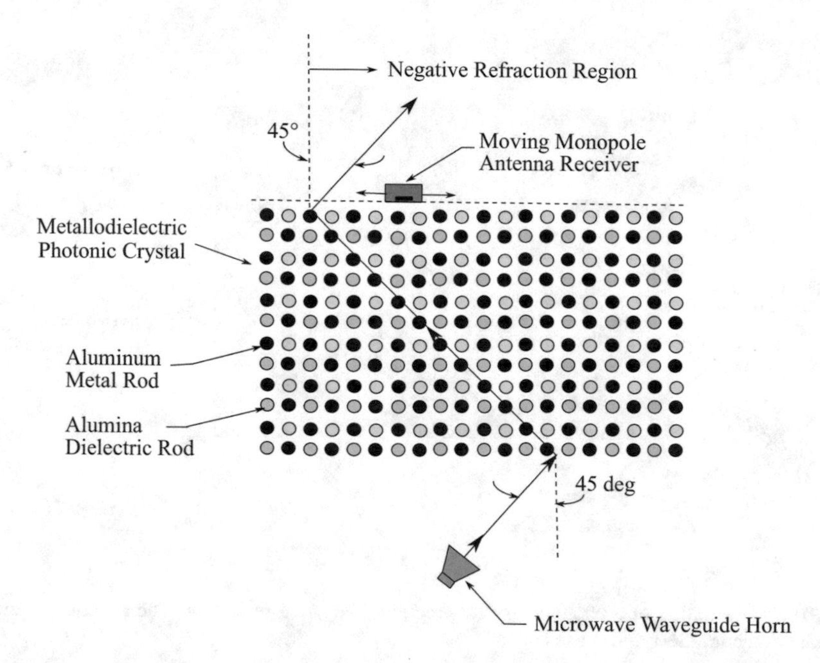

Fig. 7.20 Measurement of negative refraction using a metal-dielectric photonic crystal in the microwave region

7.5.2 *Infinite Light Speed Using a Zero Refractive Index Medium*

For a medium that could have a refractive index n close to zero, the material should have certain properties for light propagating through it. The light would have close to an infinite phase velocity v_p, but the group velocity v_g would not exceed the Einstein speed limit c. The distribution of the phase φ would be uniform, with no observable phase delay, or $\Delta\varphi = 0$. The wavelength would be close to infinite.

One method to produce a near-zero index medium involves the coupling of a negative refractive index material n_1 with a positive refractive index material n_2, such that the net phase change $\Delta\varphi = \Delta\varphi_1 - \Delta\varphi_2 = 0$, where $\Delta\varphi_1 < 0$ and $\Delta\varphi_2 > 0$. Investigators at Columbia University have fabricated alternating stacks of negative index photonic crystals (sometimes called *photonic crystal-based metamaterials*) and conventional homogeneous dielectric positive index materials to form a "superlattice" [38]. This superlattice displayed an average zero refractive index or a *zero-n gap* along the superlattice length for near-infrared light. Three superlattices were fabricated with seven, nine, and eleven unit cells, or superperiods. For the seven cell lattice the superperiod Λ was 4.564 μm, with a patterned area length d_1 = 2.564 μm and a homogeneous area length d_2 = 2 μm. Figure 7.21 illustrates the structure for three superperiods of a superlattice. Using an integrated

Mach-Zehnder interferometer, it was determined that for a cell length ratio $d_2/d_1 = 0.78$, a zero-n gap was produced at a light wavelength $\lambda \approx 1.56$ μm.

Another method of producing zero refractive index uses a nanostructure or metastructure consisting of a dielectric waveguide core enclosed in a metallic cladding. When the electric permittivity ε of the metallic waveguide approaches zero, the waveguide has refractive index $n \approx 0$, since $n = \sqrt{\varepsilon\mu}$, where μ is the permeability of the metal waveguide. This is referred to as the *epsilon-near-zero* (ENZ) condition. This condition occurs at the cutoff frequency of the waveguide (lowest frequency for which a mode will propagate), or where the propagation constant k is zero. In 2012 a team in Amsterdam and Philadelphia reported the fabrication of a rectangular waveguide that produced zero refractive index for visible light [39]. Figure 7.22 illustrates the basic construction, where an 8.5 nm high and 2 μm long SiO_2 core was encased in Ag cladding mounted on a Si substrate. A layer of Cr surrounded the Ag cladding. An external 30 keV electron beam was then tightly focused into the waveguide to initiate internal resonant guided plasmon modes. Far-field emission from the waveguide was measured using a parabolic collecting mirror and a spectrometer. It was found that no propagating modes existed beyond at a visible cutoff wavelength of ≈800 nm for a waveguide width of 190 nm. At this wavelength, the electromagnetic fields in the waveguide oscillated in phase along the entire waveguide, showing that $\varepsilon \approx 0$ and hence $n \approx 0$.

A metamaterial that exhibits ENZ over the visible spectrum was engineered in 2013 [40]. A stacked array of silver and silver nitride layers using ion and electron

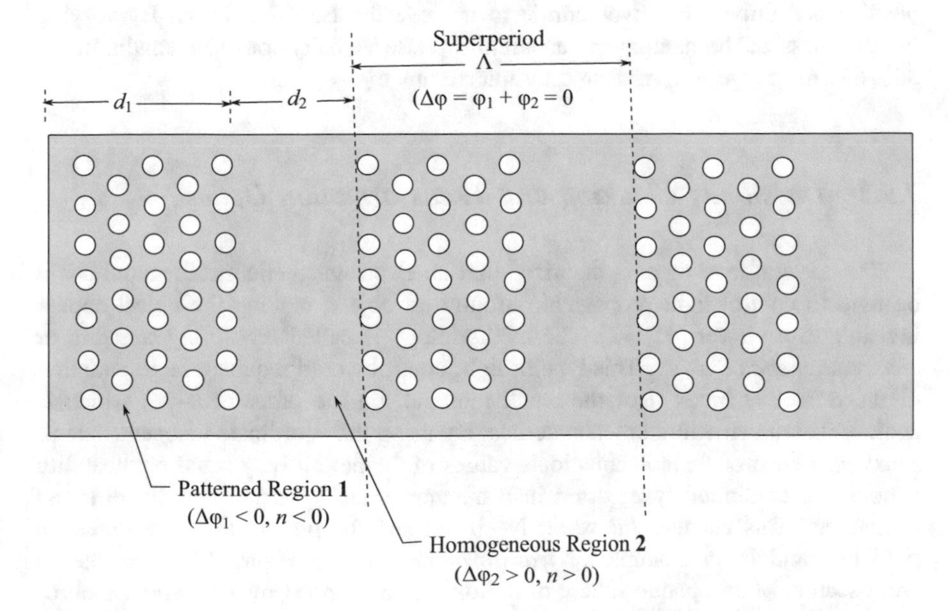

Fig. 7.21 Illustration of the structure for three superperiods of a photonic crystal superlattice producing average zero refractive index over the superlattice length

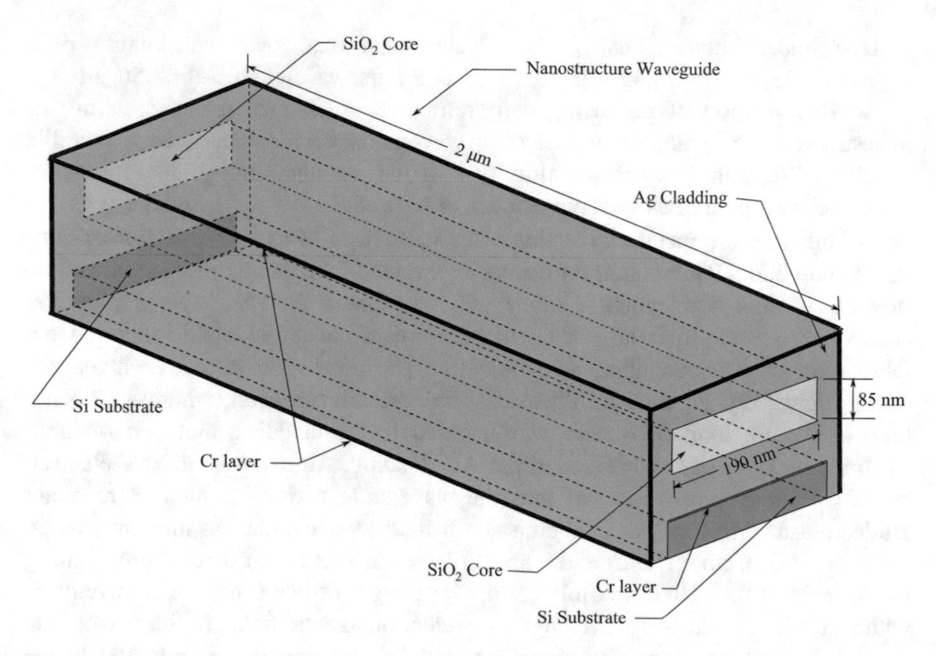

Fig. 7.22 Construction of a metallic waveguide to produce zero refractive index

beam lithography was fabricated, where the unit cell dimension was less than the wavelength of visible light. The negative permittivity of silver was balanced by the positive permittivity of silver nitride to achieve the ENZ condition. By varying the thickness of the nanolayers, an infinite phase velocity for wavelengths in the 351–633 nm range was measured by interferometry.

7.5.3 Invisibility Cloaking and Transformation Optics

In 2006 several investigators theorized that an electromagnetic metamaterial could be used to control light propagation around an object, making the object appear invisible to a viewer [41, 42]. The methodology is called *invisibility cloaking* or *electromagnetic cloaking*. This type of light control would require a large variation of the refractive index n of the cloak material, on the order of 0–36, which is achievable with structured metamaterials. By using the coordinate independence of Maxwell's electromagnetic equations, values of the permittivity ε and permeability μ could be continuously adjusted in a metamaterial by certain coordinate transformations. This changes the refractive index and the propagation direction of a light ray, and is an example of *transformation optics*. Figure 7.23 illustrates a cross-section of an opaque sphere of radius R_1, surrounded by a spherical shell of anisotropic metamaterial with inner radius $r = R_1$ and outer radius $r = R_2$. The metamaterial shell guides the propagation direction of incident light around the

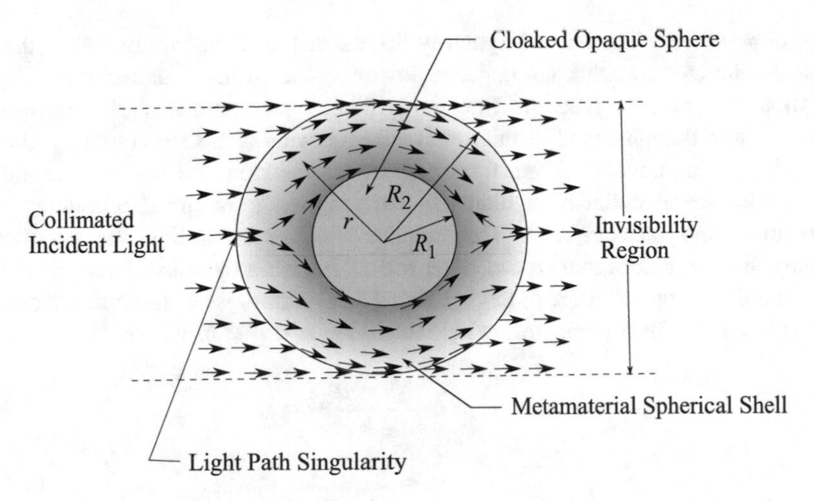

Fig. 7.23 Illustration of invisibility cloaking around an opaque sphere with a metamaterial cloaking shell having a spatially varying permittivity ε′ and μ′. A propagation direction singularity occurs at point A (Adapted with permission of AAAS from Ref. [42]; permission conveyed through Copyright Clearance Center, Inc.)

opaque sphere, making the sphere invisible to an observer. To accomplish this, all fields within the region $r < R_2$ are transformed into the region where $R_1 < r < R_2$, producing transformed spherical coordinates r', θ', φ'

$$r' = R_1 + r(R_2 - R_1)/R_2, \tag{7.8a}$$

$$\theta' = \theta, \tag{7.8b}$$

$$\varphi' = \varphi. \tag{7.8c}$$

The transformed values of permittivity ε′ and permeability μ′ in the metamaterial compression region $R_1 < r < R_2$ can take on the values

$$\varepsilon'_{r'} = \mu'_{r'} = \frac{R_2}{R_2 - R_1}\frac{(r' - R_1)^2}{r'}, \tag{7.9a}$$

$$\varepsilon'_{\theta'} = \mu'_{\theta'} = \frac{R_2}{R_2 - R_1}, \tag{7.9b}$$

$$\varepsilon'_{\varphi'} = \mu'_{\varphi'} = \frac{R_2}{R_2 - R_1}. \tag{7.9c}$$

It is assumed that $R_2 >> \lambda$, where λ is the monochromatic wavelength used in the geometric ray tracing. Use of a single frequency produces no dispersion, allowing all exiting rays to have the same phase, where the phase velocity equals the group velocity. There is a singularity for the central incident ray, where the ray could be directed to either the upper or lower region of the metamaterial.

About six months later the first experimental realization of electromagnetic cloaking was demonstrated at microwave frequencies using structured metamaterials [43]. Using developed transformation optics theories, the object to be hidden was a conducting cylinder of radius a and length coordinate z, allowing easier fabrication and measurement. The annular cloak surrounding the conducting cylinder had an inner radius a and outer radius b. Using cylindrical coordinates (r, θ, z), modifications of Eqs. (7.8a), (7.8b), (7.8c) compressed the space from the region $0 < r < b$ into a transformed cloak region $a < r' < b$, where

$$r' = a + \frac{b-a}{b} r \tag{7.10a}$$

$$\theta' = \theta \tag{7.10b}$$

$$z' = z. \tag{7.10c}$$

This transformation produced the following expressions for the permittivity and permeability components:

$$\varepsilon_r' = \mu_r' = \frac{r-a}{r}, \tag{7.11a}$$

$$\varepsilon_\theta' = \mu_\theta' = \frac{r}{r-a}, \tag{7.11b}$$

$$\varepsilon_z' = \mu_z' = \frac{r-a}{r}\left(\frac{b}{b-a}\right)^2, \tag{7.11c}$$

In order to achieve the desired dispersion in the cylindrical cloak, the cloak metamaterial was constructed to realize the dispersion produced by Eqs. (7.11a), (7.11b), (7.11c), with a radial variation of μ_r' and constant values of ε_z' and μ_θ'. This was accomplished by fabrication of a three-tier series of split-ring resonator SRR curved unit cells on ten concentric high-frequency laminate cylinders. See Fig. 7.24. Each square unit cell had about a 3.3 mm width a_θ and height a_z. The solid copper cylinder to be cloaked had a 25 mm radius, and the metamaterial cloak had an inner radius a = 27.1 mm and an outer radius b = 58.9 mm. The solid cylinder and metamaterial cloak were placed in a parallel plate waveguide and driven at an 8.5 GHz operating frequency for phase and amplitude measurements. A series of field map simulations demonstrated that the object and cloak resembled free-space, showing that the object was essentially hidden.

Although this experiment demonstrated electromagnetic cloaking, it could not achieve *optical cloaking* over a range of visible light wavelengths. The use of electromagnetic metamaterials operating at optical wavelengths would produce severe absorption losses. In 2007 scientists at Purdue University proposed a

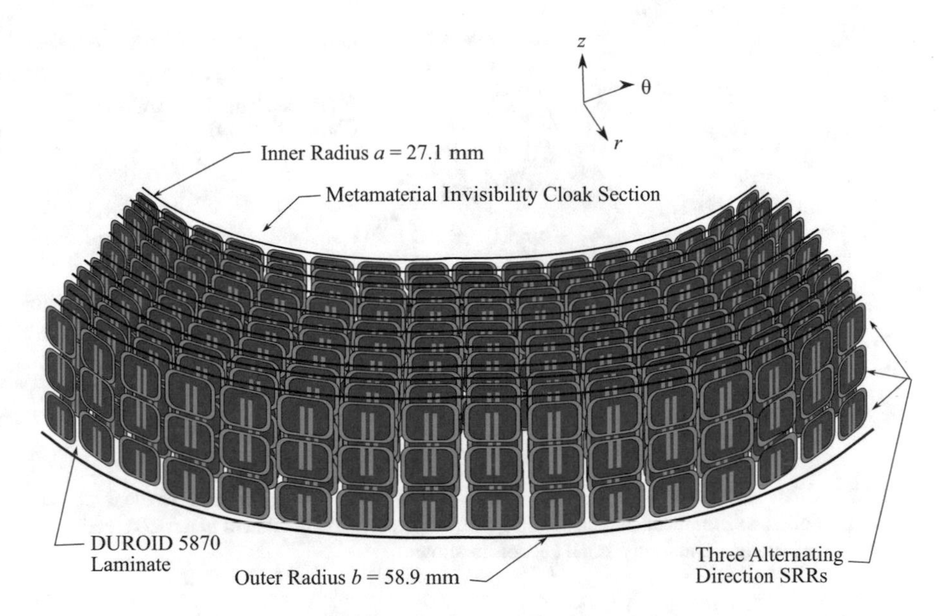

Fig. 7.24 A segment of a cylindrical invisibility cloak composed of split-ring resonators producing a radial variation of μ_r' and constant values of ε_z' and μ_θ'

variation of the cylindrical shell cloak that would operate at optical frequencies [44]. Here the incident light is TM polarized with the magnetic field polarization along the z-axis, such that only μ_z, ε_r, and ε_θ are important, and Eqs. (7.11a), (7.11b), and (7.11c) can be reduced to

$$\mu_z' = 1, \tag{7.12a}$$

$$\varepsilon_\theta' = \left(\frac{b}{b-a}\right)^2, \tag{7.12b}$$

$$\varepsilon_r' = \left(\frac{r-a}{r}\right)^2 \left(\frac{b}{b-a}\right)^2. \tag{7.12c}$$

Equations (7.12a), (7.12b), and (7.12c) produce the same wave paths as Eqs. (7.11a), (7.11b), and (7.11c), but the constant permeability $\mu_z' = 1$ value removes the effect of magnetism. The cloaking material then essentially performs as a dielectric. The construction of a dielectric cylindrical ring cloak is illustrated in Fig. 7.25. The cylindrical solid object to be cloaked is placed within this dielectric ring. The ring has an inner radius $r = a$, and an outer radius $r = b$, with the cloaked cylindrical object having a radius $\approx a$. A matrix of thin sub-wavelength metal wires is randomly distributed throughout the dielectric ring in the radial direction. For the specific construction used for a cloaking simulation, silver wires with a length-to radius ratio of 10.7 were imbedded in the cylindrical silica dielectric ring having a

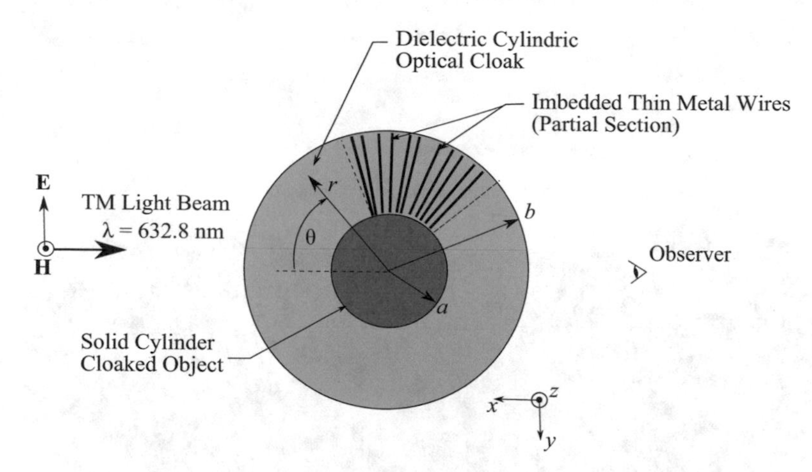

Fig. 7.25 Cross-section of a dielectric cylindrical ring cloak and cylindrical cloaked object, extending along *z*-axis. Imbedded metallic thin wires are randomly distributed in radial direction of cloak material along *z*-axis (only small section is shown)

shape factor $a/b = 0.314$. Equations (7.12a), (7.12b), and (7.12c) then give constant values $\mu'_z = 1$, $\varepsilon'_\theta = 2.10$, and a smooth variation of ε'_r from 0 to 1 between the cloak boundaries $a \leq r \leq b$. The TM polarized incident light has a He–Ne laser optical wavelength $\lambda = 632.8$ nm. This proposed system was evaluated using field-mapping computer simulations, which showed greatly reduced scattering by the cloaked object, and the capability of this non-magnetic cloak to produce invisibility.

In 2008 Li and Pendry proposed an *invisibility carpet* using isotropic dielectrics instead of magnetic resonating structures [45]. This approach would be more suitable at optical frequencies and might allow broadband performance. This carpet cloak resembles a flat plane and compresses the cloaked object to a flat sheet, rather than the point and line compressions discussed previously. Figure 7.26 illustrates a carpet cloak, where a rectangular cell grid space has been mapped to a compressed grid space with orthogonal grid intersections, called a quasi-conformal grid. In this grid, the compression anisotropy is minimized, the permittivity ε varies from about 1.5 to 4.4, and the refractive index n varies from 0.83 to 1.4, where n is high near the cloak boundary center and low toward the cloak boundary edges. The ε and n values are relative to silica glass ($\varepsilon = 2.25$, $n = 1.46$), which surrounds the entire cloak. Thus the cloak can be considered as an isotropic dielectric material with a permeability $\mu = 1.0$. The lower boundary of the cloak is coated with a highly reflective metal, and the entire cloak rests on a planar highly reflective metal conductor (ground plane). The permittivity range could be obtained by drilling a series of sub-wavelength holes of varying diameter in the cloak dielectric material, such as silicon. When a Gaussian light beam is incident at about 45°, an observer would not view the cloaked object, but only perceive a continuous flat ground plane.

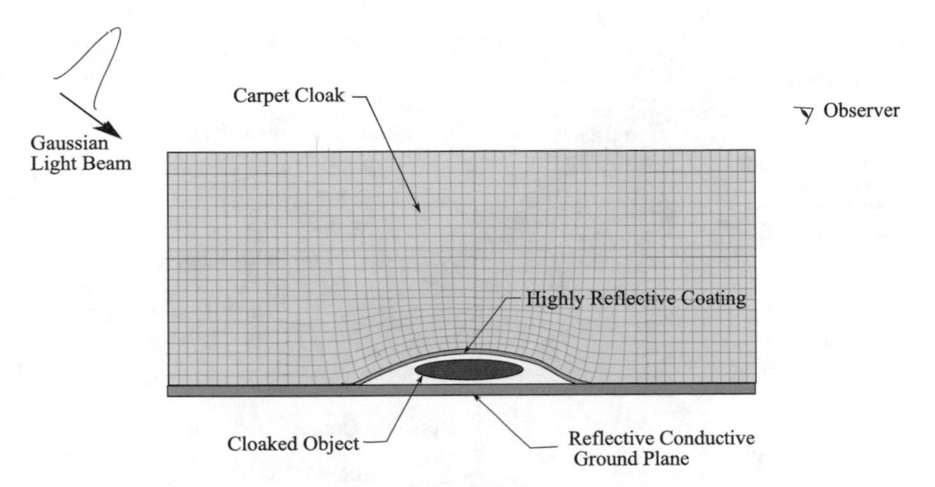

Fig. 7.26 Isotropic dielectric material carpet cloak, where a rectangular cell grid space has been mapped to a compressed grid space with orthogonal grid intersections. In this grid the permittivity ε varies from about 1.5 to 4.4, and the refractive index *n* is maximized near the cloak boundary center

Shortly thereafter several carpet cloak designs were demonstrated that operated over near-infrared wavelengths ranging from 1400 to 1800 nm. In 2009 a carpet cloak composed of dielectric materials was demonstrated that concealed a bump on a flat reflector at an operating wavelength of 1550 nm [46]. Figure 7.27 illustrates the general construction and operation. The distributed Bragg reflector, consisting of alternating layers of SiO_2 and silicon, had a reflectivity >0.999. The reflector deformation bump covered a 1.6 μm cloaked region. The cloaking device was a triangular shaped silicon-on-insulator wafer on which spatially distributed 50 nm diameter silicon nano-poles were attached. This spatial distribution produced an effective refractive index range $1.45 < n < 2.42$, where the most closely-spaced nano-poles near the bump apex produced the highest refractive index. The effectiveness of the cloak was compared using an infrared camera focused at the output edge of the cloaking device. When 1550 nm light was launched into the input edge of the cloaking device over the reflector bump, the output image was similar to that produced by a planar reflector, demonstrating the effectiveness of the cloak.

In the same year a team at the University of California, Berkeley, built and demonstrated an all dielectric carpet cloak that provided invisibility over the 1400–1800 nm wavelength range [47]. Using quasi-conformal mapping, the required refractive index changes in the rectangular cloaking region was obtained by controlling the density of 110 nm diameter nano-holes in a silicon-on-oxide (SOI) wafer. The SOI wafer consisted of a 250 nm thick silicon waveguide layer over a 3 μm silicon oxide slab over a silicon base substrate. See Fig. 7.28. The cloaking device was triangular with a constant refractive index $n = 1.58$ outside the square cloaking region. Input and output light was coupled to the cloaking device

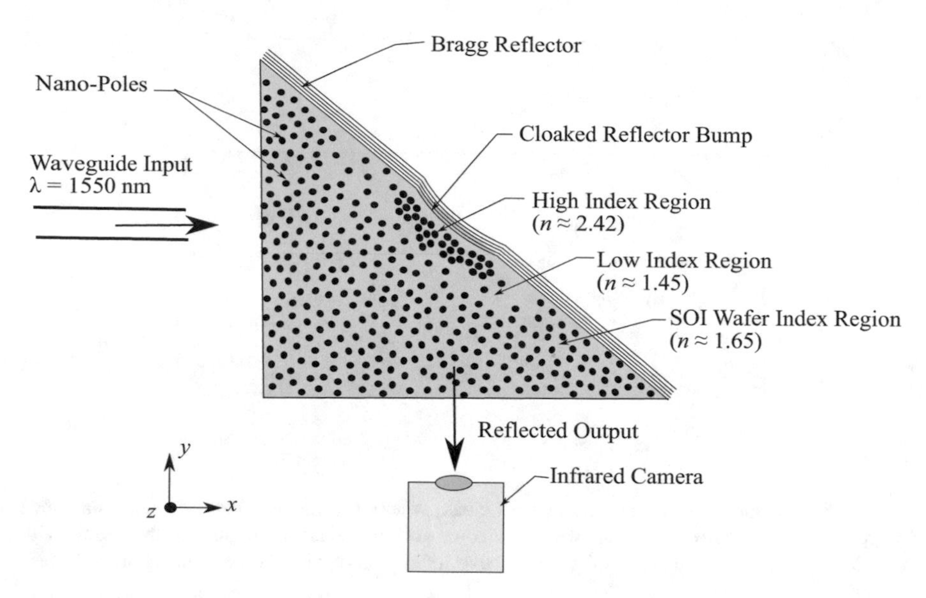

Fig. 7.27 Cross-section of a dielectric carpet cloak. The 50 nm diameter nano-poles are spatially distributed in the *x-y* plane in the *z*-axis direction. The highest nano-pole distribution density near the cloak surface produces the highest effective refractive index. 1550 nm light is launched into one edge of the cloaking device, and the reflected output is imaged by an infrared camera

by a pair of gratings. The 100 nm gold layer reflecting surface had a 3.8 μm wide by 400 nm deep bump to be concealed by the cloak. Input light from a tunable optical parametric oscillator was focused at the input grating, launching a TM wave into the silicon waveguide. A CCD camera measured the output light from the silicon waveguide at the output grating. To experimentally show the effectiveness of the cloak, Gaussian beams with wavelengths ranging from 1400 to 1800 nm were inputted, with and without the carpet cloak. With the cloak in place, the output over the wavelength range was a single Gaussian beam. However, without the cloak the output beam was split into two or three output lobes, depending on the wavelength. This result demonstrated the broadband performance of the optical cloak.

Another cloaking technique that could allow invisibility cloaking of an object in free-space was theorized in 2005 by scientists at the University of Pennsylvania and the University of Rome [48]. This method considered an isotropic material object, covered by a thin isotropic material layer having low or negative permittivity or permeability, to significantly reduce the scattering cross-section of the object. Since the cloaking material is close to its plasma resonance, this method is called *plasmonic cloaking*. The inner object (core) had permittivity ε and permeability μ with radius a, and the outer layer (cover) had permittivity ε_c and permeability μ_c with radius a_c. The objects analyzed were either spherical or cylindrical in shape, and the plasmonic metamaterial cover layer was in a free-space environment. By proper

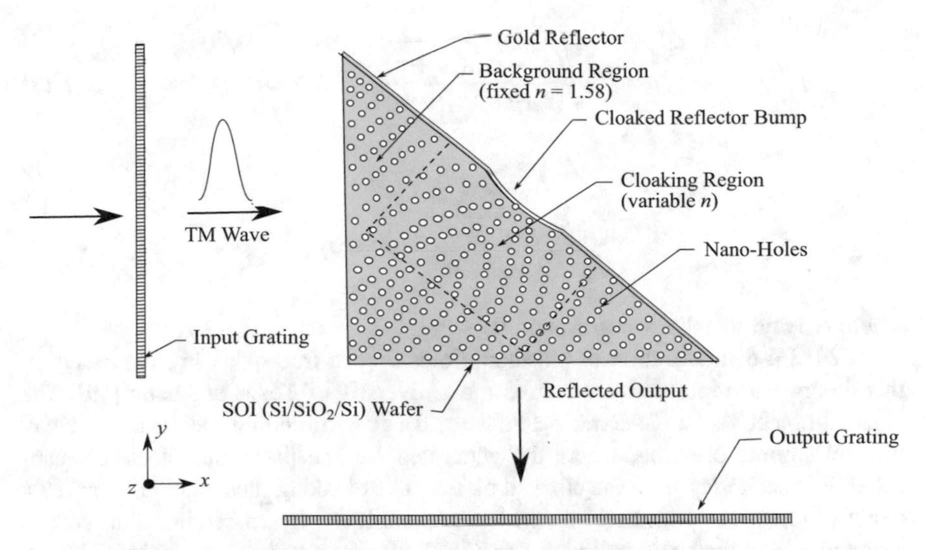

Fig. 7.28 All-dielectric carpet cloak, where the density of fixed diameter nano-holes in a triangular shaped SOI wafer control the refractive index variation in a rectangular cloaking region

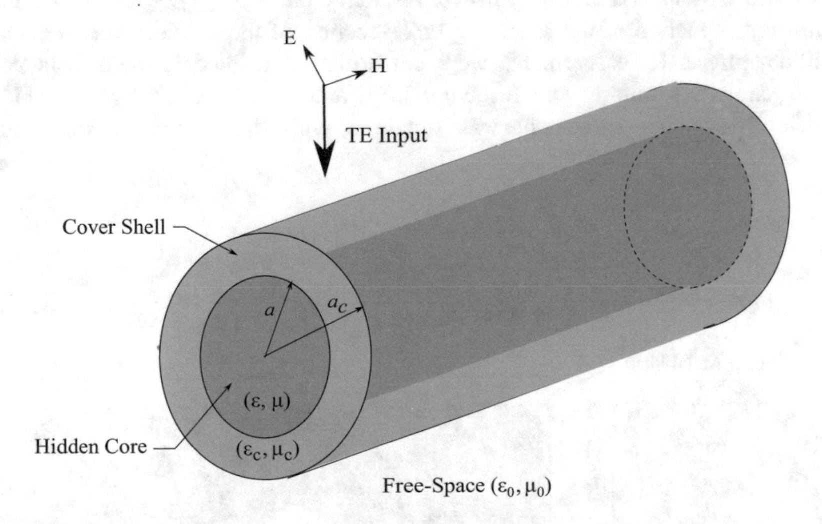

Fig. 7.29 Proposed construction of an invisibility cloak for use in free-space. The cloaked dielectric cylinder core of radius a, with permittivity ε and permeability μ, is surrounded by an outer cover layer of radius a_c, with permittivity ε_c and permeability μ_c

choice of these parameters, the scattering cross-section could be reduced to render the object almost invisible to a viewer. Figure 7.29 illustrates a design for a cylindrical geometry with a TE wave input, where the conditions for minimum cross-section scattering are given by

$$\gamma = \sqrt[2m]{\frac{(\varepsilon_c - \varepsilon_0)(\varepsilon_c + \varepsilon)}{(\varepsilon_c - \varepsilon)(\varepsilon_c + \varepsilon_0)}}, \quad (m \neq 0) \tag{7.13}$$

and

$$\gamma = \sqrt{\frac{\mu_c - \mu_0}{\mu_c - \mu}}, \quad (m = 0) \tag{7.14}$$

where m is an integer, $\gamma = a/a_c$, and $0 \leq \gamma \leq 1$.

In 2012 a demonstration of plasmonic cloaking in free-space (ε_0, μ_0) based on this design was reported by a team at the University of Texas at Austin [49]. The cloaked object was a dielectric cylinder (ε, μ). A surrounding metamaterial cloak used plasmonic cloaking, where the generated surface plasmons effectively canceled light scattering from the cloaked object. The cloaking metamaterial required a negative effective permittivity ε_c, which was achieved by embedding thin copper strips in a host dielectric material. See Fig. 7.30. Parameters were cylinder length L = 18 cm, cloaked cylinder radius a = 1.25 cm, cloak radius $a_c = 1.3a$ = 1.63 cm, $\varepsilon = 3\varepsilon_0$, $\mu = \mu_0$, and $\varepsilon_c = -13.6\varepsilon_0$. Full-wave numerical simulations were performed over an incident microwave frequency range $3 \leq f \leq 3.5$ GHz, and a significant reduction of the scattering cross-section of the cylinder was shown. In addition, physical measurements were performed at a fixed far-field distance of 1.4 m using transmitting and receiving horn antennae over a range of incident angles. Reasonable agreement was obtained with the numerical simulations,

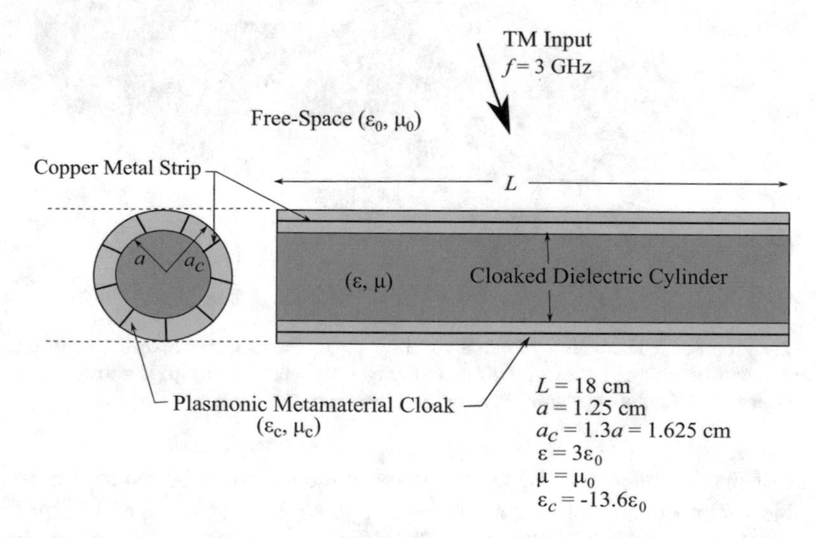

Fig. 7.30 Realization of free-space cloaking where the cloaked object is a dielectric cylinder (ε, μ). The surrounding cloaking metamaterial has a negative effective permittivity ε_c, achieved by embedding thin copper strips in a host dielectric material

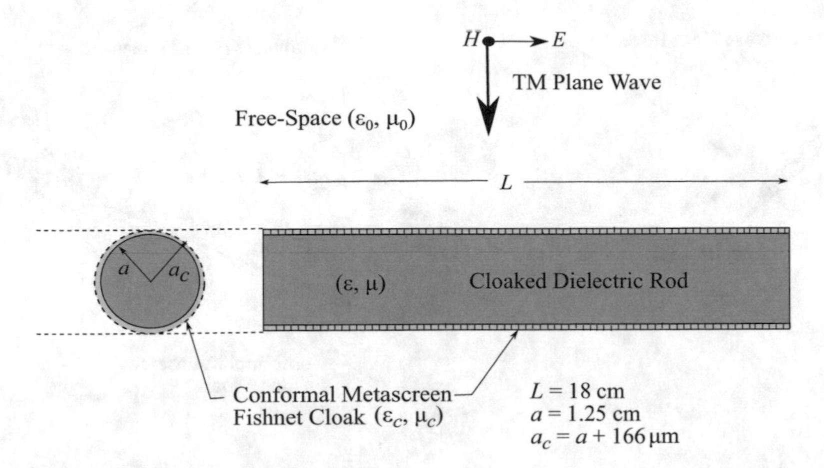

Fig. 7.31 General construction of a free-space mantel screen cloak. The screen is a "fishnet" array of 1.8 cm by 1.8 cm sub-wavelength squares, formed by 1 mm wide, 66 μm thick copper tape on a flexible 100 μm flexible polycarbonate film

showing about a 10 dB reduction in object scattering at $f = 3$ GHz for normal incidence.

The technique of *mantel cloaking* was demonstrated for a 3-D cylindrical object in 2013 [50]. To achieve this type of free-space cloaking, an ultra-thin flexible metamaterial screen was wrapped around the cylindrical object. Figure 7.31 shows the general construction of the cloak. The screen was a "fishnet" array of 1.8 cm by 1.8 cm sub-wavelength squares, formed by 1 mm wide, 66 μm thick copper tape on a flexible 100 μm flexible polycarbonate film. A series of far-field measurements were made at various elevation and azimuthal angles at a fixed distance $R = 1.75\lambda_0$, where $\lambda_0 \approx 8.2$ cm. See Fig. 7.32. Strong scattering suppression was observed over a frequency range $3.7 \leq f \leq 4.1$ GHz for this type of conformal cloak.

7.5.4 *Spacetime Hidden Event Cloaking*

Instead of transforming spatial coordinates $(x, y) \rightarrow (x', y')$ for hidden object cloaking, *spacetime cloaking* transforms spatial and time coordinates, $(x, t) \rightarrow (x', t')$. These transformed spacetime coordinates would allow an event to be hidden for a specific period of time, forming a spacetime cloak. In 2010 a theory was developed to describe how such a spacetime cloak might be realized [51]. Spacetime cloaks require temporal changes in refractive index, where sections of a transmitted light beam can slow down or speed up. Temporal index changes are achievable in certain types of electromagnetic metamaterials or non-linear optical fibers. The metamaterial must act as a medium moving through space at a velocity $v(x, t)$. Then the velocity of light $v'(x, t)$ in this medium can be calculated from a Lorentz transformation as

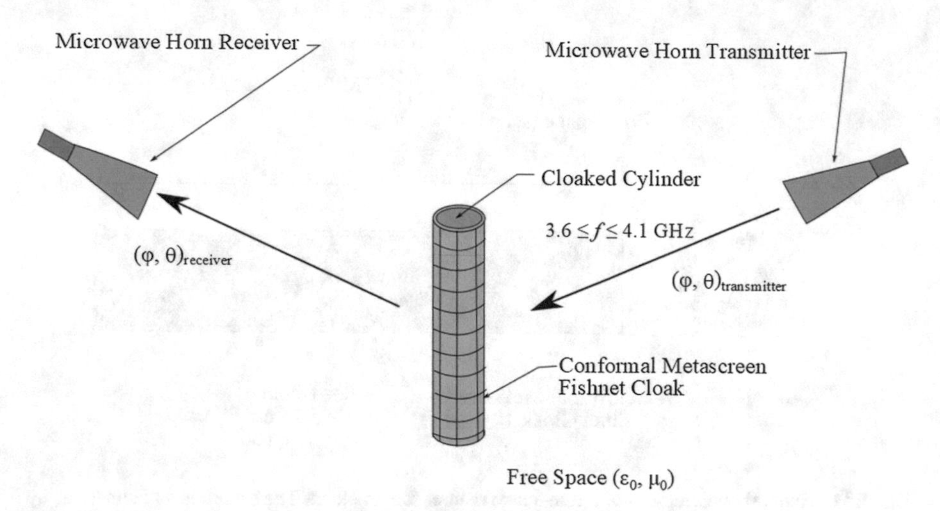

Fig. 7.32 Setup for a series of far-field measurements cloaking efficiency made at various elevation and azimuthal angles at a fixed distance $R = 1.75\lambda_0$, where $\lambda_0 \approx 8.2$ cm

$$v'(x,t) = \frac{v(x,t) + c/n}{1 + v(x,t)/(cn)}. \tag{7.15}$$

If the metamaterial is constructed such that $v(x, t)$ is negative for $x < 0$ and positive for $x > 0$ near the $x = 0$, $t = 0$ spacetime origin, then the leading point of a light beam in the medium will speed up in the $x > 0$ region, and the lagging point will slow down in the $x < 0$ region. An event cloaking region is then opened up around $t = 0$, and eventually closes by the reverse Lorentz transformation $v'(x, t) \rightarrow v(x, t)$. Thus any event occurring in the cloaking region will be perceived by an observer as occurring during a single instant, and hence remain unknown. Such a spacetime cloak is often referred to as a *history editor* or *darkness gap*. Figure 7.33 illustrates the formation of a spacetime cloak in an electromagnetic metamaterial of unspecified structure at $t = 0$, with varying x-values.

Motivated by the previous results, a team at Cornell University experimentally demonstrated spacetime cloaking using another method [52]. The technique is illustrated in Fig. 7.34a–d. Two half-lenses, oriented as shown, produce a temporal phase shift to a continuous wave incident probe beam. Dispersion through this "split time-lens" produces two adjoining light groups, one consisting of longer wavelengths and the other of shorter wavelengths. The light enters a dispersion-compensating optical fiber where the shorter wavelength group propagates faster than the longer wavelength group, resulting in a temporal gap between the two groups. This temporal gap forms a spacetime cloaking region after leaving the fiber. After the spacetime cloaking region, the light enters a single-mode optical fiber with opposite dispersion, where the shorter wavelengths propagate faster than the longer wavelengths. The temporal gap is eventually closed, and a second split

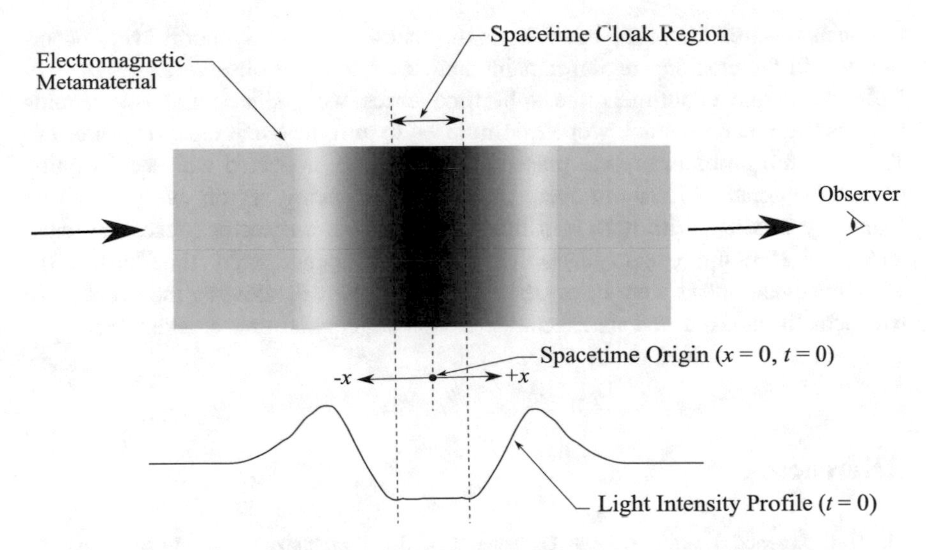

Fig. 7.33 Illustration of a spacetime cloak in an electromagnetic metamaterial of unspecified structure at $t = 0$, with varying x-values. The accompanying light intensity profile is also shown

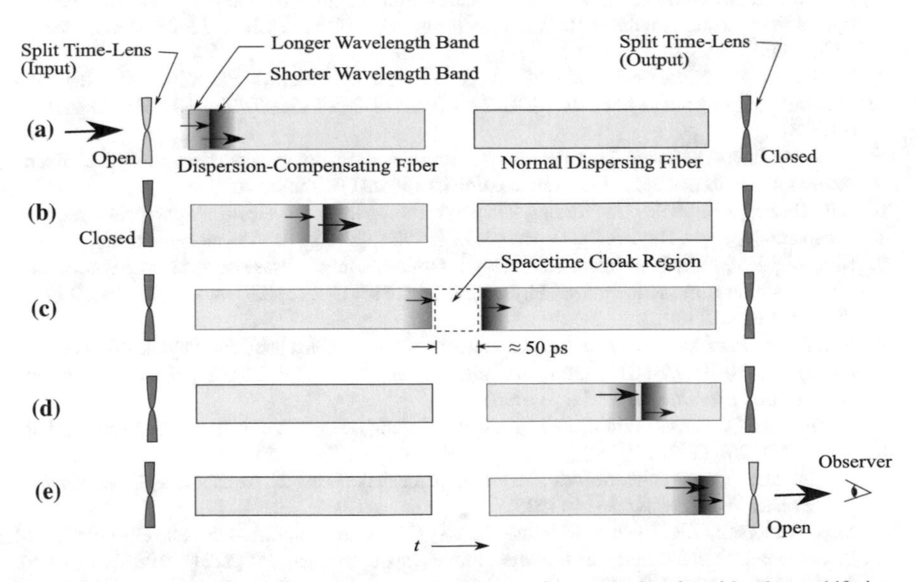

Fig. 7.34 **a** Open input split time-lens generates two frequency bands with phase shift into dispersion-compensated optical fiber. **b** Low-frequency band accelerates from low-frequency band to produce temporal gap. **c** Spacetime cloak region formed at maximum temporal gap between fibers. **d** High-frequency band accelerates in single-mode fiber, narrowing the gap between bands. **e** Frequency bands come together and are transformed into original input wavelength by open output split time-lens

time-lens restores the probe beam to its original form. Any temporal event occurring within the cloaking region remains undetectable to an observer.

In the actual experiment the split time-lenses were silicon nano-waveguide devices that could be quickly opened or closed to turn the event cloaking on or off. The 1539 nm probe beam was pulsed at 41 kHz and interacted with a 5 ps pump pulse to generate a 1539 nm event pulse in the cloaking region every 24 μs by four-wave mixing. With both split time-lenses closed, a detector (observer) registered the 1539 nm event pulses. However, for opened split time-lenses, the 1539 nm event pulses were below the detector noise level, showing the event pulse was actually cloaked and undetectable. It was anticipated that cloaked spacetime gaps could be extended to the 50 ps level.

References

1. H.A. Lorentz, Über die Beziehung zwischen der Fortpflanzung des Lichtes und der Körperdichte. Wiedermann Annalen **9**, 641–664 (1880)
2. L. Brillouin, Diffusion de la lumi`ere et des rayonnes par un corps transparent homog`ene; influence de l'agitation thermique. Annales de Physique **17**, 88–122 (1922)
3. K.Y. Song, M. Herráez, L. Thévenaz, Observation of pulse delaying and advancement in optical fibers using simulated Brillouin scattering. Opt. Express **13**(1), 82–88 (2005). doi:10.1364/OPEX.13.000082
4. Z. Zhu et al., Numerical study of all-optical slow-light delays via stimulated Brillouin scattering in an optical fiber. JOSA B **22**(11), 2378–2384 (2005). doi:10.1364/JOSAB.22.002378
5. A. Kobyakov, M. Sauer, D. Chowdhury, Stimulated Brillouin scattering in optical fibers. Adv. Opt. Photonics **2**(1), 1–59 (2010). doi:10.1364/AOP.2.000001
6. S.E. Harris, J.E. Field, A. Kasapi, Dispersive properties of electromagnetically induced transparency. Phys. Rev. A **46**(1), R29–R32 (1992). doi:10.1103/PhysRevA.46.R29
7. M.S. Bigelow, N.L. Lepeshkin, R.W. Boyd, Observation of ultraslow light propagation in a ruby crystal at room temperature. Phys. Rev. Lett. **90**(11), 113903-1–4 (2003). doi:10.1103/PhysRevLett.90.113903
8. S.N. Bose, Plancks Gesetz und Lichquantenhypothese. Zeitschrift für Physik **26**, 179–181 (1924). doi:10.1077/BF01327326]. (*Albert Einstein translated Bose's original paper into German and submitted it to this journal*)
9. A. Einstein, Quantentheorie des einatomigen idealen Gases. Sitzungsber. Kgl. Preuss. Akad. Wiss., 261–267 (1924)
10. A. Einstein, "Quantentheorie des einatomigen idealen Gases 2. Abhandlung", *Sitzungsber. Kgl. Preuss. Akad. Wiss.*, 3–14 (1925)
11. M.H. Anderson, J.R. Ensher, M.R. Matthews, C.E. Wieman, E.A. Cornell, Observation of Bose-Einstein condensation in a dilute atomic vapor. Science **269**(5221), 198–201 (1995). doi:10.1126/science.269.5221.198
12. J. Klaers, J. Schmitt, F. Vewinger, M. Weitz, Bose-Einstein condensation of photons in an optical microcavity. *Nature* **468**, 545–548 (2010). doi:10.1038/nature09567
13. L.V. Hau et al., Light speed reduction to 17 metres per second in an ultracold atomic gas. Nature **397**, 594–598 (1999). doi:10.1038/17561
14. M.M. Kash, et al., Ultraslow group velocity and enhanced nonlinear optical effects in a coherently driven hot atomic gas. Phys. Rev. Lett. **82**, 5229–5232 (1999). doi:10.1103/PhysRevLett.82.5229

15. D. Budker, et al., Nonlinear magneto-optics and reduced group velocity of light in atomic vapor with slow ground state relaxation. Phys. Rev. Lett. **83**, 1767–1770 (1999). doi:10.1103/PhysRevLett.83.1767
16. M. Notomi, et al., Extremely large group-velocity dispersion of line-defect waveguides in photonic crystal slabs. Phys. Rev. Lett. **87**(25), 235902-1–235902-4 (2001). doi:10.1103/87.PhysRevLett.87.253902
17. S. Kubo, D. Mori, T. Baba, Low-group-velocity and low-dispersion slow light in photonic crystal waveguides. Opt. Lett. **32**, 2981–2983 (2007). doi:10.1364/OL.32.002981
18. Y. Hamachi, S. Kubo, T. Baba, Slow light with low dispersion and nonlinear enhancement in a lattice-shifted photonic crystal waveguide. Opt. Lett. **34**(7), 1072–1074 (2009). doi:10.1364/OL.34.001072
19. T.F. Krauss, Slow light in photonic crystal waveguides. J. Phys. D. **40**, 2666–2670 (2007). doi:10.1088/0022-3727/40/9/S07
20. T. Baba, Slow light in photonic crystals. Nat. Photonics **2**, 465–473 (2008). doi:10.1038/nphoton.2008.146
21. C. Liu et al., Observation of coherent optical information storage in an atomic medium using halted light pulses. Nature **409**, 490–493 (2001). doi:10.1038/35054017
22. D.F. Phillips et al., Storage of light in atomic vapor. Phys. Rev. Lett. **86**, 783–786 (2001). doi:10.1103/PhysRevLett.86.783
23. J.J. Longdell et al., Stopped light with storage times greater than one second using electromagnetically induced transparency in a solid. Phys. Rev. Lett. **95**(6), 783–786 (2005). doi:10.1103/PhysRevLett.86.783
24. G. Heinze, C. Hubrich, T. Halfmann, Stopped light and image storage by electromagnetically induced transparency up the regime of one minute. Phys. Rev. Lett. **111**(3), 03360-1-5 (2013). doi:10.1103/PhysRevLett.111.033601
25. M.F. Yanik, S. Fan, Time reversal of light with linear optics and modulators. Phys. Rev. Lett. **93**(17),173903-1-4 (2004). doi:10.1103/PhysRevLett.93.173903
26. M.D. Stenner, D.J. Gauthier, M.A. Neifeld, The speed of information in a ‘fast-light’ optical medium. Nature **425**, 695–698 (2003). doi:10.1038/nature02016
27. L.J. Wang, A. Kuzmich, A. Dogariu, Gain-assisted superluminal light propagation. Nature **406**, 277–279 (2000). doi:10.1038/35018520
28. R.T. Glasser, U. Vogl, P.D. Lett, Stimulated generation of superluminal light pulses via four-wave mixing. Phys. Rev. Lett. **108**(17), 17390-2-6 (2012). doi:10.1103/PhysRevLett.108.173902
29. V.G. Veselago, The electrodynamics of substances with simultaneously negative values of ε and μ. Sov. Phys. Usp. **10**(4), 509–514 (1968). doi:10.1070/PU1968v010n04ABEH003699
30. J.B. Pendry, Negative refraction makes a perfect lens. Phys. Rev. Lett. **85**(18), 3966–3969 (2000). doi:10.1103/PhysRevLett.85.3966
31. D.R. Smith et al., Composite medium with simultaneously negative permeability and permittivity. Phys. Rev. Lett. **84**(18), 4184–4187 (2000). doi:10.1103/PhysRevLett.84.4184
32. R.A. Shelby, D.R. Smith, S. Schultz, Experimental verification of a negative index of refraction. Science **292**(77), 451–453 (2001). doi:10.1126/science.1058847
33. W.J. Padilla, D.N. Basov, D.R. Smith, Negative refractive index metamaterials. Mater. Today **9**(7, 8), 28–35 (2006)
34. C. Luo, et al., All-angle negative refraction without negative effective index. Phys. Rev. B **65**, 201104-1–201104-4 (2002). doi:10.1103/PhysRevB.65.201104
35. E. Cubukcu et al., Electromagnetic waves: negative refractive index by photonic crystals. Nature **423**, 604–605 (2003). doi:10.1038/42360b
36. I. Bulu, H. Caglayan, E. Ozbay, Negative refraction and focusing of electromagnetic waves by metallodielectric photonic crystals. Phys. Rev. B **72**, 045124-1-6 (2005). doi:10.1103/PhysRevB.72.045124
37. T. Baba, T. Matsumoto, T. Asatsuma, Negative refraction in photonic crystals. Adv. Sci. Tech. **55**, 91–100 (2008). doi:10.4028/www.scientific.net/AST.55.91

38. S. Kocaman et al., Zero phase delay in negative-refractive-index photonic crystal superlattices. Nat. Photonics **5**, 499–505 (2011). doi:10.1038/nphoton.2011.129
39. E.J.R. Vesseur, et al., Experimental verification of $n = 0$ structures for visible light. Phys. Rev. Lett. **110**, 013902-1–013902-5 (2013). doi:10.1103/PhysRevLett.110.013902
40. R. Maas et al., Experimental realization of an epsilon-near-zero metamaterial at visible wavelengths. Nat. Photonics **7**, 907–912 (2013). doi:10.1038/nphoton.2013.258
41. U. Leonhardt, Optical Conformal Mapping. Science **312**(5781), 1777–1780 (2006). doi:10.1126/science.1126493
42. J.B. Pendry, D. Schurig, D.R. Smith, Controlling electromagnetic fields. Science **312**(5781), 1780–1782 (2006). doi:10.1126/science.1125907
43. D. Schurig et al., Metamaterial electromagnetic cloak at microwave frequencies. Science **314**, 977–980 (2006). doi:10.1126/science.1133628
44. W. Cai et al., Optical cloaking with metamaterials. Nat. Photonics **1**, 224–227 (2007). doi:10.1038/nphoton.2007.28
45. J. Li, J.B. Pendry, Hiding under the carpet: a new strategy for cloaking. Phys. Rev. Lett. **101**, 2039-01-11 (2008). doi:10.1103/PhysRevLett.101.203901
46. L.H. Gabrielli et al., Silicon nanostructure cloak operating at optical frequencies. Nat. Photonics **3**, 461–463 (2009). doi:10.1038/nphoton.2009.117
47. J. Valentine, et al., An optical cloak made of dielectrics. Nat. Mater. 568–571 (2009). doi:10.1038/NMAT2461
48. A. Alu, N. Engheta, Achieving transparency with plasmonic and metamaterial coatings. Phys. Rev. E **72**, 016623-3–016623-9 (2005). doi:10.1103/PhysRevE.72.016623
49. D. Rainwater, et al., Experimental verification of three-dimensional plasmonic cloaking in free-space. New J. Phys. **14**, 0130-54-66 (2012). doi:10.1088/1367-2630/14/1/013054
50. J.C. Soric, et al., Demonstration of an ultralow profile cloak for scattering suppression of a finite-length rod in free-space. New J. Phys. **15**, 0330-37-55 (2013). doi:10.1088/1367-2630/15/3/033037
51. M.W. McCall, et al., A spacetime cloak, or a history editor. J. Opt. **3**(2), 02040-03-11 (2010). doi:10.1088/2040-8978/1.3/2/024003
52. M. Fridman et al., Demonstration of temporal cloaking. Nature **481**, 62–65 (2012). doi:10.1038/nature10695

Chapter 8
Quantum Mechanics of the Photon

8.1 Double-Slit Experiments Using Photon Particles

When the photon is treated as a subatomic quantum particle, several types of unusual effects occur, many of which are counter-intuitive. If a double-slit Young's experiment is performed using single-photons produced by a weak source of light, a photon can be directed through one slit or the other at any instant of time. In Fig. 8.1 single-photons are directed to either slit A or slit B, separated by a small time interval on the order of 1 s. The resultant intensity pattern over time can be captured on a screen, such as a photographic plate or a gain-amplified CCD array. Remarkably, an interference-type pattern is observed on the screen, implying that a single-photon passing through slit A or slit B is somehow influenced by the opposite slit. This effect is called *single-photon interference*. It also implies that the photon may have no exact location, or exhibits a wave nature.

If the slits can be observed to determine whether a photon has passed through slit A or slit B, the resultant intensity pattern at the screen changes. In Fig. 8.2a light detector has been placed close to each slit on the exit side. This allows the determination of which slit the single-photon has passed through. When the experiment is repeated, the accumulation of particles hitting the screen over time does not produce the single-photon interference pattern, but rather a pattern that exhibits the particle nature of light photons. In other words, the fact that the active slit has been observed somehow influenced the light pattern registered at the screen. If the detectors are then disabled or disconnected, the slit observation disappears and a single-photon interference pattern appears as in Fig. 8.1. This is often called the *wave-particle paradox* and is closely related to the quantum theory of light.

D.F. Vanderwerf, *The Story of Light Science*,
DOI 10.1007/978-3-319-64316-8_8

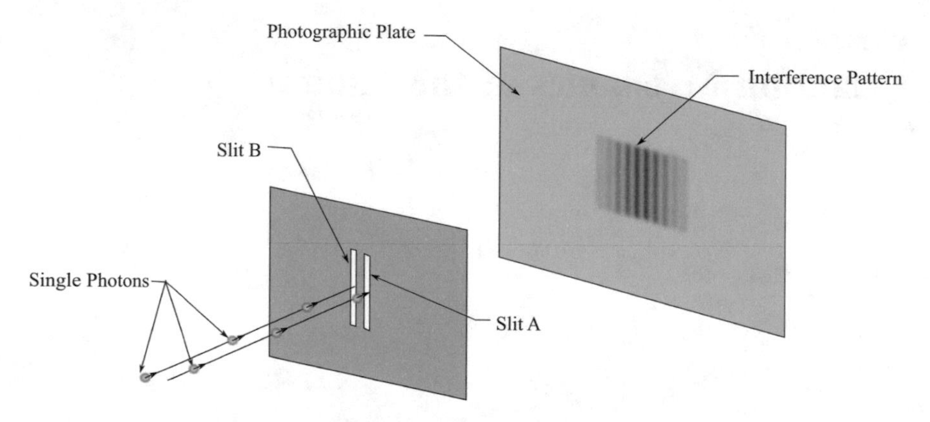

Fig. 8.1 Single photon interference producing an interference pattern on the screen, showing the wave nature of the photon

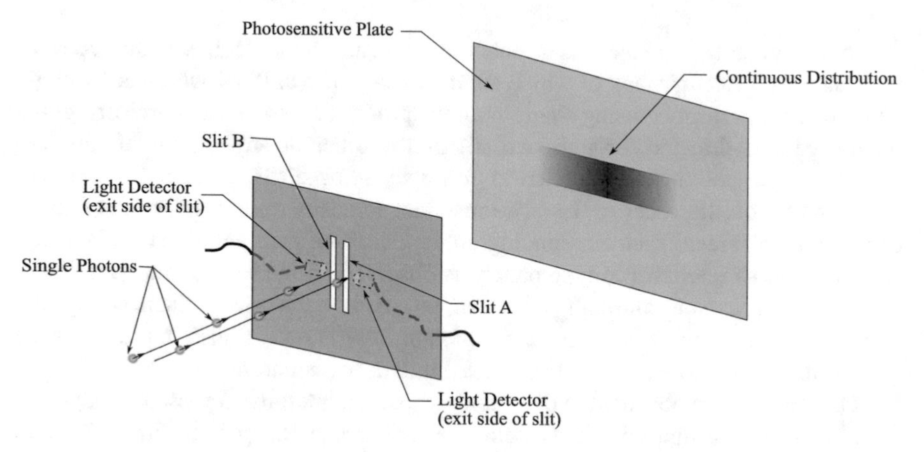

Fig. 8.2 Active photodetectors adjacent to each slit produces a continuous pattern on the screen, showing the particle nature of the photon

8.2 Delayed-Choice Experiments

8.2.1 The Delayed-Choice Experiment of Wheeler

In 1978 John Wheeler proposed a thought or *Gedanken* experiment, using a variation of the double-slit experiment [1]. Instead of the two light detectors adjacent to the slits, two light detectors A and B were positioned just behind the photosensitive screen position and focused on each slit. This produced a known distance L between the slits and the detectors, and the photon transit time could be calculated. While a single-photon was in transit between the slits and the photosensitive screen position, one could choose whether to keep the screen in place, or quickly

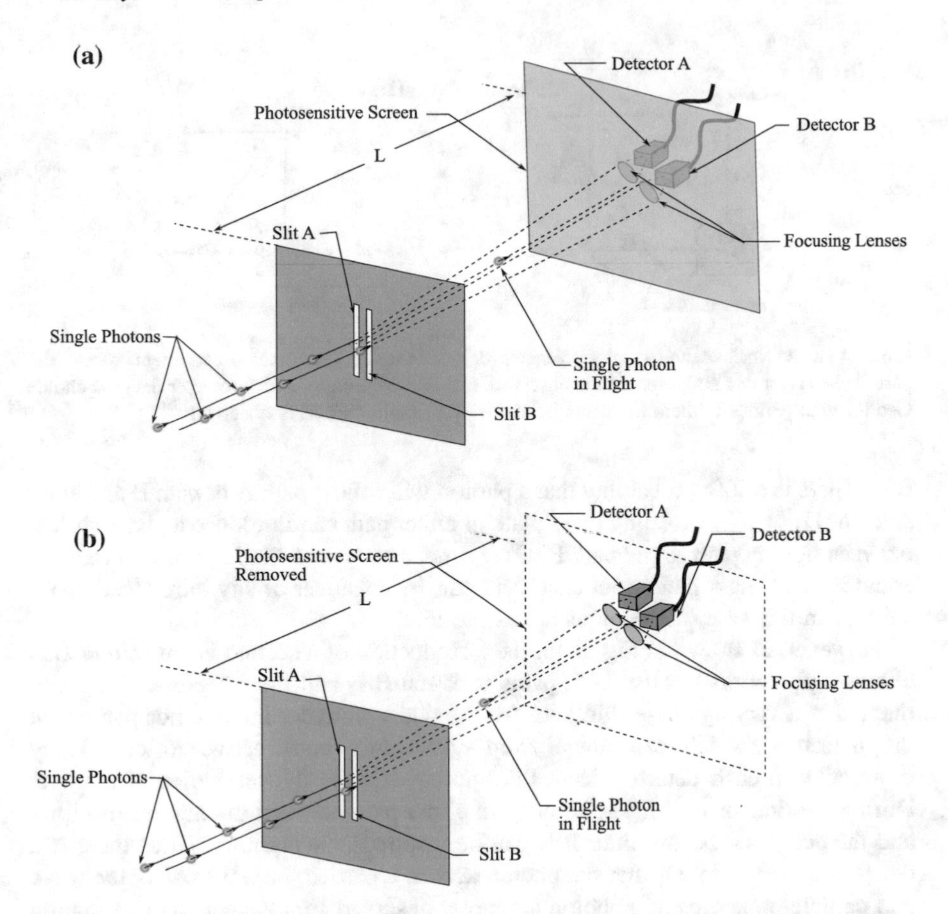

Fig. 8.3 **a** A double-slit delayed-choice Gedanken experiment where the choice is made to observe the wave behavior of a single-photon. **b** A double-slit delayed-choice Gedanken experiment where the choice is made to observe the particle behavior of a single-photon

remove it to expose the focused light detectors. Figure 8.3a shows the screen in place, and the expected interference pattern would indicate the wave nature of the photon particle. In Fig. 8.3b the screen has been removed. The photon could enter detector A or detector B, indicating which slit the photon particle entered. Thus a choice could be made on whether to leave the screen in place and make the light photon behave as a wave, or to remove the screen and make the light photon behave as a particle. This is referred to as a *delayed-choice experiment*. It seems that something done in the present could affect the nature of an event that occurred in the past!

The Mach-Zehnder interferometer is a single-pass interferometer often used for detecting optical inhomogeneities in materials. It has also been used as a platform for a delayed-choice thought experiment [2]. Figure 8.4a illustrates a stream of single-photons entering a Mach-Zehnder interferometer with a single beamsplitter

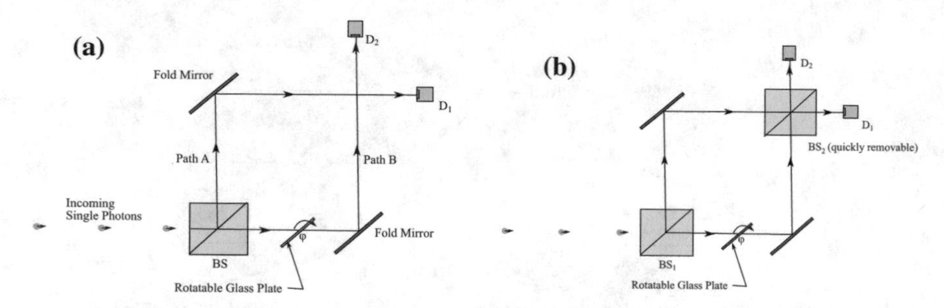

Fig. 8.4 **a** A Mach-Zehnder interferometer delayed-choice Gedanken experiment where the particle behavior of a single-photon is observed. **b** A Mach-Zehnder interferometer delayed-choice Gedanken experiment where the wave behavior of a single-photon is observed

BS_1. There is a 50% probability that a photon will follow path A or path B and enter detector D_1 or D_2. A rotating glass plate in either path can introduce a phase shift φ between light in path A and path B. Over time, a stream of single-photons produces equal intensity spot patterns at each detector, independent of any introduced phase shift φ. In this case, the photons act as particles.

However, as shown in Fig. 8.4b, the introduction of a second beam splitter BS_2 allows a photon to traverse both paths between BS_2 and the detectors. The introduction of a varying phase shift between the paths produces interference patterns at the detectors, which can range from completely constructive to completely destructive at each detector. Here the photon exhibits the properties of a wave. During the time that a single-photon is in either path between the first beamsplitter and the detectors, beamsplitter BS_2 can be inserted or withdrawn. Thus there is a delayed-choice as to whether the photon acts as a particle or wave. Also, the wave and particle properties of a photon are never observed simultaneously. In quantum mechanics, this is known as Bohr's *principle of complementarity*.

The explanation of delayed-choice has become a subject of continuous debate. In the Wheeler interpretation, Wheeler concluded that the properties of a photon in the past do not exist, except as they are recorded in the present. Niels Bohr stated that the photon sometimes acts like a wave, sometimes like a particle, and that we can measure, but never visualize or completely understand either. David Bohm used a quantum potential approach, where the photon particle is surrounded by an active energy field, forming an extended volume of influence around a core [3]. In the double-slit experiment, this allows the second slit to receive some information from a photon passing through the first slit.

8.2.2 Experimental Realization of Delayed-Choice

In 2007 a remarkable experiment was performed that closely approximated the Mach-Zehnder arrangement, and confirmed the Wheeler delayed-choice Gedanken

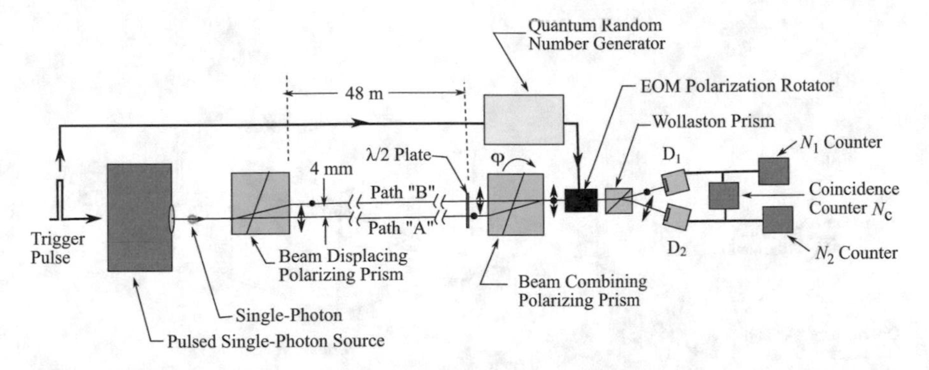

Fig. 8.5 Setup for an experimental measurement of delayed-choice for a single-photon. (Adapted with permission of AAAS from Ref. [4]; permission conveyed through Copyright Clearance Center, Inc.)

experiment [4]. The basic setup is illustrated in Fig. 8.5. A pulsed single-photon light source emitted a polarized photon into a polarizing beamsplitter prism. The resulting parallel orthogonally polarized beams had a spatial separation of about 4 mm while traversing a 48 m path length to an output assembly. The travel time was about 160 ns. The beams then entered a half-wave plate which rotated their polarization planes. These beams entered a polarization beam-combining prism, where both orthogonally polarized beams followed a single path. These overlapping beams then entered a KDP electro-optical modulator (EOM), followed by a Wollaston polarizing prism to define paths to each detector.

There was no physical removal of these output components, but by controlling the polarization modes of the combined beams entering the Wollaston prism, a choice could be made to restrict the beams to follow a single path to each detector (favoring particle behavior of photon), or to allow the photon to simultaneously follow both paths to the detectors (favoring wave behavior of photon). When no voltage was applied to the EOM, there was no change in the polarization states of photons following either path. Figure 8.6a illustrates the output components, where the photons were restricted to a single path (*open configuration*). However, when a half-wave retardance voltage $V_{\lambda/2}$ was applied to the EOM, with optical axes oriented as in Fig. 8.6b, the polarization state was rotated 45°. Beams from the paths "A" and "B" could then reach both detectors (*closed configuration*). By tilting the beam-combining prism, a phase difference was introduced between the two paths and any single-photon interference could be detected.

The voltage to the EOM could be switched from $V = 0$ to $V = V_{\lambda/2}$ within about 40 ns, and random values of these voltage values were chosen by the use of a quantum random number generator (QRNG). The QRNG was clock synchronized to the pulsed single-photon source, which emitted photons of 45 ns pulse duration at a pulse repetition rate of 238 ns. The delayed-choice decision for an open or

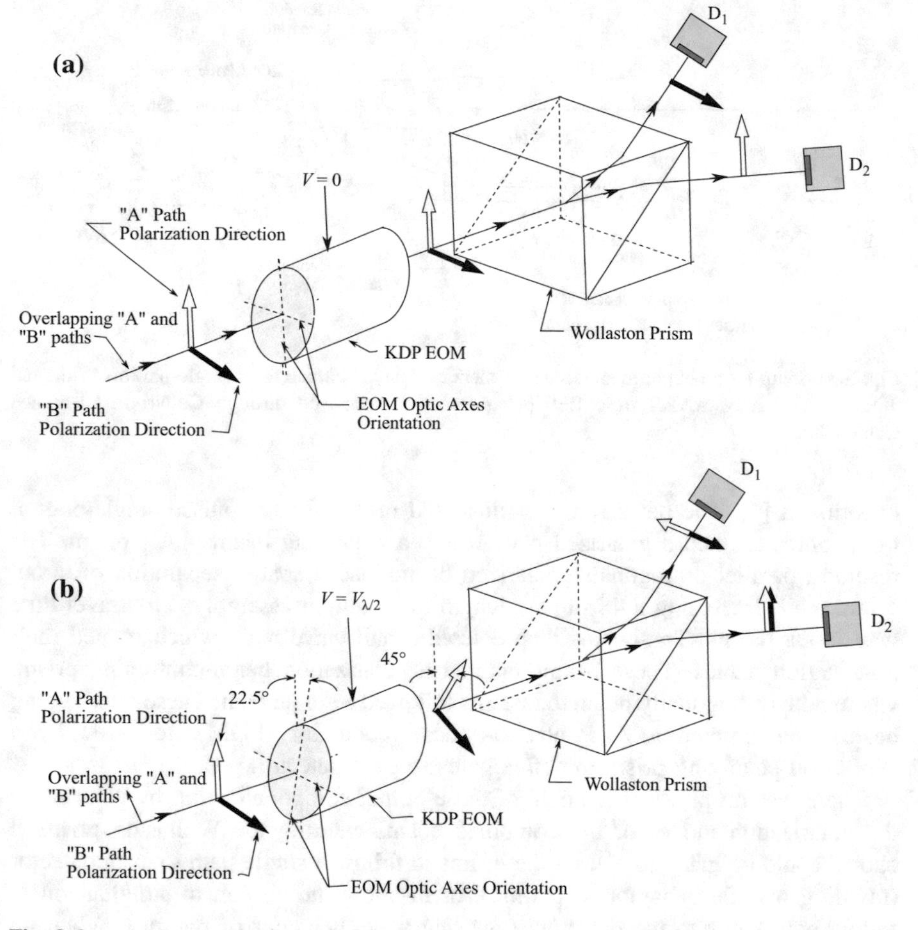

Fig. 8.6 **a** Arrangement of delayed-choice components to record the particle nature of a single-photon. **b** Arrangement of delayed-choice components to record the wave nature of a single-photon

closed interferometer was made when a single-photon was about halfway down the 48 m interferometer path. The number of photons N_1 and N_2 hitting detector D_1 or D_2 (particle behavior) and the number of photons N_c coincident on both detectors (wave behavior) were counted. Data points, consisting of about 2600 detected photons over a 1.9 s acquisition time, were then plotted against the phase shift φ. For the open configuration of the interferometer, the data points were clustered around a horizontal line with equal detection probabilities. However, for the closed configuration, the data points produced an interference pattern with a 94% visibility. Thus a delayed-choice was realized that determined the particle or wave behavior of a single-photon.

8.2.3 EPR Paradox, Particle Entanglement, and Bell's Inequalities

If a pair of spatially separated quantum particles are somehow linked such that a measurement of the state of one particle changes the state of the other, then the quantum particles are said to be *entangled*. Particles can be entangled in various types of properties, such as spin, momentum, energy, or polarization. Moreover, a determination of one of these properties on one particle would instantaneously reveal the state of the corresponding property for the other particle, irrespective of the separation of the particles. Einstein, Podolsky, and Rosen analyzed the ramifications of this effect and concluded that the condition of locality was violated and the two particles could not be described by a single wave function [5]. Also, this seems to violate Einstein's special relativity where information cannot travel faster than the speed of light. This gave rise to the famous *Einstein-Podolsky-Rosen paradox* or *EPR paradox*. These theorists believed that quantum theory was somehow incomplete, and required the introduction of some hidden or supplemental variables to explain the particle correlations.

According to quantum mechanics, the state of a quantum particle, e.g. momentum, position, or spin, cannot be determined until it is observed. In 1957 Bohm and Aharonov proposed a hypothetical experiment in which a quantum particle with spin 0 emitted two new particles, the new particles having opposite spin states of +1/2 or −1/2 [6]. Then any observation of the spin state of either spatially separated quantum particle must *instantaneously* determine the state of the second particle, with no need for further observation. *Bell's inequalities* were an attempt to describe the dependence of the correlated spins of the two particles in terms of classical probabilities [7]. Failure of this *Bell test* showed that there were no hidden variables to explain the correlation between entangled quantum particles. The mainstream view today is that entangled information is instantly communicated, that quantum mechanics displays the condition of non-locality, and that Einstein's principle is not violated for other types of information transfer.

8.2.4 Experimental Verification of Photon Entanglement

The first complete experimental verification of correlation between two spatially separated entangled photons was made in 1982 by Alain Aspect and colleagues in France [8]. The optical technique of *polarization entanglement* was used for entanglement correlation measurements. The experiment used uncorrelated time-variable polarizers during the transit of the entangled photons, and restricted any information exchange between the two photons to occur at less than, or at the speed of light. The experimental setup is illustrated in Fig. 8.7. A light source consisting of cascaded atomic transitions in calcium emitted two entangled photons in opposite direction, at wavelengths 422.7 and 551.3 nm. Each photon was

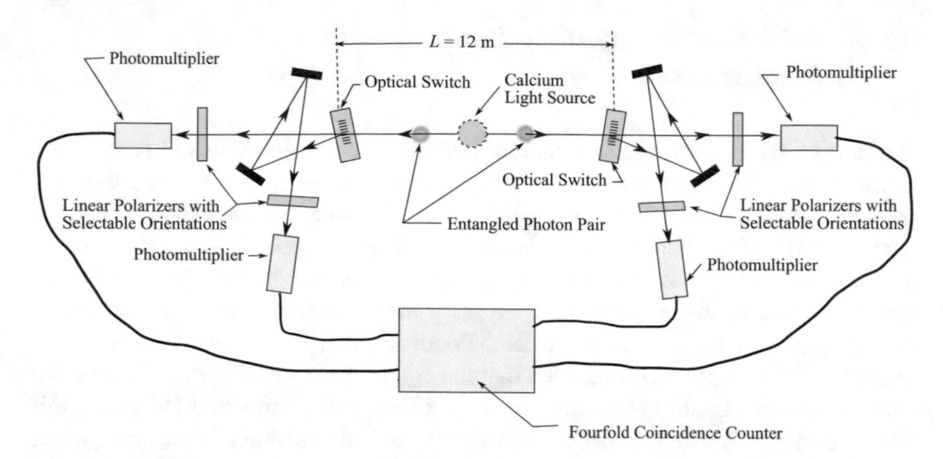

Fig. 8.7 Experimental setup to verify the correlation of entangled photon pairs. (This figure adapted with permission from Ref. [8]. Copyrighted by the American Physical Society)

incident upon an acoustic Bragg grating optical switch with a cycle time of $\approx$20 ns. There was a separation of L = 12 m between the optical switches, giving a 20 ns transit time from the light source to each switch. Each optical switch directed the light to one of two linear polarizers with selectable non-orthogonal polarization axes, with the two switches randomly switched at $\approx$50 MHz, and uncorrelated with each other. This was equivalent to changing the settings of the polarizers at a rate exceeding c/L (25 MHz). Light from each polarizer was detected by a photomultiplier and the signals were sent to a fourfold coincidence counter.

Using possible combinations of the switched linear polarizations in both channels, a form of Bell's inequalities was obtained. For a typical run of about 12,000 s, the coincidence rates were compared to the predictions of Bell's inequalities. It was found that the results agreed with the quantum mechanical predictions, but violated Bell's inequalities by a statistically significant amount.

8.2.5 A Quantum Mechanical Beamsplitter

While the classic treatment of a beamsplitter considers the partial reflection of light waves, quantum beamsplitting is described in terms of quantum mechanical probability amplitudes that determine how incident photons will be reflected or transmitted. For a 50/50 transmission/reflection beamsplitter, there is a 50% probability that the photon will be transmitted, and a 50% probability that the photon will be reflected. A beamsplitter analyzed in quantum mechanical terms is often called a *quantum beamsplitter* (QBS). The quantum state of a photon is usually designated as $|\Psi\rangle$, where Ψ represents a combination of properties of the photon particle. The bracket symbol $|\ \ \rangle$ is called a *ket*, from quantum notation introduced by Paul Dirac. Figurc 8.8 illustrates a beamsplitter with two possible entrance paths. This two-state system has input paths 1 and 2, where a photon in path 1 has reflection and transmission probability amplitudes r_1 and t_1, and a photon in path 2 has reflection and transmission probability amplitudes r_2 and t_2. To detect the output states, detectors D_1 and D_2 are positioned as shown. A unitary linear transformation matrix $\mathbf{R}$ can be defined for the beamsplitter that transforms the input state $|\Psi_{\text{in}}\rangle$ to the output state $|\Psi_{\text{out}}\rangle$, such that $|\Psi_{\text{out}}\rangle = \mathbf{R}|\Psi_{\text{in}}\rangle$, where $\mathbf{R} = \begin{pmatrix} r_1 & t_2 \\ t_1 & r_2 \end{pmatrix}$. If the beamsplitter is symmetric, then $r_1 = r_2$ and $t_1 = t_2$. However, there is a phase lag $\delta = \pi/2$ of the reflected beams from the transmitted beams, which introduces the imaginary number i into the transmission probability amplitudes. For a 50/50 symmetric beamsplitter $r = \frac{1}{\sqrt{2}}$ and $t = \frac{i}{\sqrt{2}}$, resulting in $\mathbf{R} = \frac{1}{\sqrt{2}}\begin{pmatrix} 1 & i \\ i & 1 \end{pmatrix}$. Moreover, quantum mechanics predicts that each photon will travel to either D_1 or D_2 with 50% probability, but never to both [9].

If the two photons in Fig. 8.8 are entangled and emitted simultaneously from a light source, then there is a 50% probability that both photons will be registered as a single count at D_1 or D_2. This is designated as a *symmetric exchange* of quantum spatial states, either or $|\Psi\rangle = |1_1\rangle|1_2\rangle$ or $|\Psi\rangle = |2_1\rangle|2_2\rangle$, where the subscript refers to the mode of the photon in each path. Another type of symmetric exchange occurs when $|\Psi\rangle = \frac{1}{\sqrt{2}}[|1_1\rangle|2_1\rangle + |1_2\rangle|2_1\rangle]$, and one photon goes to D_1 and the other to D_2. A *non-symmetric exchange* is described by the spatial quantum state. $|\Psi\rangle = \frac{1}{\sqrt{2}}[|1_1\rangle|2_1\rangle - |1_2\rangle|2_1\rangle]$, and again, one photon goes to D_1 and the other to D_2. The latter two exchanges can produce coincidence counts between the detectors. However, all possible photon paths are indistinguishable and there is no evidence of interference at either detector.

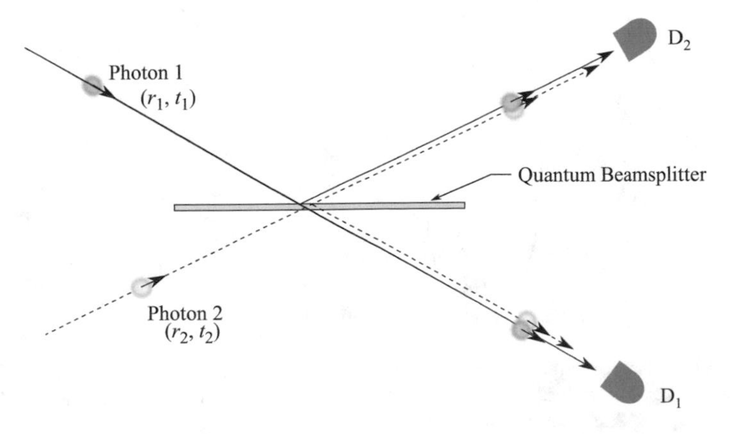

Fig. 8.8 A quantum beamsplitter with two possible entrance paths and four possible exit paths

8.2.6 Delayed-Choice Quantum Eraser

According to the principle of complementarity, the wave and particle behavior of a photon cannot be observed simultaneously for a given type of measurement. However, for an entangled photon pair, it may be possible to perform one type of observation on a *test* or *idler* photon, and another type of observation on the second *corroborative* or *signal* photon to determine which type of behavior the test photon has displayed. In 1982 a theoretical experiment was proposed and analyzed using entangled test photon pairs emitted from atoms "A" and "B", produced by a decay process initiated from incident laser pulses [10]. A configuration of such an experiment is illustrated in Fig. 8.9. The emitting atoms were at the focal points of a double elliptical cavity. The emitted photons from atom "A", designated as φ_A and γ_A, were generated from different atomic transitions and had different wavelengths. Similarly, the photons emitted from atom "B" were designated as φ_B and γ_B. The differing wavelengths allowed the φ_A and φ_B idler photons to be reflected to a φ detector positioned at the common focal points of the elliptical cavity, while the transmitted γ_A and γ_B signal photons were directed to a detector outside the cavity. This γ detector was significantly farther from the emitting atoms than the φ detector, resulting in detection of the φ_A and φ_B photons before detection of the γ_A

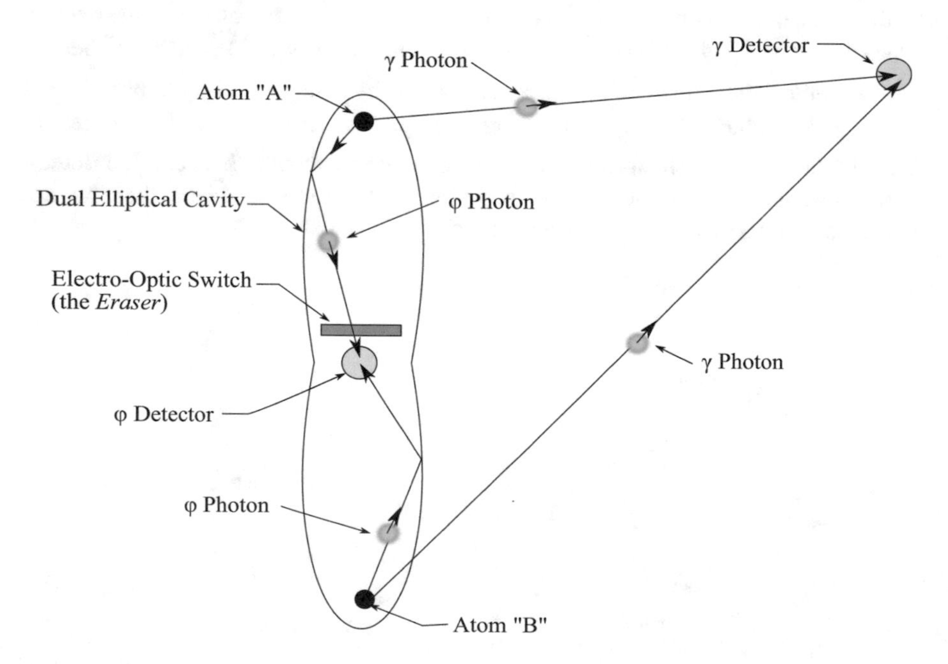

Fig. 8.9 A proposal for a quantum eraser experiment using a dual elliptical cavity, two detectors, and an electro-optic switch as the eraser. (This figure adapted with permission from Ref. [10]. Copyrighted by the American Physical Society)

and γ_B photons. Closure of an electro-optic shutter blocked φ_A photons from reaching the φ detector. When the electro-optic shutter was open, the paths of the γ photons were indistinguishable and interference patterns formed at the γ detector. However, with a closed electro-optic shutter, the φ_A photon was not registered at the φ detector, and by the entanglement of the φ_A and γ_A photons, which-path information was revealed and the interference pattern at the γ detector disappeared. The γ photons would now behave as particles. However, while these γ photons were in transit, or even at the γ detector, wave behavior of the photons could be recovered by reopening the electro-optic shutter. The shutter essentially erased the which-path information, and wave behavior of the γ photons was restored. Interference patterns then reappeared at the γ detector. This is called a *delayed-choice quantum eraser.*

Certain types of dielectric nonlinear crystals, when pumped by a laser pulse on one side, emit a pair of correlated light cones from the opposite side. The photons in these cones are weaker than the laser pump beam and have different frequencies, while total quantum energy is conserved. The frequencies of the pump beam and the down-converted signal and idler beams are related by $\omega_{pump} = \omega_{signal} + \omega_{idler}$. This is known as *parametric down-conversion* PDC (usually called *spontaneous parametric down-conversion* SPDC). In Type-II SPDC the two emitted light cones have an angular separation. Photons at the intersection points of these light cones form a superposition of two quantum states, entangled with respect to the orthogonal polarization states of the photon pair. These photon pairs remain entangled as their spatial separation increases. Polarization entanglement can be achieved using a BBO (β-BaB_2O_4) optical crystal [11]. Another type of SPDC is Type-I SPDC, where the output light cones are centered along the pump beam propagation axis, and photons in each beam are correlated with respect to a common polarization direction. However, the photons are not in a superposed, entangled state.

Figure 8.10 shows another proposed arrangement for quantum erasure [12]. Response of the test photons at either of two detectors D_3 and D_4 provides which-path information for both the test and the entangled correlated photons. However, by an arrangement of four detectors and three 50/50 quantum beam-splitter plates, it is possible for the idler photon to trigger one of two detectors that reveals which-path information or either of two other detectors D_1 and D_2 that does not reveal this information. The entangled signal photons are focused to a scanning detector D_0, where they can generate an interference pattern. However, which-path information from responses at D_3 or D_4 can extinguish this interference pattern. The which-path information can subsequently be erased by responses from D_1 or D_2, where the paths are again made indistinguishable, and interference from signal photons can occur at D_0. The decision on distinguishability of the idler photons can be performed by the type of measurement on the entangled signal photons.

In 1999 the first experimental realization of a delayed-choice quantum eraser was performed using a modification of Fig. 8.10 [12]. As shown in Fig. 8.11, a series of two sets of 702.2 mm polarization entangled photon pairs were emitted

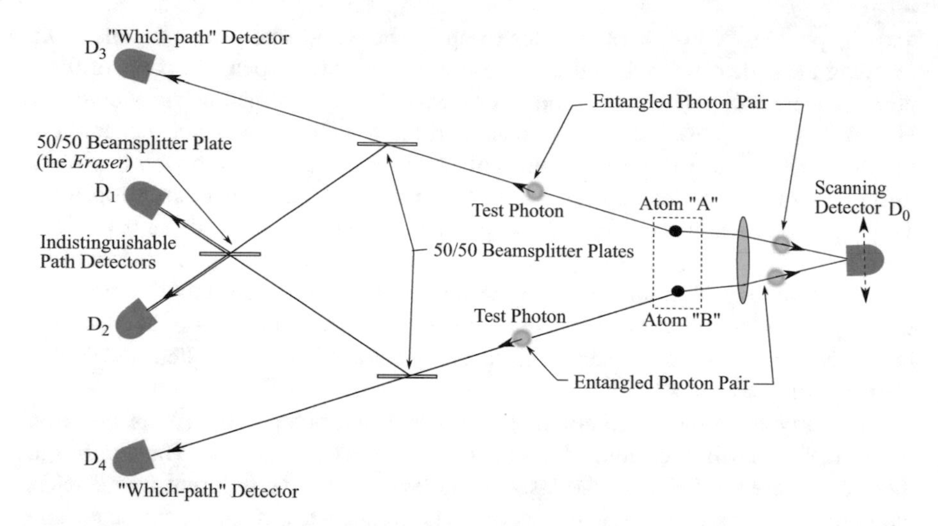

Fig. 8.10 A proposal for a quantum eraser experiment using four detectors and three 50/50 quantum beamsplitter plates. (This figure adapted with permission from Ref. [12]. Copyrighted by the American Physical Society)

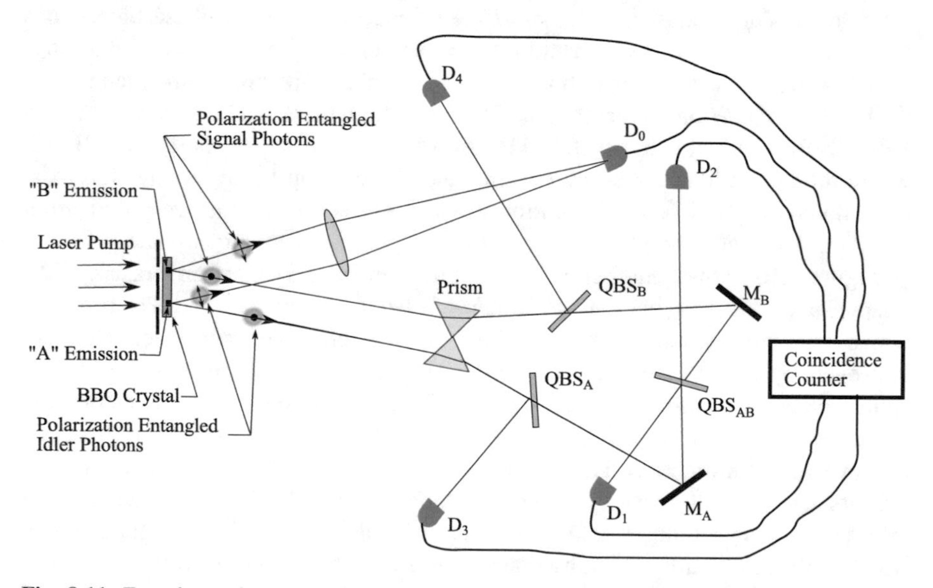

Fig. 8.11 Experimental setup to observe quantum erasure, using a polarization entangled photon pair, three quantum beamsplitters, and five detectors. (This figure adapted with permission from Ref. [12]. Copyrighted by the American Physical Society)

from regions “A” and “B” of a BBO crystal pumped by a pulsed split-beam 351.1 mm argon ion laser. One pair of entangled signal photons was directed by a refracting prism to two legs of an interferometer with quantum beamsplitters. The

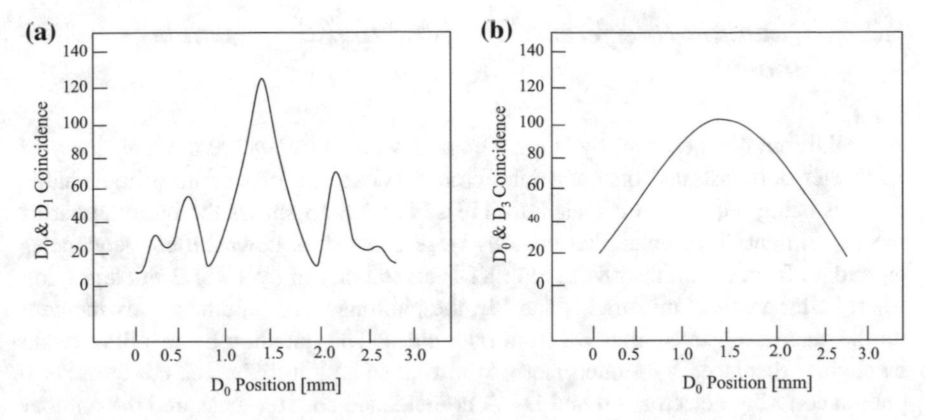

Fig. 8.12 **a** Typical curve fitted to coincidence counts between detectors D_0 and D_1 for various positions of scan detector D_0 (actual coincidence count points not shown). The interference-type pattern shows the wave behavior of the photon. (This figure adapted with permission from Ref. [12]. Copyrighted by the American Physical Society.) **b** Typical curve fitted to coincidence counts between detectors D_0 and D_3 for various positions of scan detector D_0 (actual coincidence count points not shown). The curve exhibits no interference-type pattern, showing the particle behavior of the photon. (This figure adapted with permission from Ref. [12]. Copyrighted by the American Physical Society.)

quantum beamsplitters determined which path the beam traversed in the interferometer. Four detectors, D_1, D_2, D_3, and D_4, indicated idler photon path information. The other pair of entangled signal photons was focused at detector D_0, located a shorter distance from the BBO crystal than the four idler path detectors. Detection of the signal photon at D_0 occurred at $\approx$8 ns *before* any information was received by any of the idler detectors, preventing any initial which-path information. As detector D_0 was scanned along its transverse axis, joint detection of the signal and idler photons on specific detectors indicated which-path or indistinguishable both-path information. A coincidence count between D_0 and D_1 or D_2 indicated both-path information, and a coincidence count between D_0 and D_3 or D_4 indicated which-path information. All joint detections were recorded during a single scan of D_0, with the quantum beamsplitters determining which type of path information was revealed. Figure 8.12a shows a type of interference pattern when the coincidence counts between D_0 and D_1 are plotted against the D_0 position. But in Fig. 8.12b, a similar plot of the coincidence counts between D_0 and D_3 or D_4 showed an absence of interference. The both-path information must have been erased. However, another plot of coincidence counts between D_0 and D_3 showed another interference-type pattern, caused by the erasure of the which-path information. This switching between wave and particle behavior of the photons is an intrinsic property of the system, and required no external intervention. Since the erasures occur *after* the detection of the signal photon at D_0, the system is referred to as a *delayed-choice quantum eraser*.

8.2.7 *Quantum Interference Between Indistinguishable Photons*

A breakthrough experiment by Hong, Ou, and Mandel in 1987 at the University of Rochester demonstrated quantum interference between a pair of indistinguishable photons using a quantum beamsplitter [13]. Figure 8.13 shows the basic layout of the experiment. Two entangled photons were emitted as downshifted frequencies ω_1 and ω_2 from a nonlinear 8 cm long KDP crystal driven by a 351.2 nm argon ion laser. Using a pair of mirrors M_1 and M_2, the photons were simultaneously incident on the quantum beamsplitter BS from both sides. The quantum beamsplitter could be slightly displaced by a micrometer to introduce a path difference $c\delta t$ to the two photon counting detectors D_1 and D_2. A coincidence counter measured the number of coincidences N_{coinc} between the two detectors. If the path lengths from the beamsplitter to the detectors were the same, quantum mechanics predicted that both photons would combine to strike either D_1 or D_2. However, for a significant value of $c\delta t$, the path lengths were unequal, and N_{coinc} is proportional to $(r^2 + t^2)$, where r is the beamsplitter reflectance, and t is the beamsplitter transmittance. A significant number of coincidence counts were measured. When $c\delta t$ approached zero, then N_{coinc} is proportional to $(r - t)^2$, and for a 50/50 beamsplitter ($r = t = 0.5$), few detector coincidences were observed. Both incident photons had a 50% probability of being in the same detector path. A single-photon in each path was not observed. Figure 8.14 illustrates the number of coincidences N_{coinc} versus the beamsplitter displacement $c\delta t$. The observed minimum represented destructive interference, which could only be explained by the simultaneity of the spatial quantum states of indistinguishable photons at the beamsplitter. This observed minimum is referred to as the *Hong-Ou-Mandel dip*.

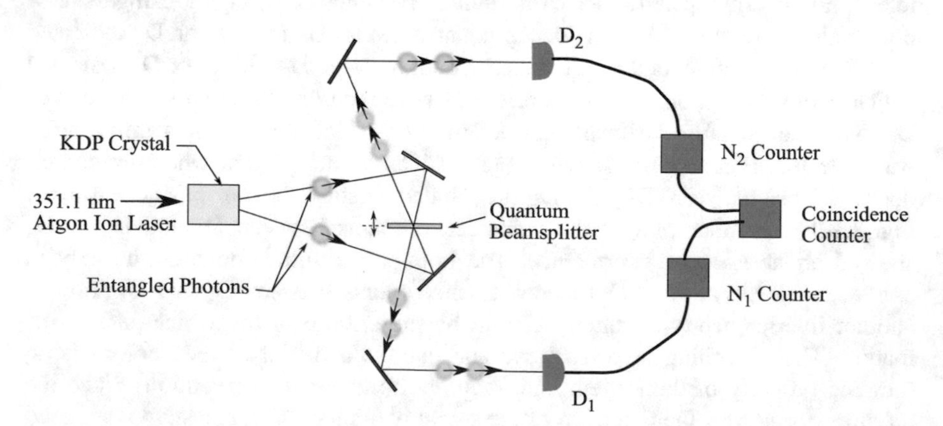

Fig. 8.13 Experimental setup to demonstrate quantum interference between a pair of indistinguishable photons using a quantum beamsplitter, as performed by Hong, Ou, and Mandel

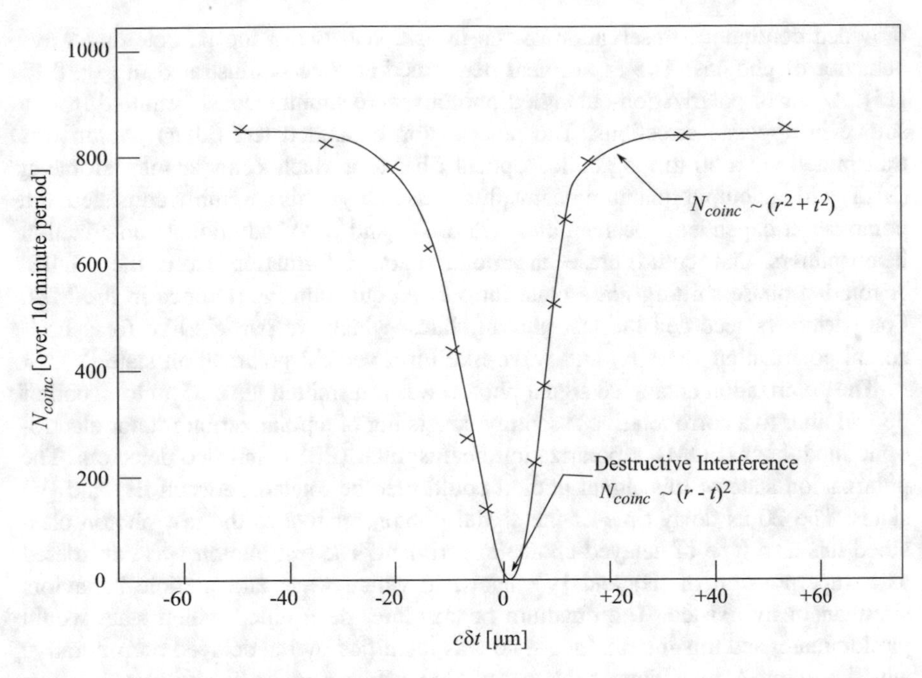

Fig. 8.14 Plot of the number of coincidences N_{coinc} versus the beamsplitter displacement $c\delta t$, showing destructive interference and the H-O-M dip

These results state that when two indistinguishable photons are simultaneously incident on a beamsplitter with an accuracy less than their coherence time, that both photons will emerge together from one possible path or the other. This outcome is referred to as the *Hong-Ou-Mandel effect* (H-O-M effect), and has numerous applications in quantum mechanical particle experiments.

8.2.8 Observation of Photon Wave-Particle Transitions

In 2011 a delayed-choice experiment using a quantum beamsplitter was proposed that could distinguish wave and particle states of a photon [14]. As a quantum-controlled device, the quantum beamsplitter treats the incident photon as a superposition of particle and wave states having equal probability. Although the complementarity principle is still valid, a properly designed delayed-choice quantum experiment could then observe the transition or morphing between the particle and wave behavior of a photon.

In 2012 independent investigators at the University of Bristol and the University of Nice used entangled photons to perform delayed-choice experiments that provided continuous observation of the transition between the particle and wave behavior of photons. The experiment performed at Nice is illustrated in Fig. 8.15 [15]. A pair of polarization-entangled photons were simultaneously emitted from a source in opposite directions. The polarization entangled test (idler) photon was transmitted via a 50 m long coiled optical fiber to a Mach-Zehnder interferometer (MZI) with an output quantum beamsplitter assembly. This assembly consisted of a polarization-dependent beamsplitter (PDBS), and two additional polarization beamsplitters that could erase the polarization information from the PDBS. A rotating phase shifting glass plate introduced quantum interference in the MZI. Four detectors recorded the test photon states, which are particle-like for a horizontal polarization state $|H\rangle$ and wave-like for a vertical polarization state $|V\rangle$.

The polarization entangled signal photon was transmitted via a 55 m long coil of optical fiber to a corroborative assembly consisting of a polarization rotator electro-optic modulator (EOM), a polarization beamsplitter (PBS), and two detectors. The polarization state of this signal photon could then be rotated between $|H\rangle$ and $|V\rangle$ states. The 20 ns delay time of the signal photon relative to the test photon classified this as a type of delayed-choice experiment. The test photon was considered as a superposition of $|H\rangle$ and $|V\rangle$ states, in which wave and particle behaviors simultaneously existed. The quantum beamsplitter determined which state would predominate, and this interference state was identified by the delayed corroborative signal photon. A transition of the test photon between wave-like and particle-like behavior was thus observed.

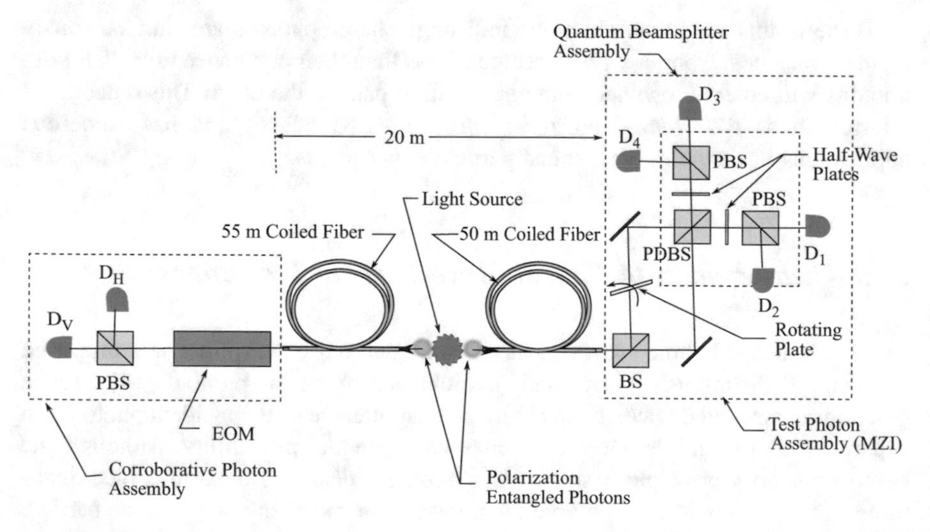

Fig. 8.15 Illustration of a delayed-choice experiment to simultaneously observe transitions between wave and particle behavior of photons. (Adapted with permission of AAAS from Ref. [15]; permission conveyed through Copyright Clearance Center, Inc.)

8.3 Quantum Electrodynamics and the Photon

The theory of quantum electrodynamics (QED) considers electromagnetic theory incorporating quantum mechanics and relativistic effects. It was first formalized in the late 1920s by Paul Dirac, and was quite useful in describing the interaction of light with matter (specifically the electrons in the matter). QED was popularized by physicist Richard Feynman, where he considered the photon *only* as a quantum particle—the "corpuscular" photon, eliminating the need for any wave-particle duality [16]. Moreover, all common optical phenomena, such as refraction, reflection, and diffraction could be explained by the use of *probability amplitudes P* for the paths of these photons.

Figure 8.16 illustrates a QED explanation of reflection from a planar mirror, as described by Feynman. All paths from the light source to the detector are considered as possible. The paths are considered as vectors, having a determined length L and direction relative to a fixed reference line. Vector length determines the possibility of a photon traveling on any particular path. The direction angle is determined by the travel time for each possible path (analogous to the final position of an ultra-fast spinning hand on a stopwatch). Each path vector defines a *probability amplitude P*, where $P = L^2$. The net effect is obtained by vector addition of all the path vectors. Figure 8.17 shows the travel time differences for each path, usually of the order of several nanoseconds, where all vector lengths are approximately the same. The resultant vector occurs at the point where the travel time is minimized and corresponds to the classic law of reflection, where the angle of incidence equals the angle of reflection.

Quantum electrodynamics describes the interaction of photons and electrons, which both travel in four-dimensional Minkowski spacetime (Sect. 3.6). In addition, electrons can emit or absorb photons. To illustrate the concepts of these interactions, Feynman describes these interactions in two-dimensional spacetime,

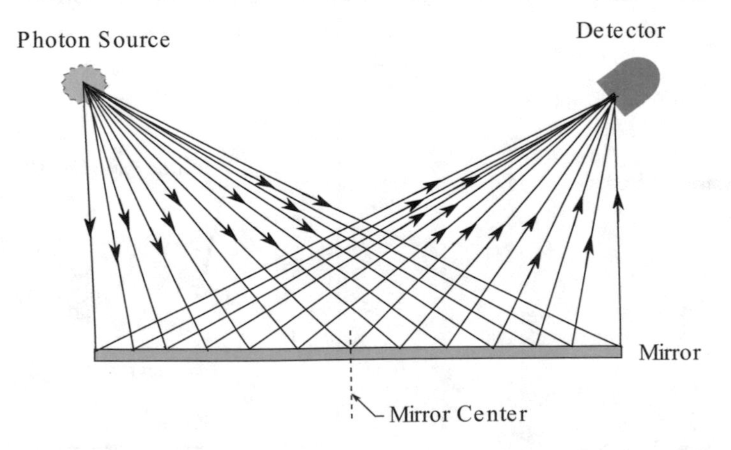

Fig. 8.16 Reflection of light from a planar mirror, as described by quantum electrodynamics

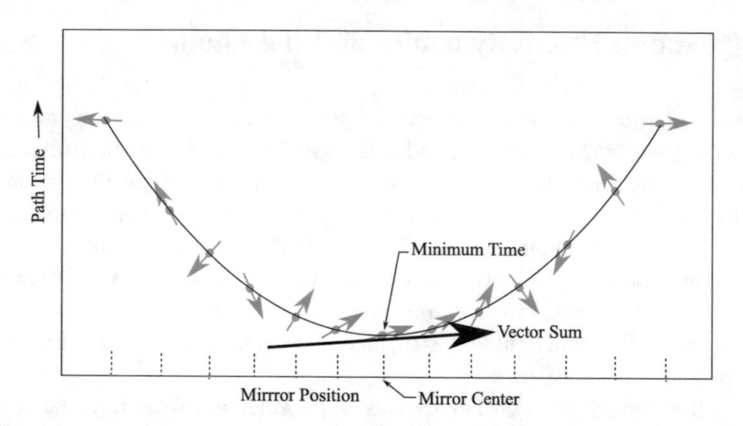

Fig. 8.17 Travel time differences for each possible path of reflection from a planar mirror

using simplified real polarization and spin numbers [17]. Motions, emissions, and absorptions are characterized by probability amplitudes *P*, using electron *P* values, and photon *P* values. Figure 8.18 illustrates the emission (or absorption) of a photon by an electron in two-dimensional spacetime. The electron movements from point A to point B and from point B to point C are represented by a straight line composed of the sum of shorter paths having various *P* values. The photon is indicated by a wavy line. This type of representation is called a *Feynman diagram*.

Figure 8.19a, b illustrate paths in which two electrons could travel with an intermediate emission and absorption of a photon, where the *P* values for absorption and emission are about the same. The photon is absorbed at a later time than it is emitted. In the first case the AC path and the DF path remain separate, while in the second case these paths cross. The end result for both cases is indistinguishable, and the photon can be considered as an *exchange particle* that is not actually observed. For this reason it is often called a *virtual photon.* A function of virtual photons is to allow information to be transferred between two electrons. As shown

Fig. 8.18 Feynman diagram showing the interacting paths of an electron and photon in one-dimensional spacetime

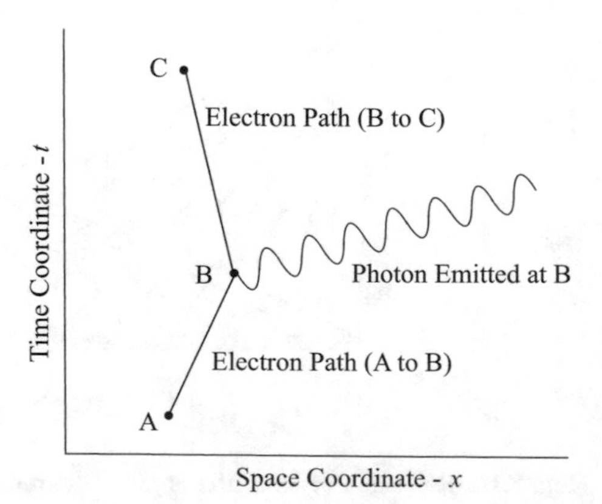

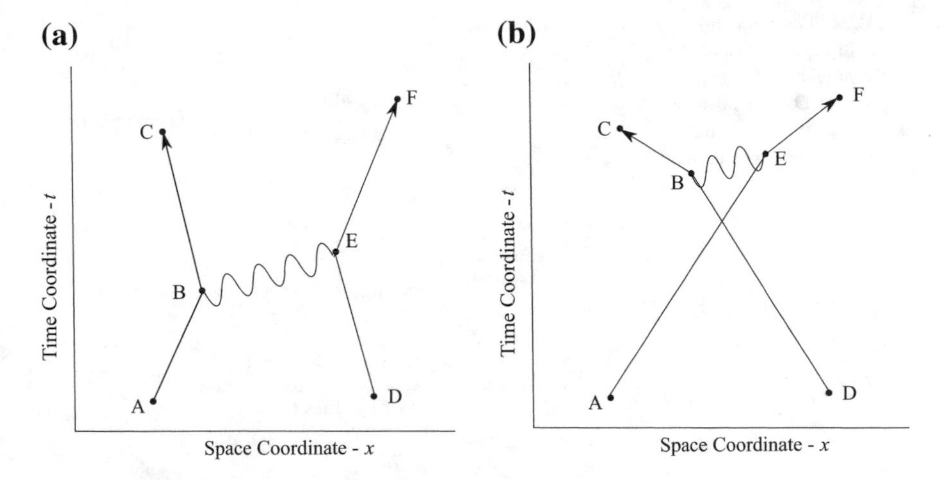

Fig. 8.19 **a** Diagram showing emission and absorption of an exchange photon between two electron paths that do not cross. **b** Diagram showing emission and absorption of an exchange photon between two electron paths that cross

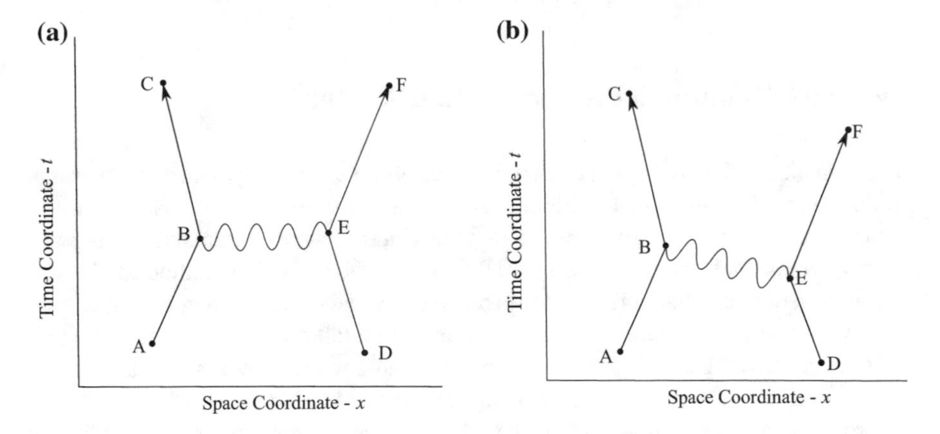

Fig. 8.20 **a** Diagram showing photon emission and absorption between electrons at B and E that occur at the same time. **b** Diagram showing photon emission from an electron at B that is absorbed by another electron at E at an earlier time, or emitted at E and absorbed by another electron at B at a later time

in Fig. 8.19a, a virtual photon can be emitted at B and absorbed at E at a later time. Emission or absorption of a photon at B or E at the same time is shown in Fig. 8.20a. Lastly, the photon could be considered to be emitted at B and absorbed at E at an earlier time, or emitted at E and absorbed at B at a later time, as shown in Fig. 8.20b.

Another type of photon-electron interaction is shown in Fig. 8.21. Here an electron moves along path AB and a photon moves along path CD. At point D the photon disintegrates to create another forward moving electron and a backward

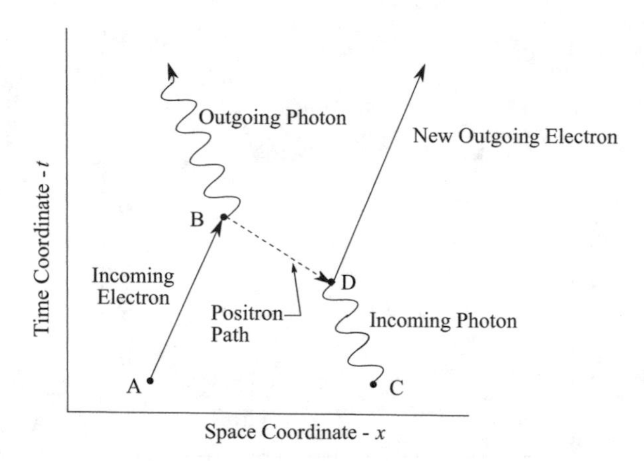

Fig. 8.21 Diagram showing disintegration of a photon to create an electron and a positron at D. The positron annihilates the electron at B and a real photon is emitted

moving *positron*, where the positively charged positron is the antiparticle of the electron. The positron annihilates the original electron at B and emits a real photon, while the electron created at D continues its path. A photon is its own antiparticle, and movement of a photon forward or backward in time cannot be distinguished.

8.4 The Casimir Effect and Virtual Photons

Quantum electrodynamics tells us that in a spacetime vacuum, quantum fluctuations produce virtual photons that randomly appear and disappear. The energy of these virtual photons moves above and below the electromagnetic quantum zero-point energy—a residual vacuum energy for all space fields. In 1948 a method to observe a consequence of this effect was proposed by Hendrick Casimir using two closely-spaced (on the order of microns) parallel metallic plates in a vacuum [18]. These conductive plates act as cavity boundaries for the wavelengths associated with the virtual photons, each electromagnetic mode having energy $\hbar\omega/2$. Certain electromagnetic modes that do not fit within this cavity are suppressed, while the modes in the free vacuum outside the cavity are unaffected. For boson particles such as photons, the resultant difference between the mode density in the cavity and the surrounding vacuum produces a small attractive force between the plates, and the plates move closer together. This is called the *Casimir effect*. The actual force that is pushing on the plates from the free vacuum side is called the *Casimir force F*, and can be calculated from

$$F = \hbar c \left(\frac{\pi^2}{240} \right) \left(\frac{A}{d^4} \right), \tag{8.1}$$

where A is the area of each plate, and d is the separation between the plates.

An illustrative configuration is shown in Fig. 8.22. The conductive plates are very thin to minimize any gravitational attraction, and grounded to eliminate electrostatic attraction. The plates are allowed to move in a direction orthogonal to the plates. Accurate observable measurements of the Casimir force for this configuration is very difficult, due to the need for accurate parallelism of the thin plates.

The first successful measurement of the Casimir force was reported by Lamoreaux at the University of Washington in 1997, using a flat conductive plate and a conductive sphere [19]. This geometry eliminated the need for precise parallelism between flat plates. The Casimir force F is given by

$$F = \frac{2\pi R}{3}\left(\frac{\pi^2}{240}\right)\left(\frac{\hbar c}{d^3}\right), \tag{8.2}$$

where R is the radius of the sphere, and d is the distance between the spherical surface and the flat plate. A torsion pendulum arrangement was used with a 2.54 cm diameter quartz plate and a spherical lens with an 11.3 cm radius of curvature. Both were vapor coated with a 0.5 μm thick layer of copper and then overcoated with a 0.5 μm thick layer of silver. The Casimir force was measured by changing the gap distance d in discrete steps between the 0.6 μm point of closest approach and about 10 μm, and measuring the restoring force at each step. By subtracting the electrostatic force, the resultant force values matched the theoretical Casimir force values to within 5%. See Fig. 8.23.

Another experiment, using a pair of parallel conducting plates as originally proposed by Casimir, was performed in 2002 in Italy [20]. The optically flat plates were 1.9 cm by 1.2 mm, about 47 μm thick with a 50 nm thick chromium

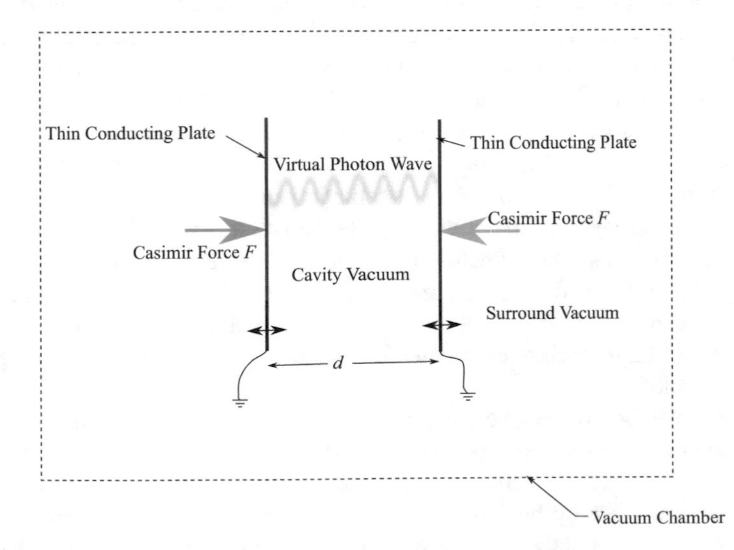

Fig. 8.22 Configuration of two parallel conducting plates in zero-point energy vacuum to produce a Casimir force

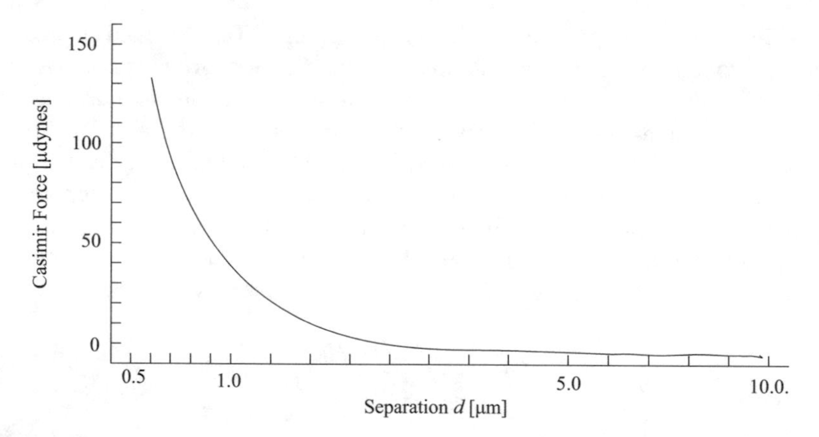

Fig. 8.23 Variation of the measured Casimir force between a conductive sphere and a conductive flat plate at various separations. The expected electrostatic attractive force was subtracted from the measured data

overcoat, and supported by a cantilever beam. The plate separation was precisely varied between 0.5 and 3.0 μm by piezoelectric transducers, with auxiliary control of the parallelism of the plates. With voltage reduction of the electrostatic force and extreme cleaning measures for the plates, a dominant Casimir force was measured. The strongest Casimir force was observed for plate separation in the 0.5–1.1 μm region. However, the experimental difficulties for this type of set-up produced measurement accuracy only on the order of 15%.

These static Casimir force experiments demonstrated the effect of virtual photon particles that dart in and out of existence in the spacetime vacuum. Around 1970 it was theorized that a rapid change in the boundary conditions of the electromagnetic modes in a vacuum zero-point energy field cavity could create real photons from these virtual photons. This could be accomplished by rapid movement of a conducting mirror in a cavity, referred to as *motion induced radiation* (MIR) [21]. In the *dynamical Casimir effect*, virtual photons are converted into observable real photons by rapid relativistic harmonic oscillation of one of the reflecting parallel plates in a direction perpendicular to the plate. However, even with a paramagnetic resonance cavity boost, the reflecting plate would require an oscillation frequency in the gigahertz range and a maximum velocity close to the speed of light to convert a sufficient number of virtual photons into measurable real photons [22, 23]. Fulfilling these requirements by mechanical motion of the plate would be practically impossible.

A novel method to achieve gigahertz oscillation of a planar mirror surface without mechanical motion was proposed in 2005 [24]. A semiconductor layer such as GaAs is placed over a conductive reflecting plate. Uniform irradiation of the semiconductor surface by high-power femtosecond laser pulses at microwave frequencies forms a conductive plasma mirror that appears and disappears. With the distance between the reflecting surfaces at about one millimeter, the plasma mirror

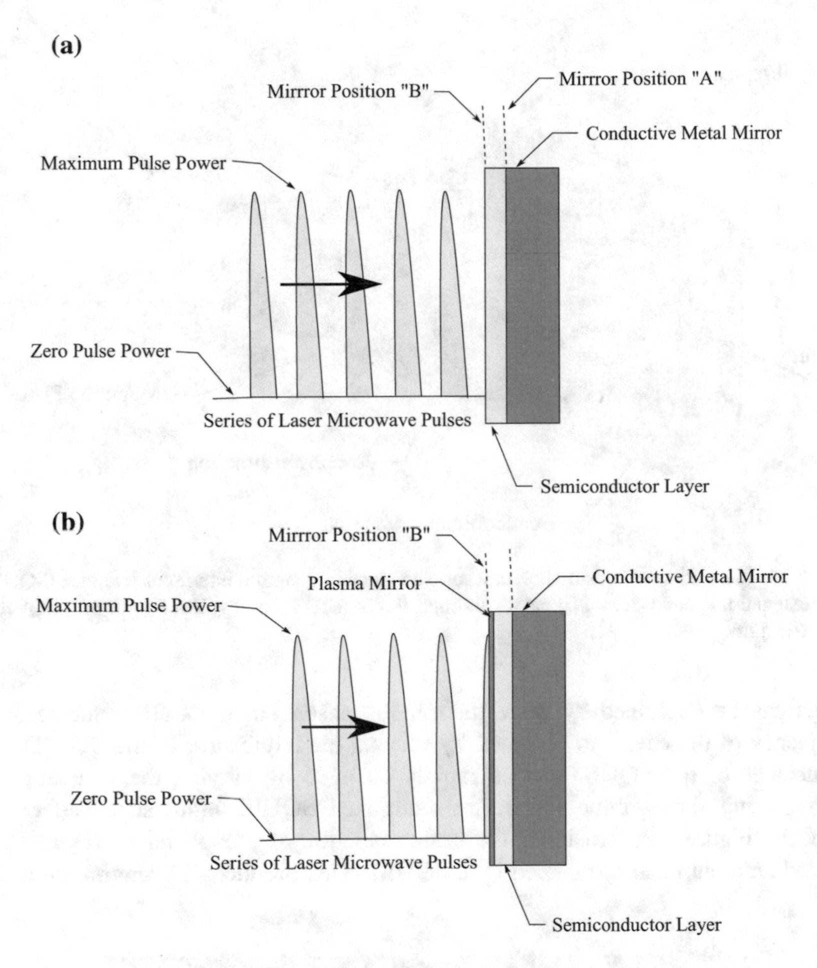

Fig. 8.24 **a** Mirror position "A" with no pulsed laser irradiation of semiconductor surface. **b** Mirror position "B" with pulsed laser irradiation of semiconductor surface

effectively oscillates between positions A and B in the gigahertz range. See Fig. 8.24a, b. If this oscillating mirror could be used as a wall in a Casimir resonant cavity, a sufficient number of real photons might be created.

Over the period 2009–2010 an international team of researchers proposed an alternate method using microwave circuits to observe real photons using the dynamical Casimir effect [25, 26]. This experiment would not use an oscillating "optical" type mirror, but make use of a superconducting quantum interference device (SQUID). The construction and operation of a SQUID is illustrated in Fig. 8.25. The SQUID is a loop formed by two superconductor sections, each joined by a barrier layer of thin insulating material known as a *Josephson junction*. Under certain electrical conditions, pairs of electron particles are able to move freely through the Josephson junction as a supercurrent. This quantum tunneling

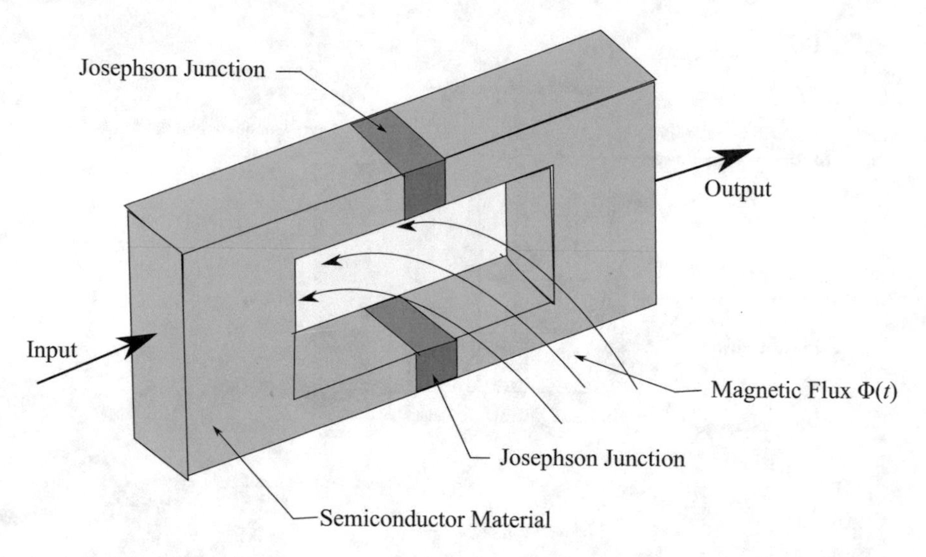

Fig. 8.25 Illustration of construction of a superconducting quantum interference device (SQUID). The external magnetic field $\Phi(t)$ passes through the SQUID loop. Typical length of a SQUID is about 13 μm

effect can be explained by wave theory. Josephson junctions allow the resonant frequency of the cavity to be tuned by varying the inductance of the SQUID. The inductance of the SQUID can be rapidly changed by varying the magnetic flux $\Phi(t)$ passing through the loop. The modulated SQUID inductance can exceed 10 GHz. Figure 8.26 illustrates the basic operation of a coplanar waveguide fabricated on a chip and terminated by a SQUID. This produces a changing boundary

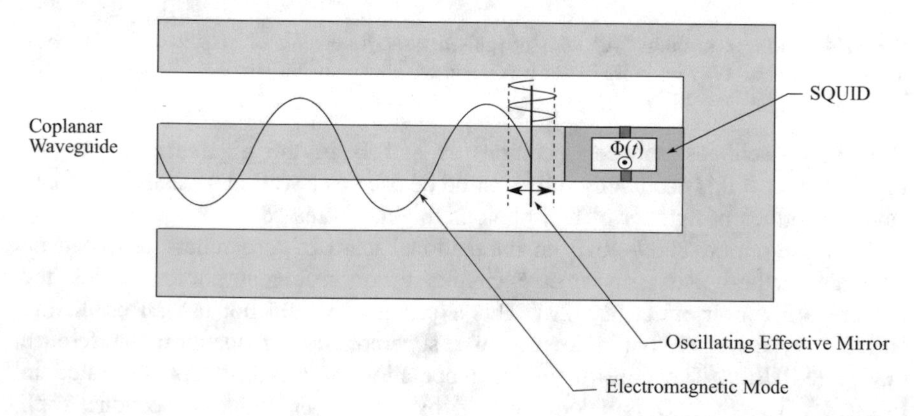

Fig. 8.26 Basic operation of a coplanar waveguide with SQUID and oscillating external magnetic field. This rapidly changes the reflective boundary position to produce dynamical Casimir effect radiation

condition, equivalent to a rapidly oscillating mirror, and real microwave photons could be generated from the virtual photons of the spacetime vacuum. Using a superconducting circuit, the generation of real photons from the spacetime vacuum was demonstrated and verified in 2011 [27]. A coplanar waveguide cavity and SQUID were fabricated on an aluminum substrate and the magnetic flux through the SQUID loop was driven at a sinusoidal frequency of about 8–12 GHz. The movement of the effective cavity boundary mirror then followed this variation. The apparatus was cryogenically cooled to less that 50 mK to eliminate spurious thermal radiation and placed in a vacuum. Real photon generation from the cavity mirror was recorded by subtracting out the expected thermal contribution. The photons covered a broad microwave band, indicating the quantum nature of the dynamical Casimir effect.

Another related method to generate real photons from the quantum vacuum was performed in Finland the same year [28]. An array of 250 SQUID elements formed a Josephson metamaterial about 4 mm long. The inductance in the Josephson metamaterial was varied by a rapidly changing external magnetic field. This caused a rapid change of refractive index in a microwave cavity transmission line. In turn, the speed of light was rapidly changed in the cavity. When cooled down to the quantum ground state at a temperature about 50 mK, correlated pairs of gigahertz frequency photons were generated by the dynamical Casimir effect.

The generation of photons (and electrons) from zero energy perturbations in the quantum vacuum state of spacetime is one of the most importance areas of study in quantum electrodynamics. In particular, the dynamical Casimir effect addresses one of the most intriguing questions of philosophy and science—“How can *something* be created from virtually *nothing*?”.

8.5 Quantum Correlations of Single and Multiple Photons

8.5.1 Single-Photon Particle Anticorrelation

In 1986 an experiment was performed by Grangier, Roger, and Aspect to detect single-photon particles incident on a beamsplitter [29]. The experimental setup is illustrated in Fig. 8.27. A train of single-state photons was generated from a two-photon radiative cascade source. One photon was directed to a photomultiplier detector PM_1, while the other photon was incident on a 50/50 plate beamsplitter with the transmitted and reflected outputs directed to photomultipliers PM_t and PM_r. Detection of a photon at PM_1 triggered a gate generator that activated PM_t and PM_r only for a short gate time τ ($\approx$9.4 ns). Photon counters on the photomultipliers recorded the counting rates N_1, N_t, and N_r. A coincidence counter recorded the counting rate of coincidence detections N_c between PM_t and PM_r.

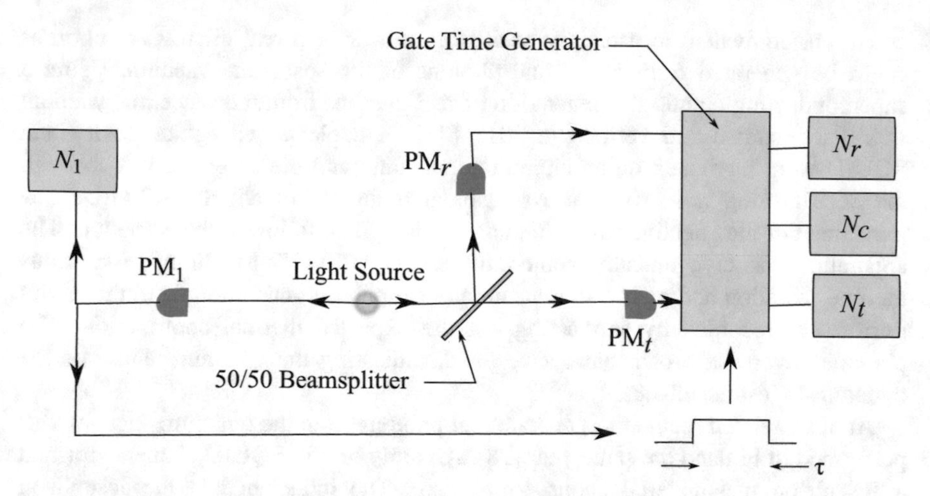

Fig. 8.27 Experiment to demonstrate single-photon particle anticorrelation. (Adapted from Ref. [29], courtesy of EDP Sciences)

The corresponding probabilities of detection during the gate time are $p_t = \frac{N_t}{N_1}$, $p_r = \frac{N_r}{N_1}$, and $p_c = \frac{N_c}{N_1}$. If the system behaved classically, then the expected inequality $p_c \geq p_r p_t$ would yield a parameter α_{CLASS} such that

$$\alpha_{CLASS} = \frac{p_c}{p_r p_t} = \frac{N_c N_1}{N_r N_t} \geq 1. \tag{8.3}$$

A measured violation of this inequality would then indicate a particle-type behavior of the photons, where N_r and N_t would dominate, and N_c would be small. Thus the transmitted and reflected photons would be uncorrelated and the anti-correlation property of single-photons between the detectors PM_t and PM_r. could be inferred. The calculated quantum mechanical value α_{QM} is

$$\alpha_{QM} = \frac{2f(\tau)N\tau + (N\tau)^2}{(f(\tau)) + N\tau)^2} < 1, \tag{8.4}$$

where $f(\tau)$ is a function relating the gate time and the cascade light source decay time, and N is the excitation rate of the light source.

The actual experiment was designed such that $f(\tau) \approx 1, N\tau << f(\tau)$, and a gate time $\tau \approx 9$ ns, to maximize the quantum mechanical effect. For a counting time of about 5 h, where $N_1 \approx 8800\ \mathrm{s}^{-1}$, a value of $\alpha_{QM} = 0.18 \pm 0.06$ was obtained. This value violated Eq. (8.3) by over 13 standard deviations. From this data, it was concluded that the transmitted and reflected light particles from the beamsplitter exhibited a quantum anticorrelation relationship, and did not support any wave-type description.

8.5.2 Single-Photon Entanglement

Entangled photon pairs were discussed in Sect. 8.2.3, and have proven useful in many delayed-choice experiments. However, when a single-photon is shared between two possible paths, it can behave as a *single entangled photon*. This may seem counterintuitive, supported by the argument that it takes "two to tangle". However, it was theoretically shown by van Enk at Bell Labs that a single-photon entangled state can be described by an equation of the form

$$|\Psi\rangle_{A,B} = \frac{1}{\sqrt{2}}[|0\rangle_A|1\rangle_B + |1\rangle_A|0\rangle_B], \tag{8.5}$$

where $|0\rangle_{A_B}$ and $|1\rangle_{A,B}$ are qubits with zero and one particle in modes A and B [30]. Qubit states in motion through space or a medium, as most photon-based qubits, are often called *flying qubits*. The state $|\Psi\rangle_{A,B}$ displays *single-particle nonlocality*. Entanglement in state $|\Psi\rangle_{A,B}$ requires that modes A and B are spatially separated and the entanglement occurs between modes A and B. This spatial separation can be achieved by passing a single-photon through a polarization beamsplitter, where A and B retain a common reference frame. An alternate form of Bell's inequality is the CHSH parameter S, derived by Clauser, Horne, Shimony, and Holt [31]. In general, when S >2 the CHSH inequality is violated and the nonlocality of a single-particle can be inferred.

An experiment that provided a robust confirmation or witness of single-photon entanglement was performed in 2013 [32]. The experimental setup is illustrated in Fig. 8.28. Single-state photons emitted from an optical parametric oscillator were incident upon a rotatable (θ = 0°–45°) half-wave plate and a polarization beamsplitter PBS to form a tunable non-localized single-photon entangled state. The state can be formulated as

$$|\Psi(\theta)\rangle = \cos(2\theta)|1\rangle|0\rangle + \sin(2\theta)|0\rangle|1\rangle, \tag{8.6}$$

which has a maximum value at θ = 22.5°. An additional coherent light beam was then combined with the initial single-photon beam. Using two additional beamsplitters, two local homodyne detector systems were formed, referred to as Alice and Bob. These homodyne detector systems measured continuous quadrature (X, P for Alice, X + P, X − P for Bob) components of discrete shared entangled states to obtain a real number S_{obs}. Random values of entangled states shared by Alice and Bob were tuned by a local oscillator driven by a coherent light source. The maximum S value for separated states $|1\rangle|0\rangle$ and $|0\rangle|1\rangle$ is $S_{sep}^{max} = \sqrt{2}(2/\pi) = 0.9$. Then if $S_{obs} > S_{sep}^{max}$, the states of Alice and Bob could be tested for entanglement. A run of about 200,000 events were used to calculated the value S_{obs}. Figure 8.29 shows the measurement results and the region of single-photon entanglement. However, since S_{obs} did not exceed the value of 2, this experiment could not validate the nonlocality of the shared entangled states.

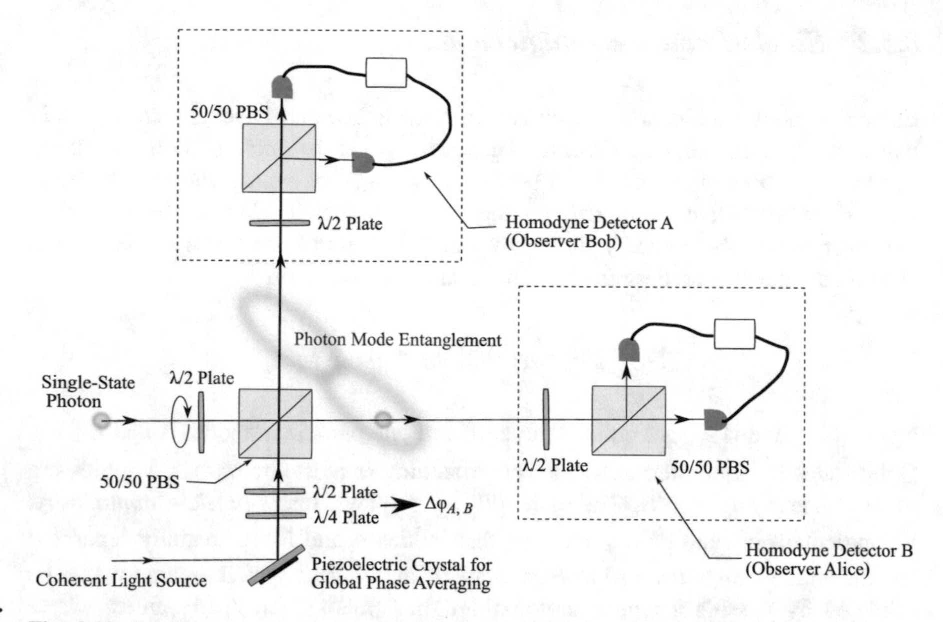

Fig. 8.28 Experiment setup to demonstrate single-photon entanglement using local homodyne measurements. (This figure adapted with permission from Ref. [32]. Copyrighted by the American Physical Society.)

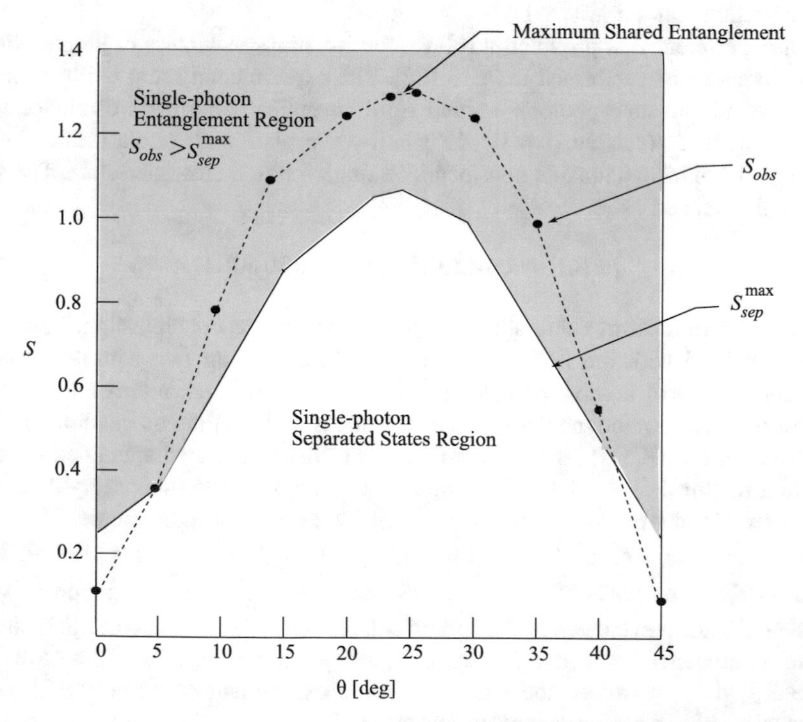

Fig. 8.29 Plot of S vs. θ, showing the region of single-photon entanglement, where $S_{obs} > S_{sep}^{max}$. (This figure adapted with permission from Ref. [32]. Copyrighted by the American Physical Society.)

8.5.3 *Validation of Single-Photon Nonlocality*

Experimental tests to confirm the nonlocality of single-photon entangled states are difficult and often controversial. Several proposals were introduced in the 1990s for experiments to demonstrated single-photon nonlocality [33, 34]. Figure 8.30 illustrates the scheme proposed by Lucien Hardy in 1994 to demonstrate single-photon nonlocality. A photon state $|1\rangle$ and a vacuum state $|0\rangle$ are incident on a 50/50 beamsplitter BS_1. The output states are sent by paths A and B to homodyne detection systems of Alice and Bob. Coherent local oscillators (e.g. lasers) are combined with these inputs at each of the Alice and Bob detection systems, where the local oscillators have independent phases φ_A and φ_B. Reported concerns with this technique are that (1) the resultant states from the BS_1 beamsplitter are not real observables, and have no utility in any particle nonlocality test, (2) the coherent reference states, not being created by a local operation, could affect the observed nonlocality of the single-particles by local hidden variables.

In 2004 an experiment performed in Sweden demonstrated single-photon nonlocality using homodyne measurements by observers Alice and Bob, and polarization beamsplitters [35]. Figure 8.31 shows the basic optical arrangement. Angular-separated signal and idler single photons were emitted from a nonlinear BBO crystal pumped by a fraction of a Ti: Sapphire laser beam transmitted by beamsplitter BS. A photodetector for the idler photon indicated the presence of the signal photon, which was polarized before entering a polarization beamsplitter PBS. The coherent laser beam (local oscillator) reflected from BS was polarized orthogonal to the signal photon, and also entered the PBS. A time delay ensured

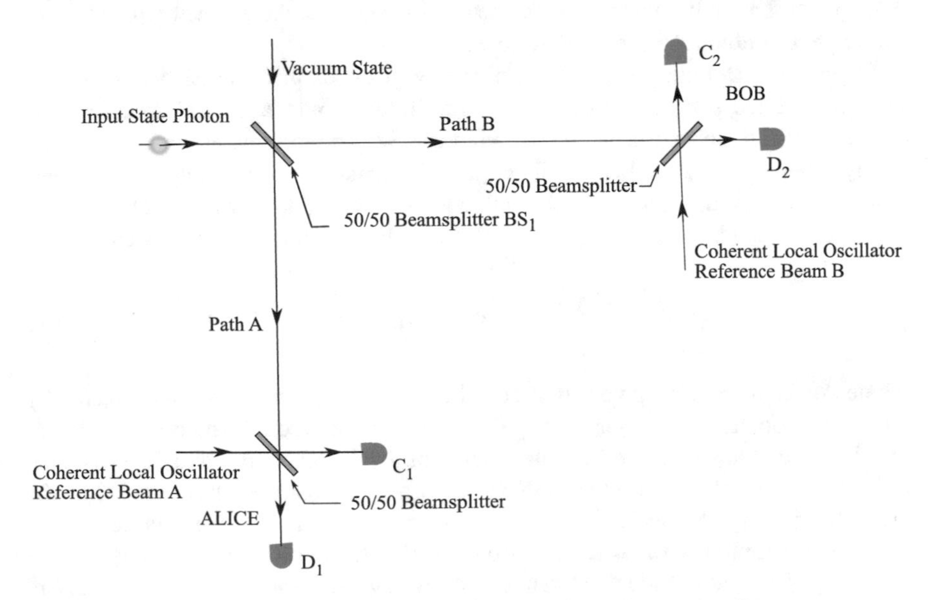

Fig. 8.30 Scheme proposed by Hardy to demonstrate single-particle nonlocality

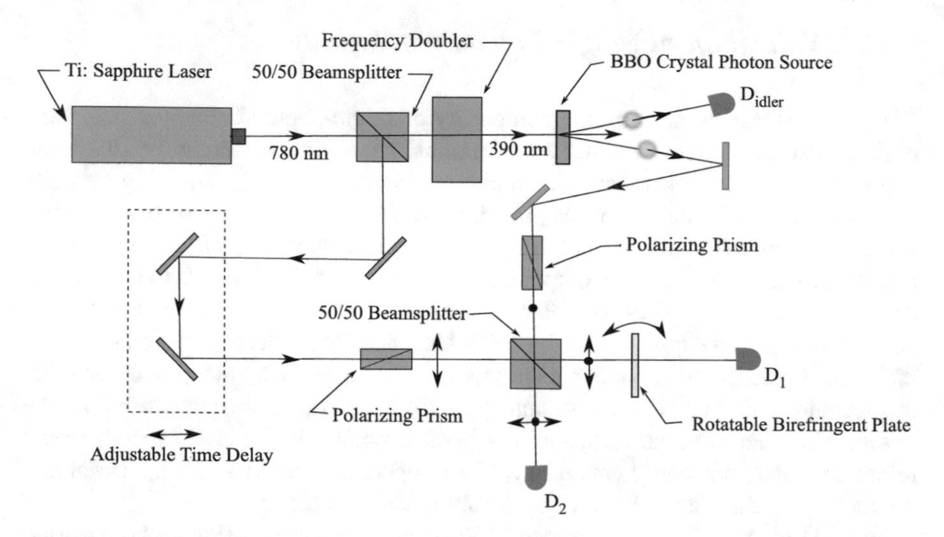

Fig. 8.31 Basic experimental setup by Hessmo and associates to witness single-particle nonlocality. (This figure adapted with permission from Ref. [35]. Copyrighted by the American Physical Society.)

that both the local oscillator beam and the signal entered the PBS at the same time. Thus coherent polarized light was co-propagated with the signal photon to each observer, with orthogonal polarizations.

The series of coincidence counts by the homodyne detectors exceeded the number that would be expected from a classical analysis. From this data, it was determined that Bell's inequality has been violated, and therefore the presence of single-photon nonlocality had been demonstrated.

Cooper and Dunningham, at the University of Leeds in Great Britain, proposed several modified schemes to observe single-photon nonlocality and address some concerns with the Hardy scheme [36]. Figure 8.32 is a modified arrangement of the Hardy scheme where the beamsplitters have instantaneously variable reflectivities, randomly chosen to eliminating the influence of local hidden variables.

A useful quantum form of the CHSH inequality for this type of system is

$$S = 2\sin^2\left(\frac{\Delta\varphi + \pi/2}{2}\right)[\sin(\xi - \eta) - \cos(\xi - \eta)] > 2, \tag{8.7}$$

where $\Delta\varphi$ is the relative phase difference between the reference beams detected by Alice and Bob, and $(\xi - \eta)$ is the angular difference between the measurement axes of Alice and Bob. By controlling the values of the variables $\Delta\varphi$ and $(\xi - \eta)$ such that $S > 2$, nonlocality of a single-photon might be demonstrated. In this scheme, the quantity $(\xi - \eta)$ can be set to a value $3\pi/4$ by changing the reflectivities of the detector beamsplitters for Alice and Bob. However, the phase relationships $\Delta\varphi = \varphi_B - \varphi_A$ of the independent reference photon states $|1\rangle$ and vacuum states $|0\rangle$ are

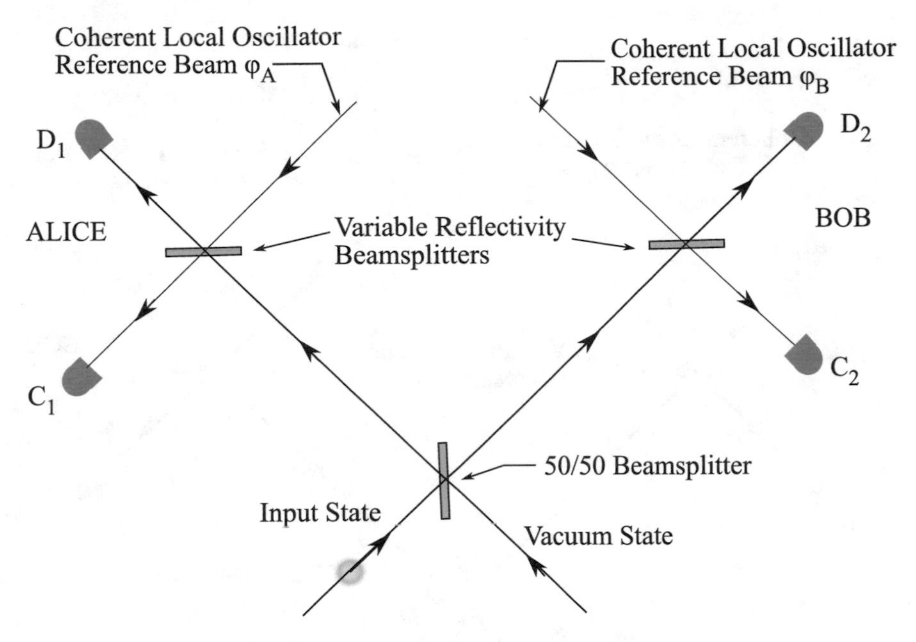

Fig. 8.32 A modified Hardy scheme proposed by Cooper and Dunningham using variable reflectivity beamsplitters

not fixed, but vary randomly between individual measurements. This randomness prevents any violation of the CHSH inequality to be observed.

A more general form of the CHSH inequality is

$$S = 2 + 2|\alpha|^2 e^{-2|\alpha|^2} \left[\sin^2\left(\frac{\Delta\varphi + \pi/2}{2}\right) |\sin(\xi - \eta) - \cos(\xi - \eta)| - 1 \right] > 2, \tag{8.8}$$

where α is a distribution parameter for coherent states, usually given the value $1\sqrt{2}$.

Figure 8.33 shows another proposed arrangement using a single reference beam incident on a 50/50 beamsplitter. The two reference beams incident on the Alice and Bob beamsplitters then have a fixed phase difference $\Delta\varphi = \pi/2$. For values of $(\xi - \eta) = 3\pi/4$, and $|\alpha| = 1\sqrt{2}$, Eq. (8.8) then reduces to

$$S = 2 + e^{-1} \left[\sin^2\left(\frac{\Delta\varphi + \pi/2}{2}\right) \sqrt{2} - 1 \right], \tag{8.9}$$

and a value $S = 2.152$ is obtained. This violates the CHSH inequality and can demonstrate nonlocality of the photon particle.

However, it could be further argued that the creation of these reference states was not strictly a local operation, and might contribute some nonlocality to the

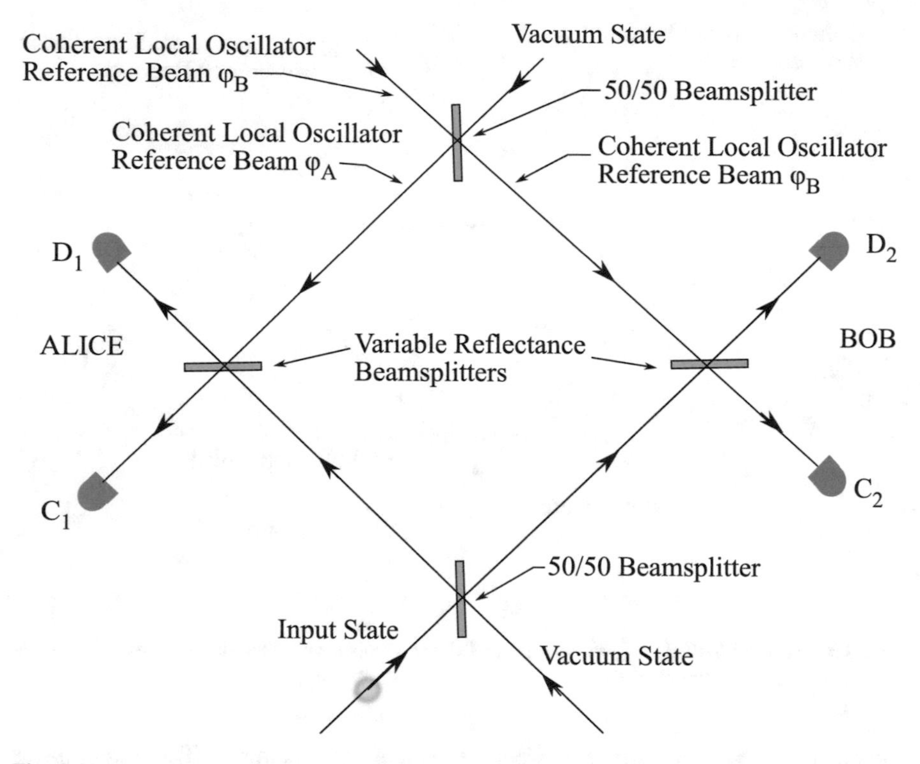

Fig. 8.33 A proposed single-particle nonlocality method with a fixed phase difference $\Delta\varphi = \pi/2$ between the reference beams. Violation of the CHSH inequality can then be evaluated

single-particle. This concern is addressed in the scheme illustrated in Fig. 8.34, where values of $\Delta\varphi = \varphi_B - \varphi_A$ are determined during each run from a local measurement using another beamsplitter and detectors C_3 and C_4. The creation of independently mixed states for Alice and Bob guarantees that all observed nonlocality is attributed entirely to the single-photon particle. The value of $\Delta\varphi$ is calculated from counting the number of particles detected at each detector, for a large number of detections. Figure 8.35 illustrates the variation of the calculated S values for various values of $\Delta\varphi$, calculated from Eq. (8.9). For the cycle illustrated, when $\Delta\varphi = 90° \pm 65.5°$, $S > 2$, and single-particle nonlocality occurs in this region.

The principles of these CHSH inequality experiments should be applicable to test the nonlocality of other types of single boson particles, and are not restricted to photons. However, an experiment of this type to demonstrate either single-particle or single-photon nonlocality has not been reported in the literature. These types of experiments could also be useful in distinguishing between *entanglement*, *nonlocality*, and *superposition* of single quantum particles, or in determining whether these properties are indeed equivalent.

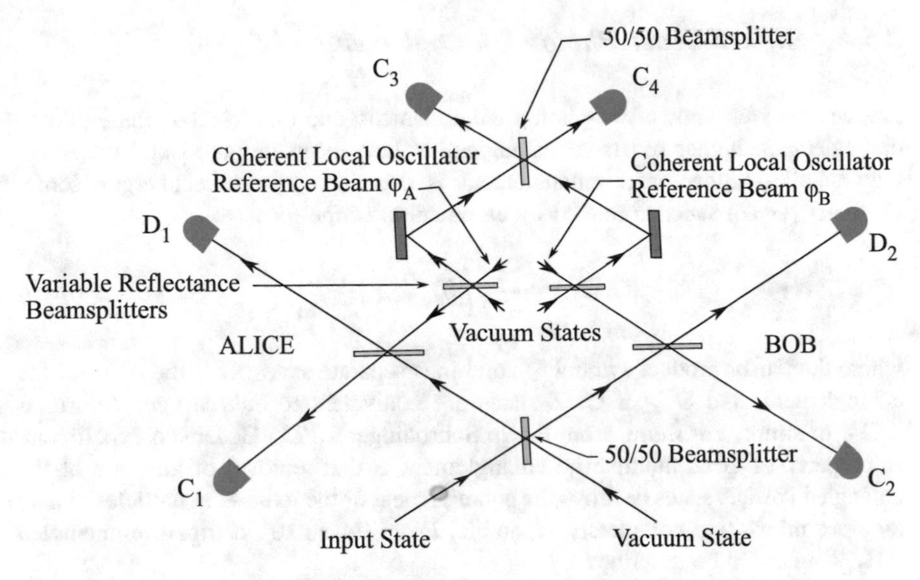

Fig. 8.34 A proposed variation of the arrangement of Fig. 8.33, where the phase difference $\Delta\varphi$ between the reference beams varies and is calculated by a local measurement

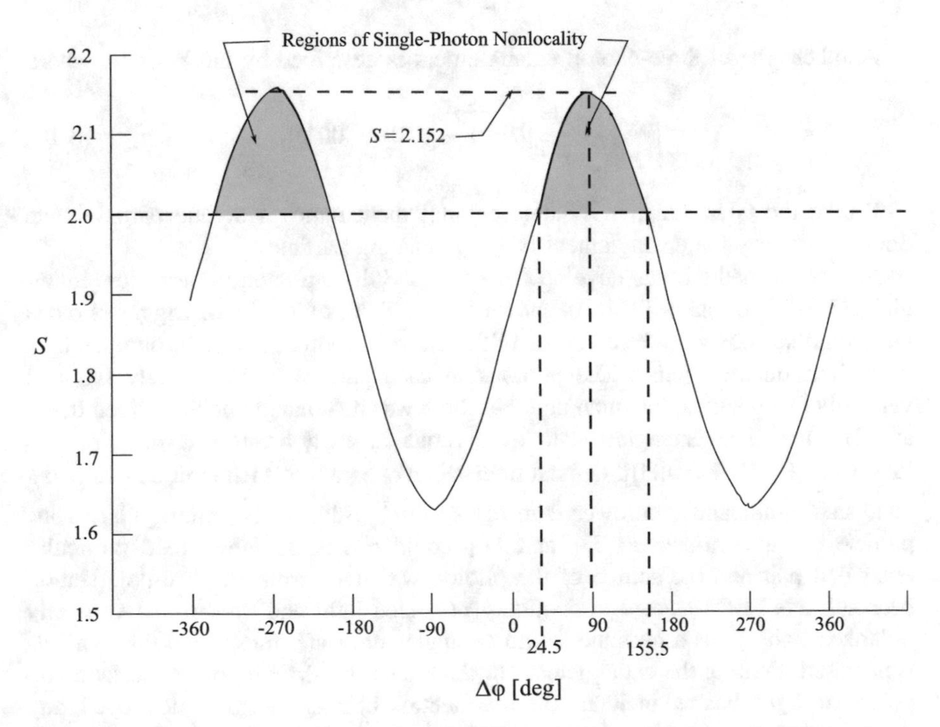

Fig. 8.35 A plot of S versus $\Delta\varphi$ using the scheme of Fig. 8.34, showing the regions where $S > 2$, indicating nonlocality of the photon particles

8.5.4 Higher-Order Photon Entanglements

Besides the well-known two-photon entanglement, and the debated single-photon entanglement, higher orders of entanglement have been proposed and observed. One type of higher order entanglement is described by a Greenberger-Horne-Zeilinger (GHZ) state, expressed by an equation of the form

$$|\mathrm{GHZ}_O\rangle = \frac{1}{\sqrt{2}}\left[|0\rangle^{\otimes O} + |1\rangle^{\otimes O}\right], \tag{8.10}$$

where the tensor product symbol $\otimes$ combines separate states, O is the order of the entanglement, and $O \geq 3$ GHZ states are equivalent to *Schrödinger cat states* $|\mathrm{SC}_O\rangle$ or simply *cat states,* from Erwin Schrödinger's 1935 Gedanken experiment. A property of GHZ multipartite entanglement is that removal of any one of the entangled particle states destroys the entanglement of the remaining particles, where they are mixed and completely separable. From Eq. (8.10), a tripartite entangled GHZ state could be described by

$$|\mathrm{GHZ}_3\rangle = \frac{1}{\sqrt{2}}[|000\rangle + |111\rangle]. \tag{8.11}$$

Another type of three-photon entanglement is described by the W-state, where

$$|\mathrm{W}_3\rangle = \frac{1}{\sqrt{3}}[|100\rangle + |010\rangle + |001\rangle]. \tag{8.12}$$

Unlike the GHZ state, in W-state entanglement removal of one particle state does not destroy the entanglement of the remaining particles.

In 1997 a method was developed to produce three-photon polarization entanglement for GHZ states [37]. As shown in Fig. 8.36, two pairs of trigger photons were simultaneously produced by two SPDC crystal sources A and B, pumped by a very short duration pulse. The photons in each pair were horizontally (H) and vertically (V) polarization entangled, but there was no entanglement between the A and B pairs. The entangled state of photons in each beam was described as $|\Psi\rangle = \frac{1}{\sqrt{2}}[|\mathrm{H}\rangle|\mathrm{V}\rangle + |\mathrm{V}\rangle|\mathrm{H}\rangle]$. Optical filters F on one path of each emitted pair had a bandpass significantly narrower than the spectral width of the pump pulse. Thus particle counts at detectors $\mathrm{D_T}$ and $\mathrm{D'_T}$ could not be assigned to a particular entangled pair and the source of the photon was undetermined. The polarization beamsplitters PBS reflected horizontally polarized light and transmitted vertically polarized light. Thus a possible fourth entanglement state transmitted through $\mathrm{PS_2}$ was erased, limiting the entanglement to three particles. Mixing of the horizontally polarized light beams incident on the neutral beamsplitter BS destroyed any identification of the light source A or B by detectors $\mathrm{D_T}$ or $\mathrm{D'_T}$.

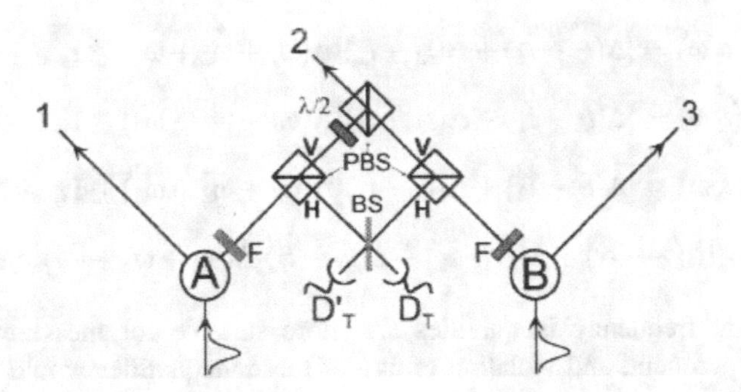

Fig. 8.36 Experimental setup to produce polarization entanglement of three photons. (This figure reproduced with permission from Ref. [37]. Copyrighted by the American Physical Society.)

Consider states that are recorded by detector D_T. If the photon originated from source A, then it would have a state $|\Psi\rangle_A \propto |V\rangle_{1A}|V\rangle_{2B}|H\rangle_{3B}$, where $|V\rangle_{1A}$ is the vertically polarized state in beam 1 originating from source A, $|V\rangle_{2B}$ is the vertically polarized state in beam 2 originating from source B, and $|H\rangle_{3B}$ is the horizontally polarized state in beam 3 originating from source B. However, if the photon originated from source B, it would have a state $|\Psi\rangle_B \propto |H\rangle_{1A}|H\rangle_{2A}|V\rangle_{3B}$, using similar nomenclature. Then the complete three-photon entangled GHZ state has the general form

$$|\Psi\rangle_{A,B} = \frac{1}{\sqrt{2}}\left[|V\rangle|V\rangle|H\rangle + e^{i\varphi}|H\rangle|H\rangle|V\rangle\right], \tag{8.10}$$

where φ is the relative phase of the states.

An experiment demonstrating three-photon energy-time entanglement was reported in 2013 by a team from the Universities of Waterloo and Calgary, Canada [38]. The entanglement test was based on continuous variables of particles having position x_i and momentum p_i, with the uncertainty relationship $\Delta x_i \Delta p_i \geq ½$ applied to each particle. They derived four general position-momentum inequalities for three particles, given by

$$f_a(x,p) = [\Delta(x_2 - x_1) + \Delta(x_3 - x_1)]\Delta(p_1 + p_2 + p_3) \geq 1, \tag{8.11a}$$

$$f_b(x,p) = [\Delta(x_2 - x_1) + \Delta(x_3 - x_2)]\Delta(p_1 + p_2 + p_3) \geq 1, \tag{8.11b}$$

$$f_c(x,p) = [\Delta(x_3 - x_2) + \Delta(x_3 - x_1)]\Delta(p_1 + p_2 + p_3) \geq 1, \tag{8.11c}$$

$$f_d(x,p) = [\Delta(x_2 - x_1) + \Delta(x_3 - x_1) + \Delta(x_3 - x_2)]\Delta(p_1 + p_2 + p_3) \geq 2. \tag{8.11d}$$

By using the relationships $x = ct$ and $p = \frac{h\omega}{2\pi c}$, these inequalities were transformed to:

$$f_{\mathrm{a}}(t,\omega) = [\Delta(t_2 - t_1) + \Delta(t_3 - t_1)]\Delta(\omega_1 + \omega_2 + \omega_3) \geq 1, \tag{8.12a}$$

$$f_{\mathrm{b}}(t,\omega) = [\Delta(t_2 - t_1) + \Delta(t_3 - t_2)]\Delta(\omega_1 + \omega_2 + \omega_3) \geq 1, \tag{8.12b}$$

$$f_{\mathrm{c}}(t,\omega) = [\Delta(t_3 - t_2) + \Delta(t_3 - t_1)]\Delta(\omega_1 + \omega_2 + \omega_3) \geq 1, \tag{8.12c}$$

$$f_{\mathrm{d}}(t,\omega) = [\Delta(t_2 - t_1) + \Delta(t_3 - t_1) + \Delta(t_3 - t_2)]\Delta(\omega_1 + \omega_2 + \omega_3) \geq 2. \tag{8.12d}$$

These time-frequency inequalities are more suitable for measurement in the designed experiment, and violation of any of these inequalities would demonstrate real time-energy entanglement between the three photons.

Figure 8.37 illustrates the experimental setup. The 404 nm 12 mW pump laser emitted about 10^6 single-photons per second into a polarization maintaining fiber (PMF). A PBS transmitted the photons to a periodically-poled potassium titanyl phosphate crystal (PPKTP), while the reflected PBS beam was sent to a Fabry-Perot interferometer to monitor the 5 MHz bandwidth of the pump laser. The PPKTP crystal down-converted the beam into orthogonally polarized 842 nm (ω_0) photons and 776 nm (ω_1) "daughter" photons. A single-mode fiber transmitted the 842 nm signal photons to a single-photon detector D_1. The 776 nm idler photon was transmitted by a polarization maintaining fiber to a periodically-poled lithium niobate waveguide. Here two additional entangled "granddaughter" photons at 1530 nm (ω_2) and 1570 nm (ω_3) were generated in an additional down-conversion—a process known as cascaded down-conversion (C-SPDC). The three entangled photons were then separated at a dichroic reflecting mirror and fiber-coupled to single-photon detectors D_2 and D_3. The arrival time differences of the three photons were measured and recorded by a time recording device. From the measured arrival time differences,

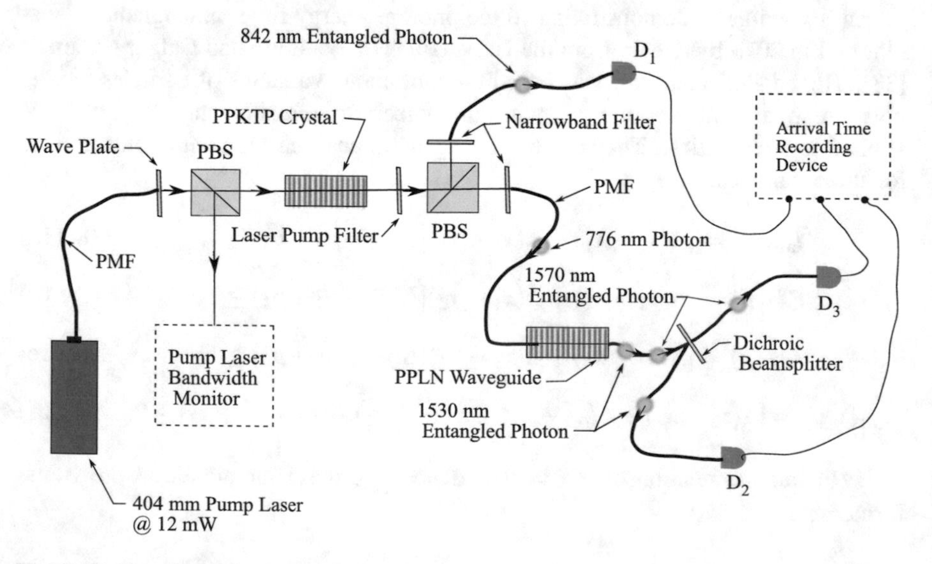

Fig. 8.37 Experimental setup to produce three photon energy-time entanglement

the inequalities in Eqs. (8.12a)–(8.12d) were calculated. It was found that $f_a(t,\omega) = 0.03 \pm 0.01$, $f_b(t,\omega) = 0.02 \pm 0.01$, $f_c(t,\omega) = 0.018 \pm 0.005$, and $f_d(t,\omega) = 0.03 \pm 0.01$. Thus all four inequalities were violated, and the existence of genuine three photon entanglement was confirmed.

About a year later some members of the same team directly produced three polarization entangled photons using cascaded down-conversion. [39] In this case, the desired three-photon entangled quantum states were of the form

$$|\mathrm{GHZ}^{\pm}\rangle_3 = \frac{1}{\sqrt{2}}\left[|\mathrm{H}\rangle_1|\mathrm{H}\rangle_2|\mathrm{H}\rangle_3 \pm e^{i\varphi}|\mathrm{V}\rangle_1|\mathrm{V}\rangle_2|\mathrm{V}\rangle_3\right]. \tag{8.13}$$

The basics of the experiment are as follows: A nonlinear PPKTP crystal was pumped by a 404 nm 25 mW laser diode to produce a polarization entangled Bell state photon pair of the form $|\Psi\rangle_2 = \frac{1}{\sqrt{2}}[|\mathrm{H}\rangle_1|\mathrm{H}\rangle_2 \pm |\mathrm{V}\rangle_1|\mathrm{V}\rangle_2]$, with photon wavelengths 776 and 842 nm. The 776 nm photons were further down-converted as in the previous experiment [38] into a polarization entangled three-photon state as in Eq. (8.13), having telecom wavelengths of 1530 and 1570 nm. To characterize and verify the three-photon state, 27 possible measurement settings were used to perform quantum state tomography. This required a detector with a high single-photon detection rate and efficiency. The detector was a superconducting nanowire single-photon detector (SNSPD), which was over 90% efficient at 1550 nm. See Sect. 9.2.7 for details on this type of detector.

8.5.5 Four-Photon Entanglement

Genuine four-photon entanglement was realized in 2010 by a team at the California Institute of Technology and University of Oregon [40]. The experiment was performed in several stages. First, entanglement was created between four supercooled caesium atom ensemble memories over a finite time period, using known quantum techniques. These quadripartite atomic entangled W-states were then transformed to mode-entangled light states a_1, b_1, c_1, d_1 in four photonic quantum channels. A strong pulsed 40 Hz laser was split into four beams that were incident on the light states. This transformation was done in a particle configuration as in Fig. 8.38a, where correlations between the single-photon detectors $D_{1\text{–}4}$ were measured through four separated fiber optic cables. This measurement yielded the photon quantum statistics (called y_c). Then the detector correlations were measured in a wave configuration setup illustrated in Fig. 8.38b, where beamsplitting junctions on fiber optic cables allowed the light states a_2, b_2, c_2, d_2 to be correlated at detectors $D_{1\text{–}4}$, called coherent Δ-states. By combining these two types of measurements using photon statistics (y_c) and coherence (Δ), the parameter space $\{\Delta, y_c\}$ of the entanglement of the four photons was determined. The values of these parameters defined the degree of non-local entanglement. The minimum values obtained were $\Delta \approx 0.07$ and $y_c \approx 0.038$, showing a strong violation of uncertainty

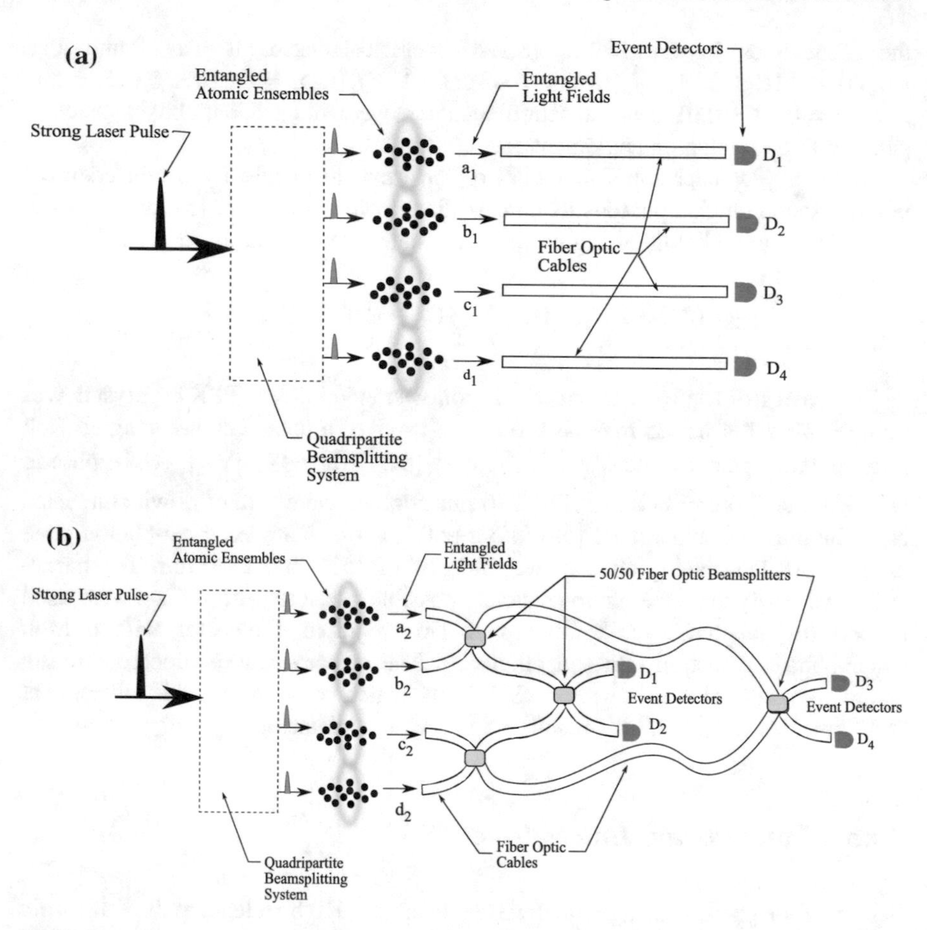

Fig. 8.38 **a** Demonstration of four-photon entanglement—Particle detection setup. **b** Demonstration of four-photon entanglement—Interferometric wave detection setup

relations. The results indicated high-fidelity four-photon entanglement that surpassed weaker forms of mode entanglement obtained from previous techniques. This experiment was reviewed in a paper by Vuletic [41].

8.5.6 Six-Photon and Eight-Photon Entanglements

There has been significant recent work on the demonstration of mode entanglement between more than four photons. For example, a scientific team from China, Austria, and Germany created a GHZ state entanglement of six photons [42]. Polarization entanglement of eight photons was demonstrated in 2011 by cooperating experimenters in China and Germany, producing an eight-photon Schrödinger cat state having genuine multipartite entanglement [43]. The entanglement takes the form

$$|\mathrm{SC}_8\rangle = \frac{1}{\sqrt{2}}\left[|\mathrm{H}\rangle^{\otimes 8} + |\mathrm{V}\rangle^{\otimes 8}\right] \tag{8.13}$$

The basic approach of the experiment is as follows: A 390 μm UV laser pumped a 120 fs pulse through four BBO crystals at a pulse rate of 76 MHz. By parametric down-conversion, four pairs of polarization entangled photons were then created, each pair having a state $|\Psi\rangle = \frac{1}{\sqrt{2}}[|\mathrm{H}\rangle|\mathrm{H}\rangle + |\mathrm{V}\rangle|\mathrm{V}\rangle]$. Using a series of polarization beamsplitters, mirrors, and retroreflecting prisms, the entangled pairs were sent to 16 single-photon avalanche diode detectors. A quarter-wave plate QWP, half-wave plate HWP, and PBS determined the polarization states of the photons for each detector pair. See Fig. 8.39. Observation of Hong-Ou-Mandel destructive interferences verified that independent photons, combined through polarization beamsplitters PBS_1, PBS_2, and PBS_3, were indistinguishable. The detector signals were sent to a coincidence logic unit, which measured the 256 possible combinations for simultaneous coincidence counts on all 16 detectors. Finally, using the experimental results, the fidelity of the eight-photon entanglement was compared to that expected from Eq. (8.13). This fidelity was calculated to be 0.708 ± 0.016.

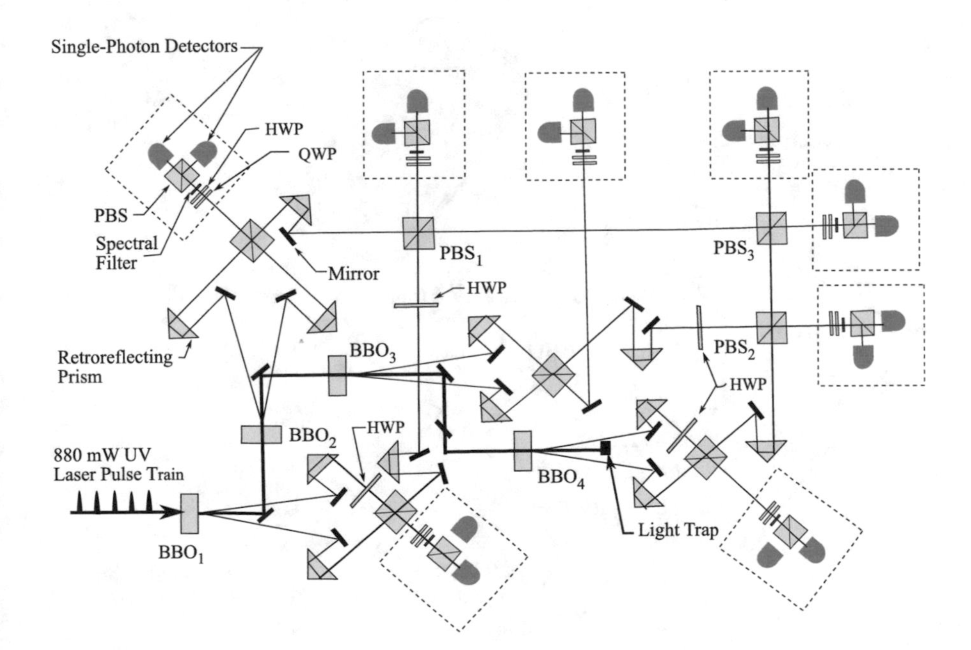

Fig. 8.39 Experiment setup to observe eight-photon entanglement

8.5.7 Entanglement of Non-coexistent Photons

At the Hebrew University of Jerusalem in 2012, photon entanglement was generated between two photons that never coexisted– the photons having temporal as well as spatial separation [44]. Simply stated, one photon of an entangled pair was detected before the other was even created! The procedure and optical setup is illustrated in Fig. 8.40. Pulses from a 400 mW Ti: Sapphire laser at 76 MHz were incident on a PDC BBO crystal. An initial pulse (pulse #1) on the BBO crystal produced emission of a pair of polarization entangled photons. One photon path was lengthened by a 31.6 m delay line, consisting of a series of high-reflection mirrors. This produced a 105 ns delay for a photon reaching a single-photon detector by the delayed path, compared to the direct "fast path". The eighth laser pulse (pulse #8), emitting a second pair of entangled photons 105 ns later than the first photon pair, was chosen for measurement comparison. Both entangled photon pairs followed the same paths, where the first emitted delayed path photon is designated as a, the second delayed path photon as a', and the first and second emitted fast path photons as b and b'. Photon a was measured before photons a' and b' were created.

Tilted crystals in both paths determined the phase φ of the state. Combined with φ, rotation angles of half-wave plates in each path produced different polarization

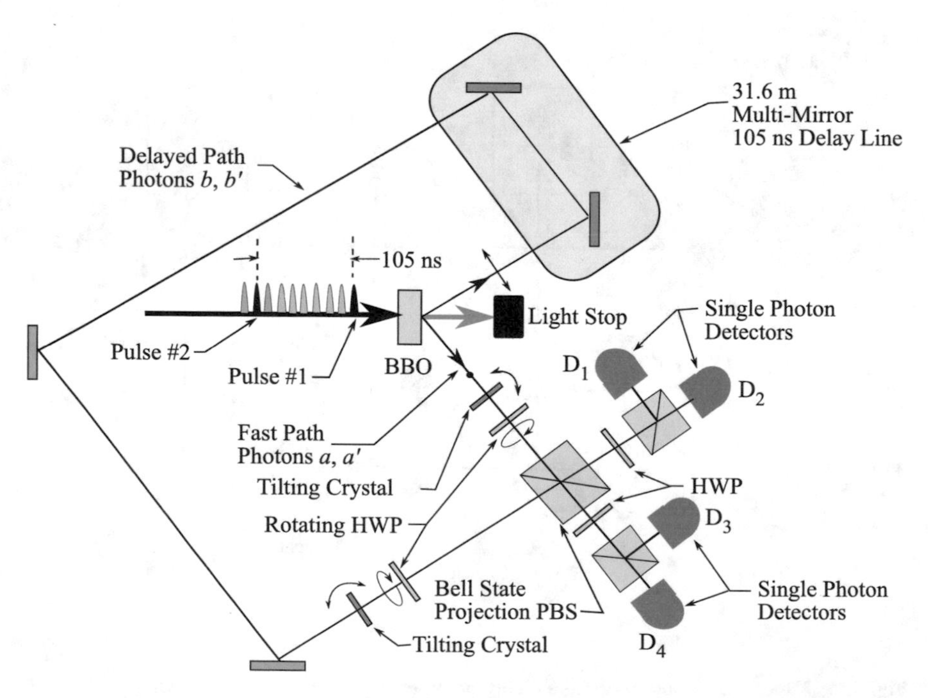

Fig. 8.40 Experimental setup to demonstrate quantum entanglement between two photons that never coexisted. (This figure adapted with permission from Ref. [44]. Copyrighted by the American Physical Society.)

states of the projected photons. Nine values of these parameters created 16 possible polarization states defining a complete quantum state description of all photons. The procedure was as follows: photon b and photon a' arrived simultaneously at the PBS and were projected into Bell state entanglement. However, this entanglements of b and a', combined with the entanglement of photons a and b produced a phenomenon called *entanglement swapping*. The result was an entanglement between the first photon a and the last photon b', which were uncorrelated and did not exist at the same time. The quantum correlations between a and b' photons were only determined after all photon measurements were performed, demonstrating both non-local and time correlations. The validity of this entanglement was tested by comparing the measured correlation with correlations from other polarization states.

8.6 Photon Tunneling and Superluminality

In the 1920 s it was theorized that the emission of alpha particles escaping from the potential well of a radioactive nucleus could be explained by wave mechanics. There was a probability that the alpha particles would tunnel through the potential barrier. This wave mechanical tunneling effect also explained the passage of electron particles through a Josephson junction, with a delay time independent of the barrier thickness. This is known as the *Hartman effect* [45].

Quantum tunneling of photon particles through a barrier is closely related to the propagation of electromagnetic wave packets or electrons through band gap materials. In 1991 an experiment was proposed to measure the tunneling time of a single-photon through a barrier with sub-picosecond resolution [46]. The experiment was performed shortly thereafter at the University of California, Berkeley using a two-photon source, where the transmission time of a photon tunneling through a barrier was compared with the other photon that encountered no barrier [47]. Figure 8.41 illustrates the experiment. A pair of polarization-entangled photons was simultaneously generated using SPDC from a nonlinear KDP crystal pumped by a UV laser at 351 mm wavelength. One photon was directed to a 7 mm thick fused silica substrate, where half of the incident face was coated with a multilayer dielectric stack having high reflectance and thickness $d = 1.1\ \mu m$. This barrier of stacked dielectric layers acted as a photonic band gap for tunneling photon wave packets. The substrate was moveable such that the beam could pass either through the reflective coating or the clear substrate. A moveable retroreflecting prism allowed small changes in the path length to be made. This photon was combined with the other reference photon at the beamsplitter of an H-O-M interferometer, where each detector coincidence count indicated a single-photon tunneling event. The time delay resolution was in the sub-femtosecond range. A series of one hour runs were made, adjusting the retroreflecting prism position while sliding the tunneling barrier in and out of the beam. The average delay time $\Delta\tau$ of the tunneling photons relative to the non-tunneling photons was $\Delta\tau = -1.47 \pm 0.19$ fs. If the tunneling photon

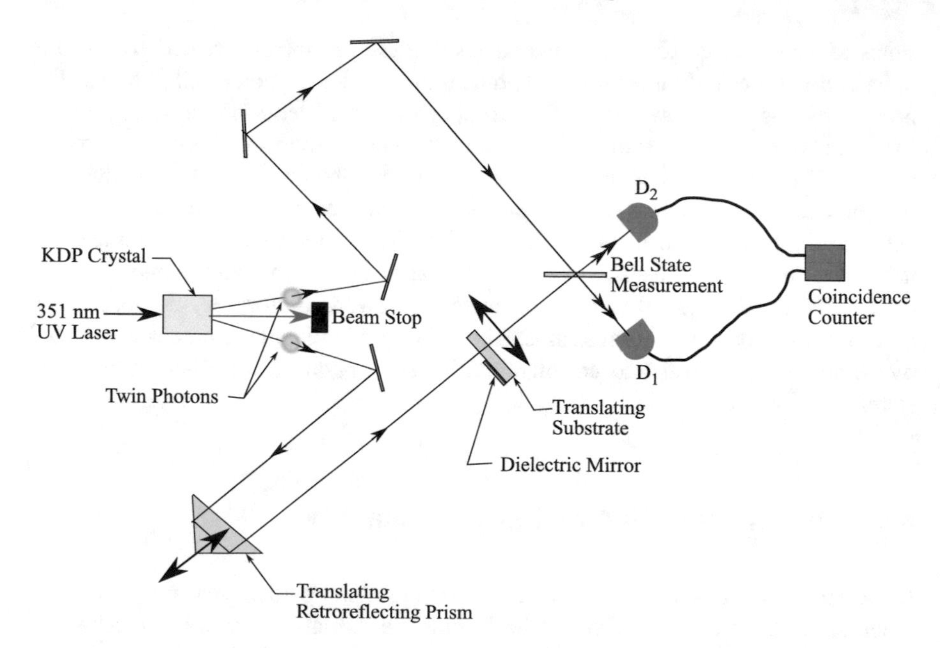

Fig. 8.41 Experimental setup to measure the tunneling time of a single-photon through a barrier

traveled at the speed of light c through the barrier, then the expected travel time τ would be $\tau = \frac{d}{c} = \frac{1.1 \times 10^{-6}\,\text{m}}{2.99 \times 10^{8}\,\text{m/s}} = 3.68\,\text{fs}$. However, the measured travel time was $3.68 - 1.47 = 2.21$ fs, yielding a speed of light $c' = \frac{1.1 \times 10^{-6}\,\text{m}}{2.2 \times 10^{-15}\,\text{s}} = 5 \times 10^{8}\text{m/s} \approx 1.7\,c$. This indicated an apparent superluminal wave-packet velocity through the barrier which was not interpreted as a signal velocity, and Einstein's principle of causality was not violated.

In addition to a stacked dielectric layer barrier, quantum tunneling of electromagnetic wave packets has been studied using evanescent modes in microwave wave guides, where superluminal movement of the wave packet has been observed [48, 49]. Spielmann and collaborators have measured the propagation of optical wave packets through photonic band gap barriers, and stated that the very short transit time indicated possible superluminal tunneling [50]. However, they concluded that causality was not violated since the transmitted signals were strongly attenuated.

In 2001 Haibel and Nimtz in Germany measured the tunneling time through an evanescent mode barrier using a prism pair [51]. Figure 8.42a shows a single right-angle glass prism (n = 1.5) in air with an internal angle of incidence I = 45° on surface C. The critical angle of total internal reflection I_{crit} = 41.8°, so the incident ray undergoes total internal reflection (TIR) at this surface. When TIR occurs, a thin evanescent light field is formed at the exterior of surface C, where the field drops off very rapidly (exponentially) with distance from the surface. However, no energy is transferred to this layer for the single prism. In Fig. 8.42b a

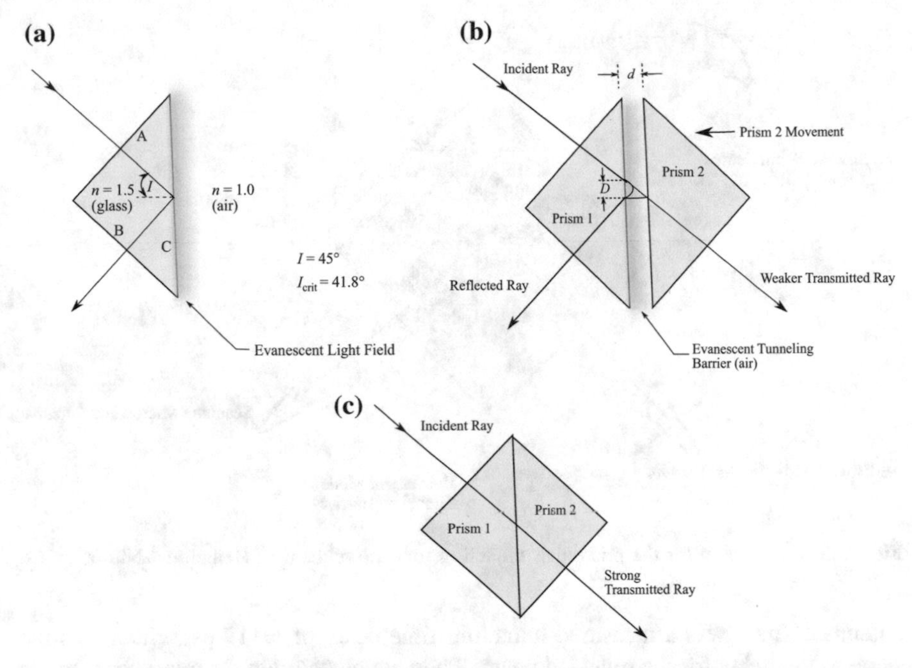

Fig. 8.42 **a** Formation of an external evanescent light field layer for total internal reflection of a right-angle prism. **b** Evanescent coupling between two right-angle prisms separated by a small distance. **c** Two right-angle prisms in contact, with the disappearance of TIR and the evanescent layer

second prism is moved toward the first prism. When this prism enters the evanescent mode region, *evanescent coupling* occurs and a small fraction of energy can be transferred through the evanescent layer from the first prism to the second prism. This is a type of *frustrated total internal reflection* (FTIR) and the evanescent layer can be considered as a tunneling barrier. Another effect which occurs is a small lateral separation D of the incident and totally internally reflected rays at surface C. The shift occurs only for an incident beam of finite width and is called the *Goos-Hänchen shift*. This shift results from wave interferences between the components of the light that are parallel to surface C in the evanescent region. The magnitude of D is on the order of the incident light wavelength. Figure 8.42c shows the two prisms in perfect contact, where TIR and the evanescent layer disappear, and ordinary refraction occurs.

The basic setup of the prism pair tunneling experiment is shown in Fig. 8.43. The separated prisms were fabricated of optical glass with $n = 1.605$ and $I_{crit} = 38.68°$. The prism pair gap d was varied between zero and 50 mm. A 190 mm microwave beam diameter, generated from a klystron oscillator and a parabolic reflector, was incident on a 400 mm long sloped face of the first prism. With $I = 45°$ at the TIR surface, the intensities of the reflected beam and the tunneling transmitted beam were measured using scanning microwave horn

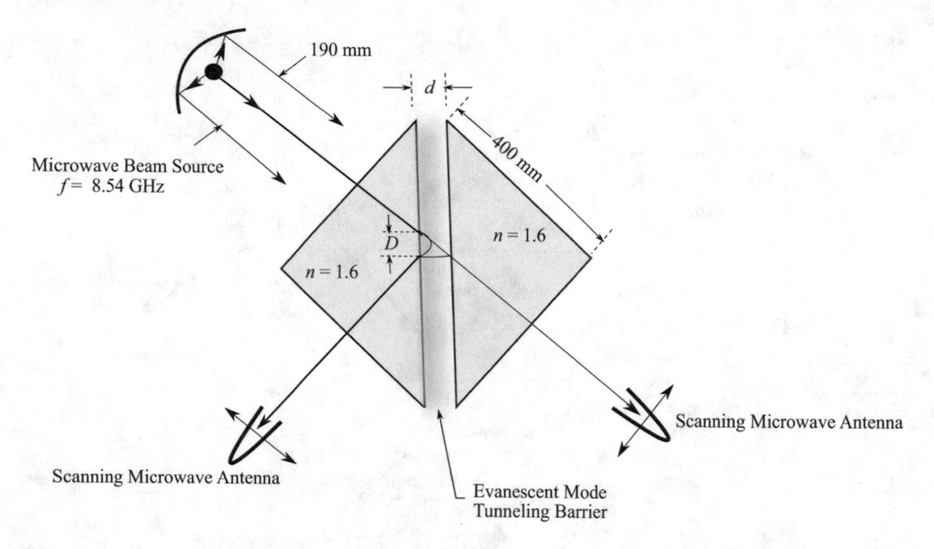

Fig. 8.43 Basic setup for the prism pair tunneling time experiment of Haibel and Nimtz

antennas. There was a measured tunneling time delay of $\approx$117 ps, which was the same as the delay for the reflected beam. They concluded that the evanescent mode tunneling time was of the same order of magnitude as previous measured tunneling times in evanescent wave guides or photonic dielectric stacks. Also, they stated that the tunneling time was close to the value of the reciprocal of the incident radiation frequency f.

The question of whether superluminality occurs for a light beam traversing a tunneling barrier is a controversial one, especially regarding any violation of Einstein relativity. Many researchers have reported differing explanations. For example, Winful at the University of Michigan has put forth arguments that superluminal velocity through a tunneling barrier is not possible [52, 53]. These arguments address the difference between *group delay* in the tunneling barrier and *transit time* through the barrier. Some of the arguments are:

1. Group delay measurements are not measurements of transit times within the tunneling barrier. Group delay is the lifetime of stored energy or photon lifetime in the barrier, and is not relevant to group velocity calculations using transit times. The tunneling group delay is shorter than in free-space and shorter than the photon transit time in the barrier.
2. This stored energy in the barrier is all contained in a $1/e$ exponential fall-off distance from the incident surface. Therefore the Hartman effect saturates at a given distance in the barrier, and essentially disappears for increased barrier length. Further increase of the tunneling distance beyond this saturation distance will not increase the tunneling velocity or produce superluminal light. However, it does increase the barrier group delay time, when the transmitted pulse peak exits the barrier.

3. For tunneling to occur, the pulse lengths must be much longer than the barrier length, and the transmitted (and reflected) pulse shapes will not be shortened or distorted to produce a superluminal effect.
4. The experiment of Steinberg, Kwait, and Chiao [47] used interference correlation to measure a mirror group delay time shift, which was equated to a tunneling transit time in the barrier. This group delay was then used to predict a superluminal velocity of the photon within the barrier. Also, the time shift cannot be attributed to pulse-reshaping, for the reason stated previously.
5. The tunneling experiment of Spielmann [50] using photonic band gaps also misinterpreted the measured group delay time as a transit time through the barrier gap.

In 2007, using a FTIR prism pair arrangement with an evanescent mode tunneling barrier gap, Nimtz and Stahlhofen reported superluminal light velocities with prism gap separations on the order of a meter [54]. The experiment was similar to the 2001 experiment [51]. Both the transmitted barrier photons and the reflected photons were received at the detectors at the same time, irrespective of prism separation, and Einstein special relativity was violated at a macroscopic level. Additionally, they identified the invisible evanescent modes as virtual photons. Winful subsequently put forth additional arguments that this reported superluminality could not possibly occur [55]. Without being judgmental on any specific arguments or conclusions, we simply state that observation of superluminality through barrier tunneling remains a controversial topic.

8.7 Two-Photon Interactions

Common knowledge says that photons, being massless and carrying no charge, do not interact with each other. Beams of light propagating through air, vacuum, or ordinary optical materials, pass through each other without any interacting effect. However, certain highly nonlinear media can produce interactions between single-photons at the quantum level. One such medium was produced in 2012 as a super-cold rubidium atomic gas with excited *Rydberg states* [56, 57]. As illustrated in Fig. 8.44, an elongated dense ensemble of $\approx 6 \times 10^6$ rubidium atoms was formed in a magneto-optical trap every 300 ms, and laser-cooled to a temperature $T = 35$ μK over a time period of about 10 ms. A strong control laser beam and narrow probe laser beam were co-propagated with the probe beam focused into the atomic ensemble with a beam waist $w \approx 4.5$ μm. In addition, a 3.6 gauss magnetic field was applied along the propagation direction. The components were enclosed in a vacuum. The strong control field coupled the probe field to the atomic Rydberg levels, and the atoms and photons form unified Rydberg polaritrons moving at a reduced speed. In addition, the control field produced a short-lived electromagnetically induced transparency (EIT) in the medium.

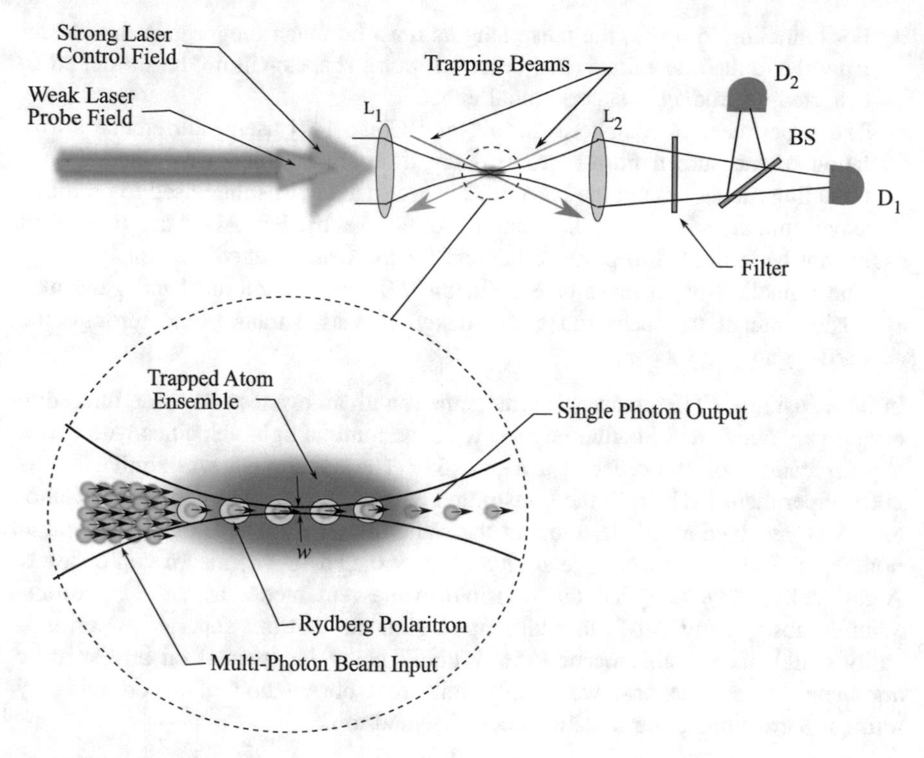

Fig. 8.44 Experimental setup to suppress photon pair interaction and produce a single-photon light source

There is a strong attraction between Rydberg atoms in the medium. However, two Rydberg atoms with a separation less than $r_b \approx 10$ μm cannot be simultaneously excited. The Rydberg polaritron radius r_b then acted as a block for other approaching probe photons (*Rydberg blockade*), and they are absorbed in the dissipative medium as the EIT condition disappears. Thus only single-photons were transmitted, since a second interacting photon was immediately absorbed, and the medium acted as a single-photon source. All photon pair transmission was suppressed over an ensemble length ≈40 μm with a coherence time ≈500 ns. Filters separated the control photons from the probe photons, and the intensity correlation of the probe photons were measured by a beamsplitter and two single-photon detectors.

A related experiment, performed at the Harvard-MIT Center for Ultracold Atoms, demonstrated photons that traveled as mass-like particles and exhibited strong attraction to produced photon pairs [58]. In this experiment a 1.6 μs^{-1} train of single linear polarized photons was incident on the super-cooled dense rubidium atomic gas. A control photon beam produced excited Rydberg states where an EIT condition was produced by slight detuning of the Rydberg resonance levels. The

Rydberg atoms allowed coherent probe photon interactions to dominate the photon propagation. The Rydberg blockade saturated the medium, making it extremely nonlinear and dispersive. The propagation of photons in the gas was slowed and two sequential probe photons coupled into a two-photon state with an effective mass $m = \frac{1000hf}{c^2}$. This polarization entangled two-photon state then exited the medium and coincidence counts were analyzed by an H-O-M interferometer. The results indicated that a close-coupled two-photon state has been created from two independent state photons.

Another photon interaction experiment showed that two independent continuous laser beams with different frequencies could modulate each other at the photon level [59]. A super-cooled dense cesium atomic ensemble (35 μK, 6×10^6 atoms) having two ground states designated as $|f\rangle$ and $|c\rangle$, and two excited Rydberg levels $|d\rangle$ and $|e\rangle$, was contained in an optical cavity. Orthogonal uncorrelated photon beams were sent through the atomic ensemble as shown in Fig. 8.45. One beam was focused in the atomic ensemble as the *cavity mode*. Another uncorrelated free-space signal beam traveled orthogonally through the atomic ensemble as the *signal mode*. A coupling laser connected the signal mode transition $|f\rangle \rightarrow |d\rangle$ with the cavity mode transition $|c\rangle \rightarrow |e\rangle$. A photon in the signal mode moved as a slow-light polaritrons through the medium, and could block the transmission of a cavity photon when the photons arrived simultaneously. Also, a photon in the cavity mode could block the transmission of a simultaneously arriving photon in the signal mode. The measured cross-correlation of the two photon beams using detectors D_1 and D_2 showed photon-photon interaction where the transmission in either mode was significantly reduced by the simultaneous presence of a photon in the other mode.

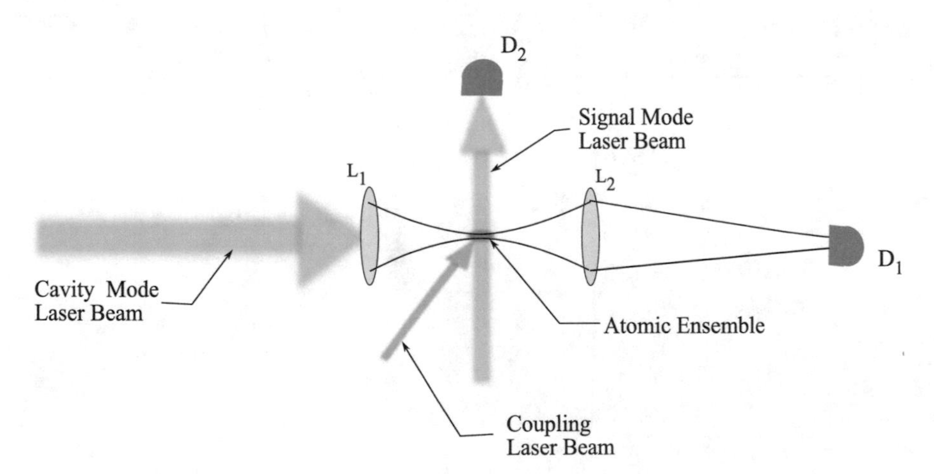

Fig. 8.45 Experimental setup to produce interaction between two orthogonal uncorrelated photon beams

8.8 Squeezed Light, Squeezed Vacuum, and Gravitational Wave Detection

By squeezing a light field component in one direction the quantum noise of that component can be reduced. Heisenberg's uncertainty principle $\Delta x \Delta p \geq \hbar/2$ for position x and momentum p can be analogously used to determine the uncertainties ΔE and ΔB in measured values of the electric and magnetic components of electromagnetic radiation at the quantum level, where $\Delta E \Delta B \geq (\text{const.})\hbar/2$. Thus a more precise measurement of the electric field E can be achieved by squeezing the uncertainty region ΔE in one direction. However, this increases the ΔB value, producing less accuracy in the ΔB measurement. The measurement accuracies can be exchanged by squeezing in the direction of ΔB, with less accuracy in the ΔE direction. Illustrative examples are shown in Fig. 8.46a–c, where the squeezed uncertainty regions have an elliptical shape.

A preferred representation for *squeezed light* uses the complex amplitude **A** and phase φ of the light displayed in dynamic phase space, the so-called phasor diagram [60, 61]. Measurements of the light vector are made in reference to quadrature axes,

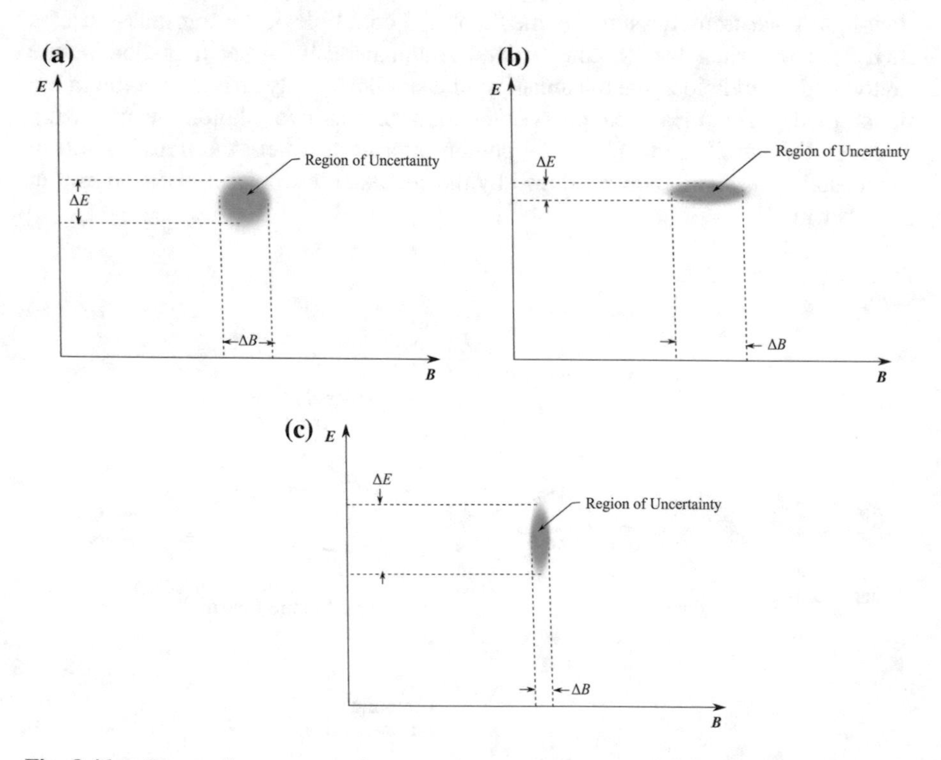

Fig. 8.46 **a** Symmetric quantum noise uncertainty region for electromagnetic and magnetic field measurements, where $\Delta E \approx \Delta B$. **b** Squeezed quantum noise uncertainty region for electromagnetic and magnetic field measurements, where $\Delta E < \Delta B$. **c** Squeezed quantum noise uncertainty region for electromagnetic and magnetic field measurements, where $\Delta E > \Delta B$

these axes representing a 90° phase shift. These quadrature-squeezed light measurements can be produced as *amplitude-squeezed light* or *phase-squeezed light*. Quantum noise can be reduced for *either* light amplitude *or* phase measurements by appropriate squeezing. The phase space diagrams Fig. 8.47a–b illustrate the uncertainty regions for amplitude and phase squeezing of light.

The quantum zero-point energy of virtual light in a spacetime vacuum (Sect. 8.4) can also be squeezed. This is referred to as *squeezed vacuum* light, where the average light amplitude is zero. Figure 8.48a–b illustrate the uncertainty regions for

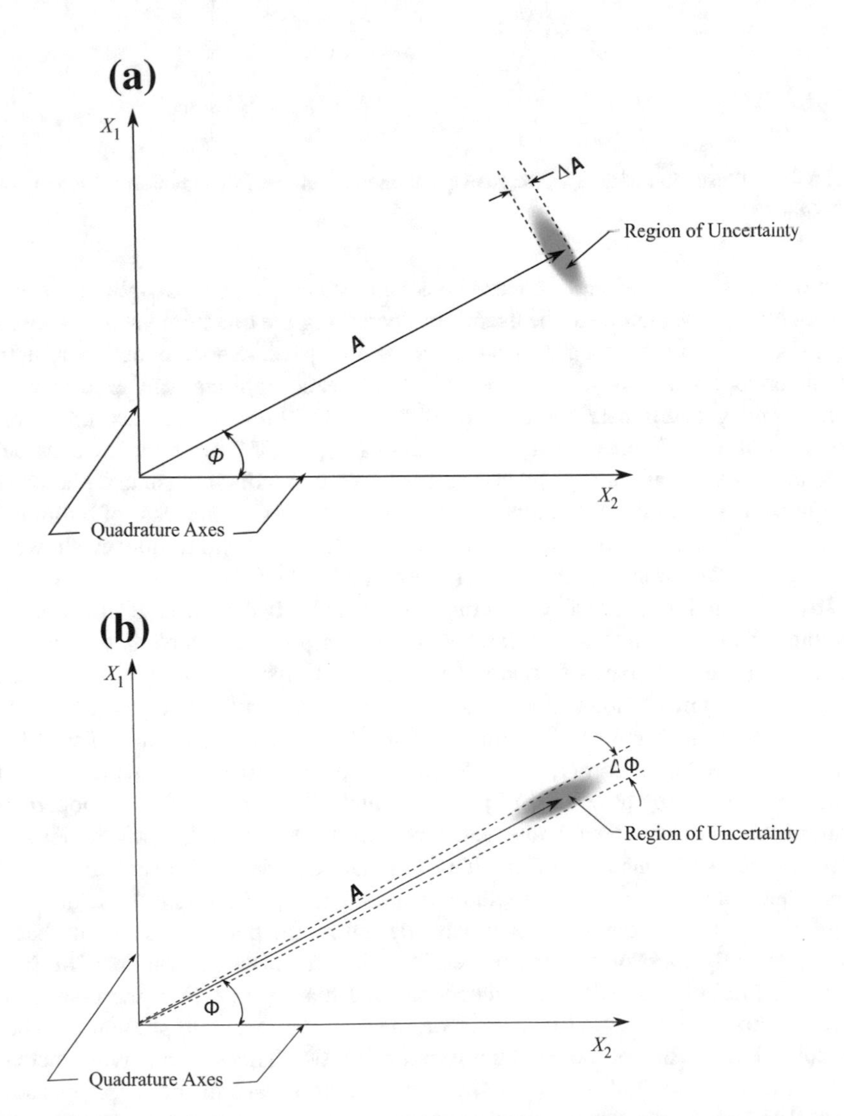

Fig. 8.47 **a** Phasor diagram showing reduction of quantum noise for light amplitude measurements. **b** Phasor diagram showing reduction of quantum noise for light phase measurements

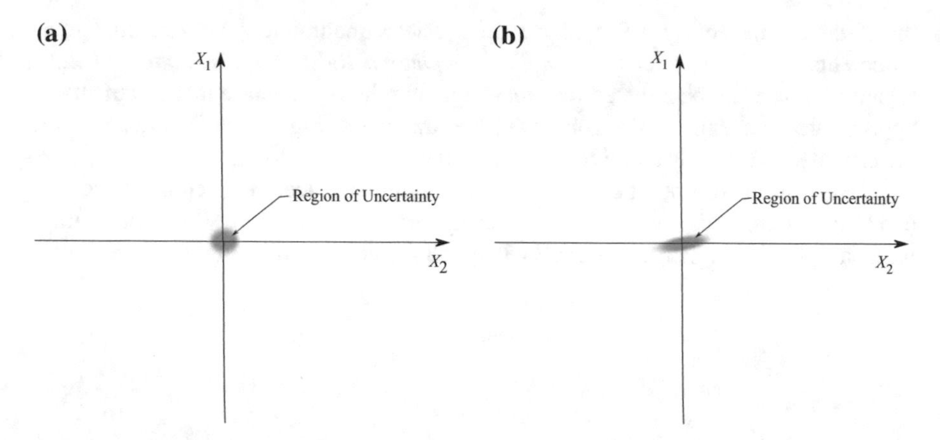

Fig. 8.48 **a** Illustration of the vacuum state quantum noise uncertainty region. **b** Illustration of a squeezed vacuum state

rotationally symmetric and squeezed vacuum states. Squeezed light, especially phase-squeezed vacuum, can be useful for decreasing the quantum vacuum noise in ultra-sensitive interferometric measurements in space. Spontaneous parametric down-conversion can be used to create both squeezed light and squeezed vacuum. A squeezed vacuum state has been produced at 1550 nm with a measured noise reduction of 12.3 ± 0.5 dB [62]. A continuous wave 1550 nm fiber laser incident on a non-linear potassium titanyl crystal generated two indistinguishable photons in a squeezed vacuum single-mode state. The measured variances of amplitude quadratures X_1 and phase quadratures X_2 at a 5 MHz sideband frequency showed a 12.3 dB quantum noise reduction over the range 80–18 kHz.

In 1997 an international consortium of over 1000 scientists from about 80 institutions in 15 countries was formed for the purpose of detecting gravitational waves in space. One type of gravitational wave detector was designated the Laser Interferometer Gravitational Wave Observatory (LIGO), and two of these detectors were situated in Hanford, Washington and Livingston, Louisiana. The LIGO detectors were Michelson-type interferometers having two beam paths of 4 km length, and operating at ≈100 kW power in ultra-high vacuum. Other cooperating interferometric gravitational wave detectors were located near Hannover, Germany (GEO600 detector), and near Pisa, Italy (VIRGO detector). The sensitivity of all these detectors is limited by vacuum quantum noise. In order to achieve the required sensitivity for gravitational wave detection, the detector sensitivity had to be significantly increased over the best detector sensitivities available. In 2008, using the GEO600 detector, researchers showed that by injecting phase-squeezed vacuum into the light beam to the detector, the important vacuum phase noise could be reduced below that of the quantum vacuum [63, 64]. The accompanying increase in the vacuum amplitude noise was relatively unimportant in the target measurement. The GEO600 squeezed vacuum system was adapted to the LIGO detectors, and combined with several other changes, the sensitivity of this Advanced LIGO was increased by a factor of four [65]. The application of phase-squeezed vacuum allowed detection of distance changes of $\approx 10^{-20}$ m between the interferometer arms.

On September 14, 2015, an historic experimental achievement occurred when the Hanford and Livingston Advanced LIGO detectors observed the first gravitational wave signal produced in space [66]. The gravitational waves were created by the coalescence of two black holes of solar mass about 29 and 36 into a larger black hole with solar mass about 62. Energy conversion of the 3 solar mass loss ($E = mc^2$) of the newly formed black hole produced an intense burst of power. This caused a transient distortion in the spacetime fabric, emitting a gravitational wave as predicted by Einstein in his general theory of relativity. The gravitational wave was created about 1.3 billion years ago and the resultant wave signal, designated GW151226, was observed on Earth by both Advanced LIGO detectors situated at Hanford and Livingston. There was about a 7 ms time difference between the observations due to separation of the locations.

8.9 Non-destructive Observation of the Photon

Observation of photons using photon detectors destroys the photon when it is absorbed and recorded. However, techniques have been developed that allow detection of a photon without destroying it, using *quantum non-demolition* (QND) measurements. In the early 1990s a method was proposed by the Serge Haroche group at the Collège de France to accomplish this by measuring atomic phase shifts in a microwave cavity using atomic interferometry [67, 68].

Using this method, in 1999 a quantum electrodynamics experiment was reported that nondestructively observed a single microwave photon using QND in a cavity [69]. See Fig. 8.49. A microwave superconducting cavity with highly reflective end mirrors acted as a trap for microwave photons, which bounced between the cavity mirrors for a finite time ($\approx$0.1–0.5 s) before absorption. An orthogonal beam of

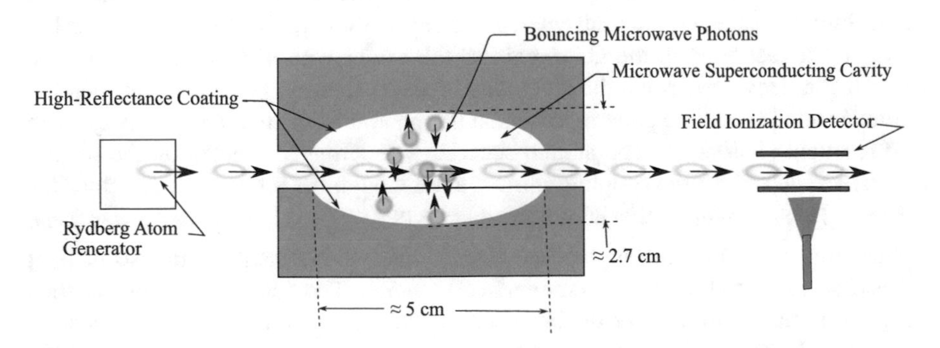

Fig. 8.49 Quantum non-destructive detection of photons using a microwave superconducting cavity and Rydberg atoms. (The interferometer surrounding the cavity for atomic phase shift measurement is not shown)

Rydberg atoms passed through the bouncing photons. The cavity apparatus was in a vacuum at a temperature a few degrees above absolute zero. A phase change occurred in a Rydberg atom if it passed through a photon—without destroying the photon. Atomic interferometry measurement of the phase change of Rydberg atoms exiting the cavity then indicated the presence of a photon in flight. Additionally, this type of experiment could detect spontaneous quantum jumps of a single-photon caused by absorption or emission from the walls. This was accomplished by repetitive interferometric phase measurement of hundreds of correlated state Rydberg atoms to record the birth, life, and death of a single-photon [70].

In an independent series of experiments at NIST in Colorado, a group headed by David Wineland used precisely controlled photon (laser) beams to study the properties of atoms at the quantum level. In one experiment a single beryllium ion was confined by an electromagnetic trap and laser-cooled in vacuum to a quantum zero-point energy temperature near absolute zero. The beryllium ion became almost motionless, with its outer electron in a quantum superposition of two “up” and “down” states. Another precisely controlled laser pulse nudged these two states apart to a spatial separation of about 83 nm, with an observable Schrödinger cat entanglement [71]. For their related work in “measuring and manipulation of individual quantum systems”, David Wineland and Serge Haroche shared the 2012 Nobel Prize in Physics [72, 73].

The QND experiments of the Haroche research group involved microwave photons. Non-destructive detection of a *visible* photon was reported in Germany in 2013 [74]. As shown in Fig. 8.50a–b, a resonant cavity was formed using a high-reflectance mirror, and a semi-transparent coupling mirror. A single ^{87}Rb atom was trapped at the center of the cavity, and brought into a superposition of two ground states $|1\rangle$ and $|2\rangle$ as an atomic state $|\Psi\rangle = \frac{1}{\sqrt{2}}[|1\rangle + |2\rangle]$ by optical pumping from a series of laser pulses. There was also an excited state $|3\rangle$, where incoming photons were in resonance with the empty cavity and the strongly coupled $|2\rangle \leftrightarrow |3\rangle$ transition frequency. The atom state $|1\rangle$ was detuned from both the cavity and any incoming photons. Using a weak 25 μs pulsed laser source, single visible photons impinged on the coupling mirror. When the atom was in the detuned or non-resonant state $|1\rangle$, photons entered the cavity through the coupling mirror but had no interaction with the atom, exiting the cavity with a phase shift of π. See Fig. 8.50a. However, when the atom was in state $|2\rangle$, strong coupling between the atomic state $|2\rangle$ and the photon prevented the photon from entering the cavity, with no resultant phase shift. The atomic state $|2\rangle$ was strongly affected by the photon directly reflected off the coupling mirror, even though the photon did not enter the cavity. This coupling produced a flipped state $|\Psi\rangle = \frac{1}{\sqrt{2}}[|1\rangle - |2\rangle]$ and a rotational phase shift of $\pi/2$ for the superposed atomic state. Subsequently, fluorescent light was emitted from the cavity, as shown in Fig. 8.50b. The fluorescent light was then used as a marker to observe the presence of a photon in a nondestructive manner. A conventional single-photon detector determined the actual number of photon reflections from the coupling mirror. Using this technique, a single-photon non-destructive detection efficiency of about 74% was obtained.

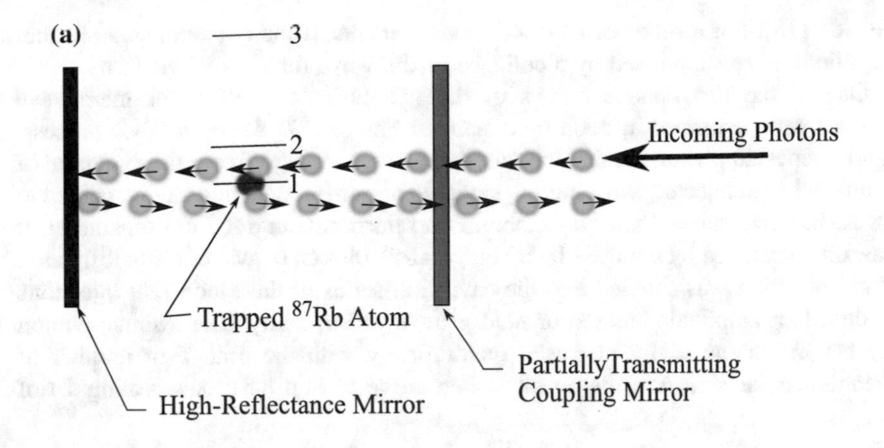

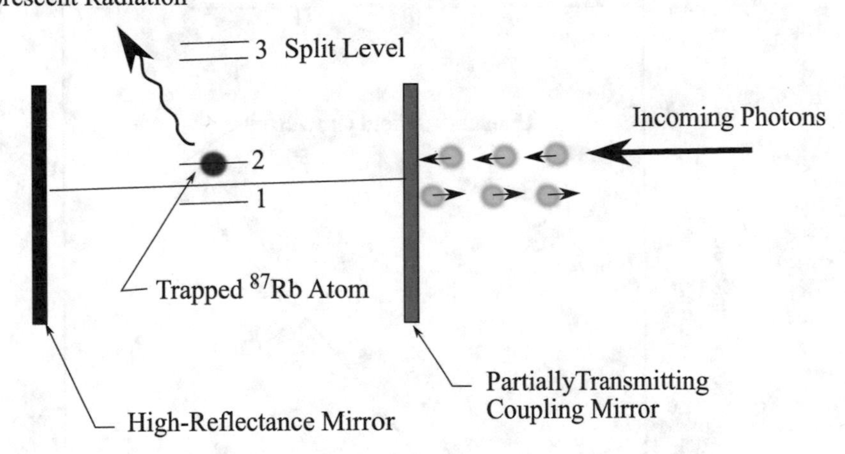

Fig. 8.50 **a** A resonant cavity and coupling mirror with the Rb atom in state $|1\rangle$ and tuned out of resonance with incident photons. Photons enter and leave cavity with no cavity atom interaction. **b** A resonant cavity and coupling mirror with the Rb atom in state $|2\rangle$ and tuned in resonance with incident photons. Photons do not enter cavity, yet react with cavity atom and are nondestructively detected

8.10 Quantum Zeno Effect for Photons

For unstable quantum particles, quantum theory predicts that the particle decay could be "frozen" if it was continuously observed. This is known as the *quantum Zeno effect* or *Zeno's paradox*. The name arose from the Greek philosopher Zeno of Eleo (ca. 490 BC–430 BC) who postulated that a flying arrow, being stationary at any given instant of time, could never progress to its final destination. A theoretical proof of the quantum Zeno effect was first reported by Misra and Sudarsjan [75]. Observation of the Zeno effect was realized in 1990 by researchers in Wineland's Colorado group, using transitions between two ground-state hyperfine levels in a

Be^+ ion [76]. For a series of repeated measurements of the quantum state, further transitions were suppressed by a collapse of the wave function of the state.

One of the first demonstrations of the quantum Zeno effect for microwave photons was performed in 2008 by a team in France [77]. Using a QND process, rapidly repeated photon number measurements were made to freeze the evolution of a coherent field injected into a high-Q cavity. The superconducting cavity, cooled to 0.8 K, had two superconducting niobium end mirrors with a 2.7 cm separation. It was side-irradiated by a pulsed light source at 51.099 GHz, where a small fraction of the radiation was coupled into the cavity. A series of these coherent injections produced an amplitude buildup or field growth in the cavity. The average photon number $\langle n \rangle$ in the cavity increased quadratically with the time T or number of injections N, as shown in the upper dashed curve of Fig. 8.51. A repeating set of

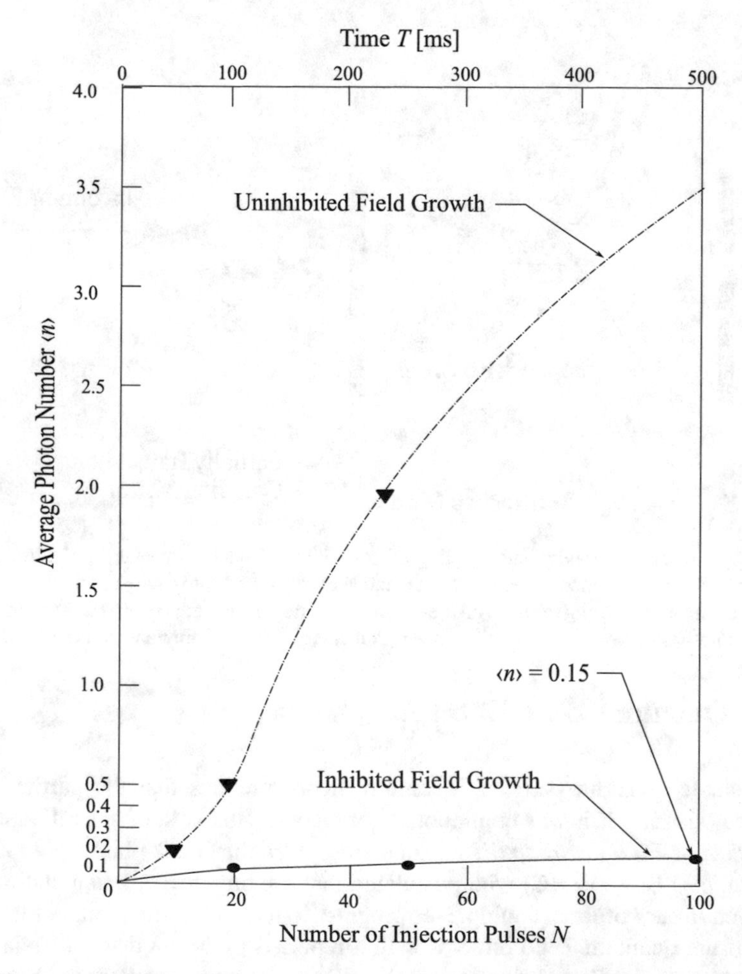

Fig. 8.51 Average photon number $\langle n \rangle$ for uninhibited field growth and inhibited quantum Zeno effect field growth versus time T and number of injection pulses N

QND field intensity measurements using rubidium Rydberg atoms was then performed in the short time interval between the coherent injections, resulting in almost complete inhibition of the original field growth, as shown in the lower solid curve of Fig. 8.51. The photon number measurements jumbled the field phase information, preventing any field amplitude buildup. A small residual growth of the field achieved an average photon number $\langle n \rangle = 0.15$ over the 500 ms time T of a complete sequence. This was a remarkable demonstration of the quantum Zeno effect for this average photon number.

8.11 Observation of Photon Trajectories

Heisenberg's uncertainty principle states that a precise measurement of a photon's spatial position is impossible due to the influence of the photon's momentum. For example, it is impossible to know the path of a photon in the double-slit experiment. (Section 8.1), or determine which slit the photon passed through. However, if the momentum component could be reduced to an insignificant yet known value, a set of paths for an ensemble of photons might be determined or observed. In 2011 researchers at the University of Toronto reported a two-slit experiment where average trajectories of single-photons could be observed [78]. Combined with *weak measurements* of photon momentum, precise measurements of the photon position in a series of planes could be used to construct a series of average photon trajectories.

The experimental setup to construct the average photon trajectories is illustrated in Fig. 8.52. A laser pumped liquid helium cooled InGaAs quantum dot source emitted single-photons at $\lambda = 943$ nm into a single-mode fiber, where they encountered a 50/50 fiber beamsplitter. The outputs were made coaxial to the propagation z-axis by right-angle mirrors, which functioned as a two-slit source having Gaussian beam profiles. A single-photon entering the split fiber would then exit either one "slit" or the other. A rotated polarizer prepared the photon into a diagonal state $|\Psi\rangle = \frac{1}{\sqrt{2}}[|H\rangle|V\rangle + |V\rangle|H\rangle]$ in the transverse x-y plane. A weak momentum value for the photon was obtained by insertion of a 0.7 mm thick birefringent calcite plate to slightly rotate the polarization state. Using a quarter-wave plate to produce circular polarization and a beam-displacement prism polarizer, the right-handed circular component of $|\Psi\rangle$ was undeviated, and the left-handed circular component of $|\Psi\rangle$ was displaced $\approx$2 mm. The separated photon beams built up two separated interference patterns over a 15 s exposure time (about 31,000 photons) at a cooled CCD. The transverse momentum was extracted from the difference in the modulation between these interference patterns. This weak momentum value was then postselected from an associated x-position strong measurement (precise to 26 μm pixel width accuracy) on the CCD in four transverse planes with z-values of 3.2, 4.5, 5.6, and 7.7 m. Over many measurements, it was possible to construct trajectories representing the average paths of a single-photon ensemble. A series of 80 reconstructed single-photon trajectories is shown in Fig. 8.53, using weak momentum values from 41 imaging planes and the

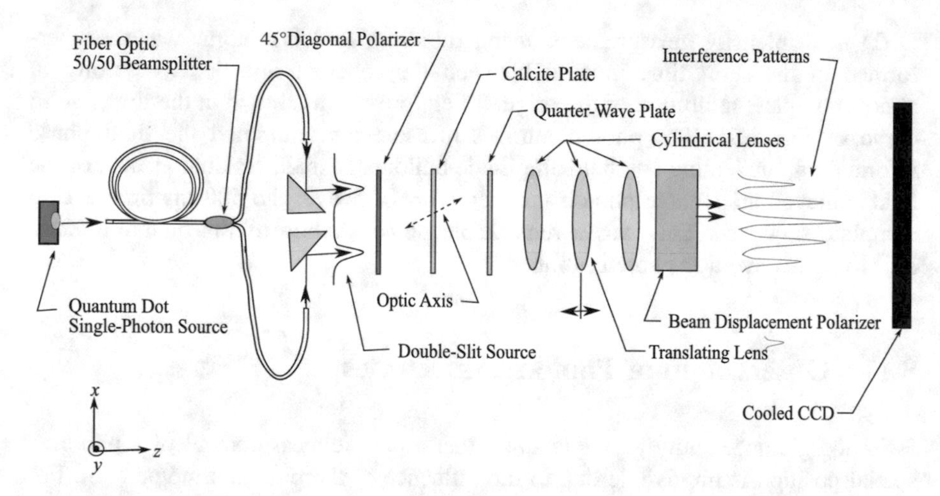

Fig. 8.52 Experimental setup to measure average single-photon trajectories. (Adapted with permission of AAAS from Ref. [78]; permission conveyed through Copyright Clearance Center, Inc.)

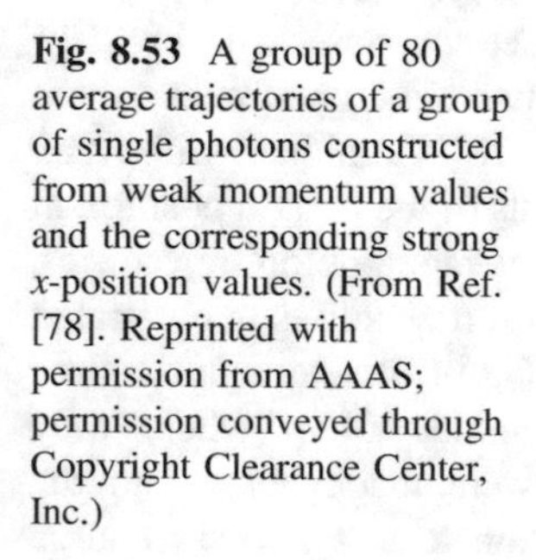
Fig. 8.53 A group of 80 average trajectories of a group of single photons constructed from weak momentum values and the corresponding strong x-position values. (From Ref. [78]. Reprinted with permission from AAAS; permission conveyed through Copyright Clearance Center, Inc.)

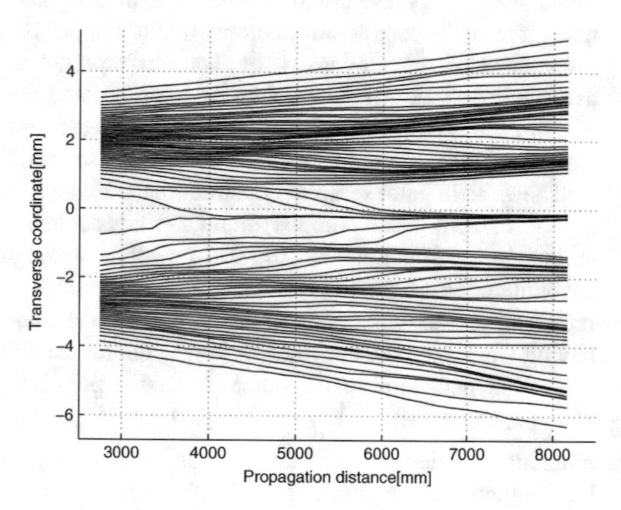

corresponding calculated x-position values. It was concluded that trajectories originating from one slit did not reach the opposite side of the interference pattern, and trajectories near the edge of a bright fringe moved over to other bright fringes, forming the observed interference pattern.

8.12 Creation of Matter by Colliding Photons

In 1934 Breit and Wheeler theoretically proposed a process whereby light could be transformed into matter [79]. This process is illustrated in Fig. 8.54, where a collision of two high-energy photons produces a positron-electron pair, known as a

Breit-Wheeler pair. Because of the high energy requirement of colliding gamma ray photons to produce a Breit-Wheeler pair, the idea lay dormant for about 80 years.

However, in 2014 a team at the Imperial College London proposed a type of photon-photon collider that could achieve the required energy levels [80]. The apparatus is illustrated in Fig. 8.55. An ultra-relativistic electron beam, created by a high-energy laser, impacts a 5 mm thick gold slab. The slab subsequently emits an

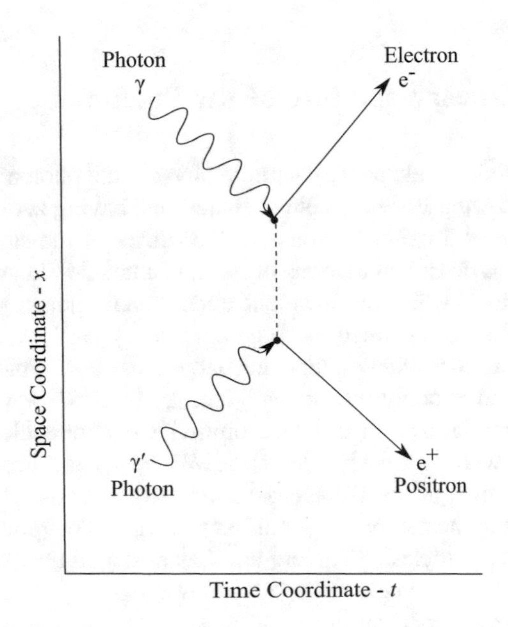

Fig. 8.54 Diagram illustrating production of an electron-positron pair (Breit-Wheeler pair) by the collision of photon particles

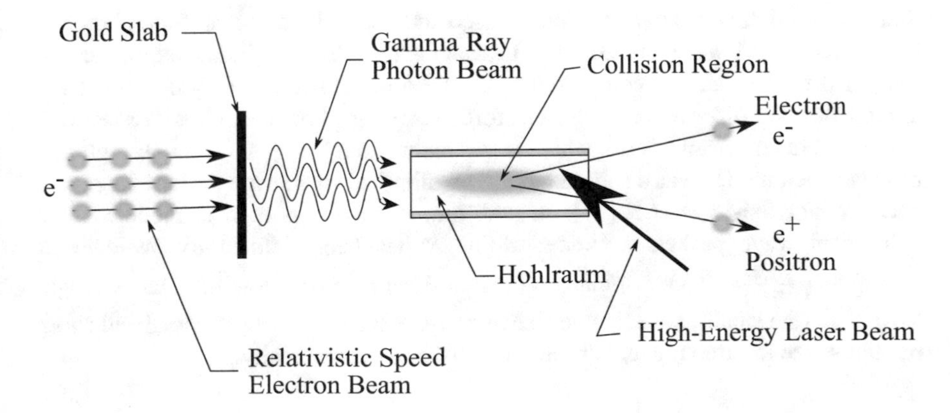

Fig. 8.55 Conceptual diagram of a proposed experiment to created matter in the form of electrons and positrons from the collision of photons in a hohlraum

ultra-high energy gamma ray photon beam by the *bremsstrahlung* process. A high-energy thermal radiation photon field (blackbody) is created throughout a small gold cylinder (called a *hohlraum*) by aiming a high-energy laser at the inner wall. The gamma ray photon beam is axially directed into the hohlraum, where the collision of the high-energy photons with the high-temperature radiation would create matter in the form of detectable electrons and positrons. An actual experiment confirming this process has yet to be reported.

8.13 Gauge Theory and Size of the Photon

In the Standard Model of elementary particle physics, the photon γ is classified as a boson particle following Bose-Einstein statistics, and having zero mass, a spin of 1, and the dimension of a point. Photons are the carrier of the strong electromotive force. The photon is described in terms of its fundamental field and is referred to as a *gauge* boson. Fields which are invariant under certain *gauge transformations* to other fields are called *gauge invariant*. The photon is gauge invariant. Other gauge invariant bosons are the weakly interacting large-mass boson particles, called W^+, W^-, and Z, and the strongly interacting gluon *g*. In 1967 Steven Weinberg and Abdus Salam formulated a theory that unified the strong electromotive force-carrying photon γ with the weakly interacting W^+, W^-, and Z bosons. The resulting gauge theory is known as the *Weinberg-Salam theory*, or the *electroweak theory*. For "contributions to the theory of the unified weak and electromagnetic interaction between elementary particles" Weinberg and Salam shared the 1979 Nobel Prize in Physics (also shared with Sheldon Lee Glashow).

Photons can be assigned an energy $E = hc/\lambda$ and momentum $p = hf/c$, but cannot be precisely localized to determine a position x. For this reason, Heisenberg's uncertainty principle $\Delta x \Delta p \geq \hbar/2$ is not applicable to a photon. However, the subject often arises as to whether the photon can be assigned a size or dimension. In the Standard Model, elementary particles are considered to be a point. If the photon is considered to be a wave, one could argue that the non-zero amplitude determines the size, or the size could be considered to be the finite coherence length. Some have argued that the distance at which the force-carrying photon exhibits "resistance" could be a determinant of size, which is typically about 0.5×10^{-15} m. The photon has also been modeled as a soliton having an elliptical shape with major axis $a = \lambda$, and a minor axis $b = \lambda/\pi$ [81]. In addition, when the photon is modeled as a cylindrical wave packet, a photon radius R has been defined by Weidner as $R = \frac{3\sqrt{2}c}{4\pi f} = \frac{3\sqrt{2}\lambda}{4\pi} = 0.338\lambda$. Finally, at dimensions near or below the Planck length $\approx 10^{-35}$ m (the smallest observable size of matter), little can be conceptualized about photon's size or structure, either theoretically or experimentally.

References

1. J.A. Wheeler, The 'Past' and the 'Delayed Choice' Double-Slit Experiment" in *Mathematical Foundations of Quantum Theory*, ed. by A. R. Marlow (Academic Press, New York, 1978), pp. 9–48
2. W.A. Miller, J.A. Wheeler, "Delayed-choice experiments and Bohr's elementary quantum phenomenon", *Proc. Int. Symp. Foundation of Quantum Mechanics (Tokyo)*, ed. by S. Kamefuchi et al. (Physical Society of Japan, 1983), pp. 140–151
3. D.J. Bohm, B.H. Hiley, C. Dewdney, A quantum potential approach to the Wheeler delayed-choice experiment. Nature **315**, 294–297 (1985). doi:10.1038/315294a0
4. V. Jacques et al., Experimental realization of Wheeler's delayed-choice gedanken experiment. Science **315**(5814), 966–968 (2007). doi:10.1126/science.1136303
5. A. Einstein, B. Podolsky, N. Rosen, Can quantum mechanical description of physical reality be considered complete? Phys. Rev. **41**(10), 777–780 (1935). doi:10.1103/PhysRev.47.777
6. D.J. Bohm, Y. Aharonov, Discussion of experimental proof for the paradox of Einstein, Rosen, and Podolsky. Phys. Rev. **108**(4), 1070–1076 (1957). doi:10.1103/PhysRev.108.1070
7. J.S. Bell, On the Einstein Podolsky Rosen paradox. Physics **1**(3), 195–200 (1964)
8. A. Aspect, J. Dalibard, G. Roger, Experimental test of Bell's inequalities using time-varying analyzers. Phys. Rev. Lett. **49**(25), 1804–1807 (1982). doi:10.1103/PhysRevLett.49.1804
9. C.H. Holbrow, E. Galvez, M.E. Parks, Photon quantum mechanics and beam splitters. Am. J. Physics **70**(7), 260–264 (2002). doi:10.1119/1.1432972
10. M.O. Scully, K. Drühl, Quantum eraser—A proposed photon correlation experiment concerning observation and 'delayed choice' in quantum mechanics. Phys. Rev. A **25**, 2208–2213 (1982). doi:10.1103/PhysRevA.25.2208
11. G. Kwiat et al., New high-intensity source of polarization-entangled photon pairs. Phys. Rev. Lett. **75**(24), 4337–4332 (1995). doi:10.1103/PhysRevLett.75.4337
12. Y.-H. Kim et al., A delayed choice quantum eraser. Phys. Rev. Lett. **84**, 1–5 (2000). doi:10.1103/PhysRevLett.84.1
13. C.K. Hong, Z.Y. Ou, L. Mandel, Measurement of subpicosecond time intervals between two photons by interference. Phys. Rev. Lett. **59**(18), 2044–2046 (1987). doi:10.1103/Phys.Rev.Lett.59.2044
14. R. Ionicioiu, D.R. Terno, Proposal for a quantum-delayed choice experiment", *Phys. Rev. Lett*. **107** (23), 2304-06-10 (2011). doi:10.1103/PhysRevLett.107.230406
15. F. Kaiser et al., Entanglement-enabled delayed-choice experiment. Science **338**, 637–640 (2012). doi:10.1126/science.1226755
16. R.P. Feyman, Photons: particles of light. Chapter 2 in *QED—the Strange Theory of Light and Matter*, (Princeton University Press, Princeton, New Jersey, 1985)
17. R.P. Feyman, Electrons and their interactions. Chapter 3 in *QED—the Strange Theory of Light and Matter*, (Princeton University Press, Princeton, New Jersey, 1985)
18. H.G.B. Casimir, On the attraction between two perfectly conducting plates. Proc. of the Koninklijke Nederlandse Akademie van Wetenschappen, **51**, 793–795 (1948)
19. S.K. Lamoreaux, Demonstration of the Casimir force in the 0.6 to 6 μm range. Phys. Rev. Lett. **78**(1), 5–8 (1997). doi:10.1103/Phys.Rev.Lett.78.5
20. G. Bressi et al., Measurement of the Casimir force between parallel metallic surfaces. Phys. Rev. Lett. **88**(4), 041804-1-4 (2002). doi:10.1103/PhysRevLett.88.041804
21. G.T. Moore, Quantum theory of the electromagnetic field in a variable-length one-dimensional cavity. J. Math. Phys. **11**, 2679 (1970). doi:10.1063/1.1665432
22. V.V. Dodonov, A.B. Klimov, Generation and detection of photons in a cavity with a resonantly oscillating boundary. Phys. Rev. A **53**(4), 2664–2682 (1996). doi:10.1103/PhysRevA.53.2664
23. J. Jeong-Young et al., Production of photons by the paramagnetic resonance in the dynamical Casimir effect. Phys. Rev. A **56**(6), 4440–4444 (1997). doi:10.1103/PhysRevA.56.4440

24. C. Braggio et al., A novel experimental approach for the detection of the dynamic Casimir effect. Europhys. Lett. **70**(6), 754–760 (2005). doi:10.1209/epl/i2005-10048-8
25. J.R. Johansson et al., Dynamical Casimir effect in a superconducting coplanar waveguide. Phys. Rev. Lett. **103**, 147003-1-4 (2009). doi:10.1103/PhysRevLett.103.147003
26. C.M Wilson et al., Photon generation in an electromagnetic cavity with a time-dependent boundary. Phys. Rev. Lett. **105**, 233907-1-4 (2010). doi:10.11103/ PhysRevLett.105.233907
27. C.M. Wilson et al., Observation of the dynamical Casimir effect in a superconducting circuit. Nature **479**, 376–379 (2011). doi:10.1038/nature10561
28. P. Lähteenmäki et al., Dynamical Casimir effect in a Josephson metamaterial. Proc. Nat. Acad. Sci. **110**(11), 4234–4238 (2013). doi:10.1073/pnas.1212705110
29. P. Grangier, G. Roger, A. Aspect, Experimental evidence for a photon anticorrelation effect on a beamsplitter: a new light on single photon interferences. Europhys. Lett. **1**(4), 173–179 (1986). doi:10.1209/0295-5075/1/4/004
30. S.J. van Enk, Single particle entanglement. Phys. Rev. A **72**, 064306-1-3 (2005). doi:10.1103/ PhysRevA.72.064306
31. J.F. Clauser et al., Proposed experiment to test local hidden-variable theories. Phys. Rev. Lett. **23**(15), 880–884 (1970). doi:10.1103/PhysRevLett.23.880
32. O. Morin et al., Witnessing trustworthy single-photon entanglement with local homodyne measurements. Phys. Rev. Lett. **110**, 13040-1-10 (2013). doi:10.1103/PhysRevLett.110. 130401
33. S.M. Tan, D.F. Walls, M.J. Collett, Nonlocality of a single photon. Phys. Rev. Lett. **66**, 252–255 (1991). doi:10.1103/PhysRevLett.66.252
34. L. Hardy, Nonlocality of a single photon revisited. Phys. Rev. Lett. **73**(17), 2279–2283 (1994). doi:10.1103/PhysRevLett.73.2279
35. B. Hessmo et al., Nonlocality of a single particle. Phys. Rev. Lett. **92**, 18040-1-4 (2004). doi:10.1103/PhysRevLett.92.180401
36. J.J. Cooper and J.A. Dunningham, "Single particle nonlocality with completely independent reference states", *NJP* **10**, 11302-4-18 (2008). doi:10.1088/1367-2630/10/11/113024
37. A. Zeilinger et al., Three-particle entanglement from two entangled pairs. Phys. Rev. Lett. **78** (16), 3013–3034 (1997). doi:10.1103/Phys.Rev.Lett.78.3031
38. L.K. Shalm, et al., "Three-photon energy-time entanglement", *Nat. Phys.*, 19–22, (2013). doi:10.1038/nphys2492
39. D.R. Hamel et al., Direct generation of three-photon polarization entanglement. Nat. Photonics **8**, 801–807 (2014). doi:10.1038/nphoton.2014.218
40. K.S. Choi et al., Entanglement of spin waves among four quantum memories. Nature **468**, 412–416 (2010). doi:10.1038/nature09568
41. V. Vuletic, Quantum physics: entangled quartet. Nature **468,** 384–385 (2011). doi:http://dx. doi.org/1038/468384a
42. C.-Y. Lu et al., Experimental entanglement of six photons in graph states. Nat. Phys. **3**, 91–95 (2007). doi:10.1038/nphys507
43. X.-C. Yao et al., Observation of eight-photon entanglement. Nat. Photonics **6**, 225–228 (2012). doi:10.1038/nphoton.2011.354
44. E. Megidish et al., Entanglement between photons that have never coexisted. Phys. Rev. Lett. **110**, 21040-3-6 (2013). doi:10.1103/PhysRevLett.110.210403
45. T.E. Hartman, Tunneling of a wave packet. J. Appl. Phys. **33**, 3427–3433 (1962). doi:http:// dx.doi.org/10.1063/1.1702424
46. R.Y. Chaio, P.G. Kwait, A.M. Steinberg, Analogies between electron and photon tunneling: a proposed experiment to measure photon tunneling times. Physica B **75**(1–3), 257–262 (1991). doi:10.1016/0921-4526(91)90724-S
47. A.M. Steinberg, P.G. Kwiat, R.Y. Chaio, Measurement of the single-photon tunneling time. Phys. Rev. Lett. **71**(5), 708–711 (1993). doi:10.1103/PhysRevLett.71.708
48. A. Enders, G. Nimtz, On superluminal barrier traversal. J. Phys. I France **2**(9), 1693–1698 (1992). doi:10.1051/jp1:1992236

49. A. Enders, G. Nimtz, Evanescent-mode propagation and quantum tunneling. Phys. Rev. E **48**, 632–634 (1993). doi:10.1103/PhysRevE.48.632
50. C. Spielmann et al., Tunneling of optical pulses through photonic band gaps. Phys. Rev. Lett. **73**(17), 2308–2311 (1994). doi:10.1103/Phys.Rev.Lett.73.2308
51. A. Haibel, G. Nimtz, Universal relationship of time and frequency in photonic tunneling. Annalan der Physik **10**(8), 707–712 (2001). doi:10.1002/1521-3889(200108)10:8<707
52. H.G. Winful, Tunneling time, the Hartman effect, and superluminality: a proposed resolution of an old paradox. Phys. Rep. **436**(1–2), 1–69 (2006). doi:10.1016/j.physrep.2006.09.002
53. H.G. Winful, Do single photons tunnel faster than light? Proc. SPIE **6664**, 6664C (2007). doi:10.1117/12.740086
54. G. Nimtz, A.A. Stahlhofen, Macroscopic violation of special relativity. arXiv:0708.0681v1 (2007)
55. H. Winful, Comment on 'Macroscopic violation of special relativity' by Nimtz and Stahlhofen. arXiv:0709.2736v1 (2007)
56. T. Peyronel et al., Quantum nonlinear optics with single photons enabled by strongly interacting atoms. Nature **488**, 57–60 (2012). doi:10.1038/nature11361
57. T.G. Walker, Strongly interacting photons. Nature **488**, 39–40 (2012). doi:10.1038/nature11384
58. O. Firstenberg et al., Attractive photons in a quantum nonlinear medium. Nature **502**, 71–75 (2013). doi:10.1038/nature12512
59. K.M. Beck et al., Cross modulation of two laser beams at the individual photon level. Phys. Rev. Lett. **113,** 11360-3-7 (2013). doi:10.1103/PhysRevLett.113.113603
60. D.F. Walls, Squeezed states of light. Nature **306**, 141–146 (1983). doi:10.1038/306141a0
61. M.C. Teich, B.E. Saleh, Squeezed states of light. Quantum Opt. **1**, 153–191 (1989)
62. M. Mehmet et al., Squeezed light at 1550 nm with a quantum noise reduction of 12.3 dB. Opt. Express **19**(25), 25763–25772 (2011). doi:10.1364/OE.19.025763
63. R. Schnabel, Gravitational wave detectors: squeezing up the sensitivity. Nat. Phys. **4**, 440–441 (2008). doi:10.1038/nphys990
64. K. Goda et al., A quantum-enhanced prototype gravitational-wave detector. Nat. Phys. **4**, 472–476 (2008). doi:10.1038/nphys920
65. J. Aasi et al., Enhanced sensitivity of the LIGO gravitational wave detector by using squeezed states of light. Nat. Photonics **7**, 613–619 (2013). doi:10.1038/nphoton.2013.177
66. B.P. Abbott et al., Observational of gravitational waves from a binary black hole merger. Phys. Rev. Lett. **116**, 061102-1- (2016). doi:10.1103/PhysRevLett.116.2061102
67. M. Brune et al., Quantum nondemolition measurement of small photon numbers by Rydberg-atom phase-sensitive detection. Phys. Rev. Lett. **65**(8), 976–979 (1990). doi:10.1103/PhysRevLett.65.976
68. M. Brune et al., Manipulation of photons in a cavity by dispersive atom-field coupling: quantum-nondemolition measurements and generation of 'Schrödinger cat' states. Phys. Rev. A **45**(7), 5193–5214 (1992). doi:10.1103/PhysRevA.45.5193
69. G. Nogues et al., Seeing a single photon without destroying it. Nature **400**, 239–442 (1999). doi:10.1038/22275
70. S. Gleyzes et al., Quantum jumps of light recording the birth and death of a photon in a cavity. Nature **446**, 297–300 (2007). doi:10.1038/nature05589
71. C. Monroe, A 'Schrödinger cat' superposition state of an atom. Science **272**(5265), 1131–1136 (1996). doi:10.1126/science.272.5265.1131
72. D.J. Wineland, Nobel Lecture: Superposition, entanglement, and raising Schröedinger's cat. Rev. Mod. Phys. **85**, 1103–1114 (2013). doi:1103/RevModPhys.85.1103
73. S. Haroche, Nobel lecture: controlling photons in a box and exploring the quantum to classical boundary. Rev. Mod. Phys. **85**, 1083–1102 (2013). doi:1103/RevModPhys.85.1083
74. A. Reiserer, S. Ritter, G. Rempe, Nondestructive detection of an optical photon. Science **342** (6164), 1349–1351 (2013). doi:10.1126/science.1246164
75. B. Misra, E.C.G. Sudarshan, The Zeno's paradox in quantum theory. J. Math. Phys. **18**, 756–763 (1977). doi:10.1063/1.523304

76. W.M. Itano et al., Quantum zeno effect. Phys. Rev. A **41**, 2295–2300 (1990). doi:10.1103/PhysRevA.41.2295
77. J. Bernu et al., Freezing a coherent field growth in a cavity by quantum Zeno effect. Phys. Rev. Lett. **101**, 18040-2-5 (2008). doi:10.1103/PhysRevLett.101.180402
78. S. Kocsis et al., Observing the average trajectories of single photons in a two-slit interferometer. Science **332**(6034), 1170–1173 (2011). doi:10.1126/science.1202218
79. G. Breit, J.A. Wheeler, Collision of two light quanta. Phys. Rev. **46**, 1087–1094 (1934). doi:10.1103/PhysRev.46.1087
80. O.J. Pike et al., A photon-photon collider in a vacuum hohlraum. Nat. Photonics **8**, 434–436 (2014). doi:10.1038/nphoton.2014.95
81. G. Hunter, M. Kowalski, C. Alexandrescu, The Bohr Model of the Photon. Chapter 13 in "*The Nature of Light: What is a Photon?*" 1st edn. (CRC Press, Boca Raton, Florida, 2008)

Chapter 9
Quantum Applications of the Photon

9.1 Entanglement-Enhanced Quantum Communication

9.1.1 Theoretical Foundation

For classic particles with well-defined states (bits 0 or 1), information is sent at a rate of one bit at a time. However, in 1992 Charles Bennett and Stephen Wiesner theorized that by using entangled particles in quantum channels, two bits of information could be encoded in one qubit [1]. This procedure is called *dense coding*, and the necessary system for this coding could be realized by the use of a single entangled pair of polarized photons.

9.1.2 Experimental Quantum Communication Using Dense Coding

In 1995 researchers in Austria performed the first experiment designed for two bit quantum communication using a pair of polarization entangled photons [2]. The entanglement allowed two bits of information to be transmitted by only one photon of the pair, with no direct encoding of the second photon. Figure 9.1 shows the general layout of the experiment. A nonlinear BBO crystal, driven by a 353 nm UV pulsed laser, produced a pair of orthogonally polarized entangled photons in the $|\Psi^+\rangle_{12}$ Bell-state

$$|\Psi^+\rangle_{12} = \frac{1}{\sqrt{2}}[|\mathrm{H}\rangle_1|\mathrm{V}\rangle_2 + |\mathrm{V}\rangle_1|\mathrm{H}\rangle_2], \tag{9.1}$$

where $|\mathrm{H}\rangle$ denotes a horizontally polarized qubit, and $|\mathrm{V}\rangle$ denotes a vertically polarized qubit. Bob received photon 1 of the pair and could create three other

D.F. Vanderwerf, *The Story of Light Science*,
DOI 10.1007/978-3-319-64316-8_9

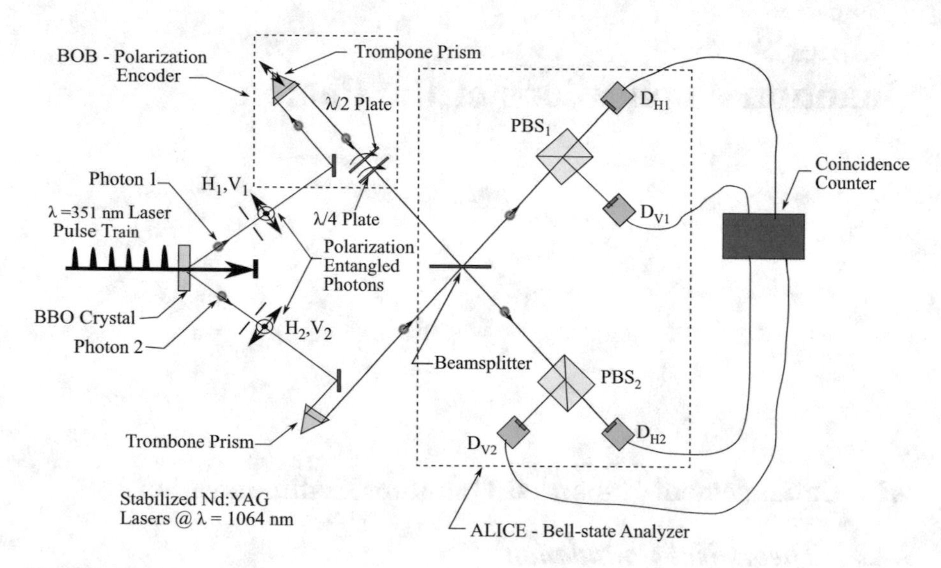

Fig. 9.1 Experimental setup for quantum communication between Bob and Alice using polarized entangled photons with quantum dense coding (This figure adapted with permission from Ref. [2]. Copyrighted by the American Physical Society)

maximally entangled states by using rotatable half- and quarter-waveplates to effect polarization axis rotation and phase shift: $\lambda/2 = 0°$ and $\lambda/4 = 90° \rightarrow |\Psi^-\rangle_{12}$ state, $\lambda/2 = 45°$ and $\lambda/4 = 0° \rightarrow |\psi^+\rangle_{12}$ state, $\lambda/2 = 45°$ and $\lambda/4 = 90° \rightarrow |\psi^-\rangle_{12}$ state. These additional Bell states are defined as

$$|\Psi^-\rangle_{12} = \frac{1}{\sqrt{2}}[|H\rangle_1|V\rangle_2 - |V\rangle_1|H\rangle_2], \tag{9.2}$$

$$|\psi^+\rangle_{12} = \frac{1}{\sqrt{2}}[|H\rangle_1|H\rangle_2 + |V\rangle_1|V\rangle_2], \tag{9.3}$$

$$|\psi^-\rangle_{12} = \frac{1}{\sqrt{2}}[|H\rangle_1|H\rangle_2 - |V\rangle_1|V\rangle_2]. \tag{9.4}$$

The original state $|\Psi^+\rangle_{12}$ was obtained with $\lambda/2 = 0°$ and $\lambda/4 = 0°$. Thus there were four maximum entangled two-qubit Bell states. These four states, entangled between photon 1 and photon 2, allowed the transmission of two classical bits of information to Alice by Bob's switching the states of photon 1 between $|\Psi^+\rangle_{12}$, $|\Psi^-\rangle_{12}$, $|\psi^+\rangle_{12}$, and $|\psi^-\rangle_{12}$.

Initially Bob sent out the $|\Psi^+\rangle_{12}$ state. Alice then received the various Bell states of photon 1 set by Bob, along with the second entangled photon 2, and both were combined at a beamsplitter for Alice's Bell-state measurements. Optical trombones allowed the path differences between photon 1 and photon from the source to the beam splitter to be set close to zero, where constructive interference occurred at the beamsplitter. The detection combinations recorded by coincidence counters allowed the states of Bob to be identified by Alice, and encoded information could be read. An additional beamsplitter with coincidence detectors (not shown in Fig. 9.1) allowed two photon states simultaneously incident on a detector

Table 9.1 Twenty-bit stream of polarized photons sent from Alice to Bob, producing a quantum key distribution 0011001001 using the BB84 protocol*

Alice sends and measures a polarized photon	Bob's basis for polarization measurement	Bob receives and measures polarization of the photon	Correct basis type (●)	Photon bit stream key (QKD)
↔	○	↻		
↔	+	↔	●	0
↔	○	↻		
↺	○	↺	●	0
↻	+	↔		
↔	○	↻		
↕	+	↕	●	1
↻	○	↻	●	1
↺	○	↺	●	0
↺	+	↔		
↺	○	↺	●	0
↕	○	↻		
↕	+	↕	●	1
↔	○	↺		
↻	+	↔		
↔	+	↔	●	0
↕	○	↻		
↺	○	↺	●	0
↻	+	↕		
↕	+	↕	●	1

*This data was generated by a BB84 encryption simulation algorithm written by Fred Henle at Dartmouth College. At the time of this writing, it was accessible at the web site http://fredhenle.net/bb84/

to be distinguished. Three messages were sent, corresponding to the ASCII-codes of "KM°", encoded in 15 trits rather than the classical requirement of 24 bits. For this proof-of-concept experiment, the channel capacity was increased by a factor of 1.58 by the use of dense coding.

9.2 Encrypted Quantum Communication Using Photons

9.2.1 Quantum Cryptography Protocols

One of the important applications, using the quantum properties of a photon, is ultrafast and secure transmission of data over optical fibers. This has been a subject of intensive investigation over the past decades and has necessitated the development of several related technologies.

To ensure that the transmitted data from a transmitter (Alice) to a receiver (Bob), several methods of secure encryption have been developed. One cryptographic method consists of using polarized photons to create a secret key that is used for encryption and decryption. The first secret key protocol using quantum techniques was proposed by Bennett and Brassard in 1984, known as the BB84 protocol [3]. Alice sends and measures a random polarized bit (0 or 1) in either a rectilinear or circular polarized mode over a one-way *quantum channel* to Bob. Bob then measures the photons using a randomly chosen rectilinear or circular polarized measurement setup (basis) and records the type of polarization detected. Due to the quantum nature of the photon, it consists of a superposition of all polarization modes, and Bob's determination of the mode is known only after he measures it. This is determined by the basis that Bob chooses. For example, if Alice sends a horizontally polarized photon, Bob will measure it as circular polarized if he uses a circular polarization basis. This is repeated until a series of common polarizations is obtained, and the corresponding types of measurement basis are shared between Alice and Bob over a separate two-way *public channel*. The actual measurement results are withheld and bits from non-common bases of Alice and Bob are discarded. The retained series of bits forms a secret key that is that can be used to encrypt a message, referred to as *quantum key distribution* (QKD).

An example of BB84 protocol quantum encryption using a series of twenty randomly polarized photons sent from Alice to Bob is shown in Table 9.1. By convention, a QKD bit = 0 corresponds to a horizontal or left-circular polarized photon, and a QKD bit = 1 corresponds to a vertical or right-circular polarized photon, as sent by Alice. For this example, the secret key is 0011001001, using matches for 50% of the sent bits. In practice, the QKD consists of a much longer string of bits. An eavesdropper can access the public channel without being detected, but can only obtain the measurement *bases*, not the measurement *results*.

Thus the eavesdropper cannot recreate the secret key. An eavesdropper on the quantum channel could steal some of Alice's polarized photons from a pulse of photons, but then must randomly guess Bob's measurement type.

Therefore the secret key cannot be reproduced with any certainty. Also, techniques have been developed to determine the presence of eavesdropping on the quantum channel.

Another quantum encryption technique using entangled photons, was first proposed by Artur Ekert at Oxford University in 1991 [4]. A light source emits a pair of entangled particles, one having a spin up number +½, and the other having a spin down number −½. These particles are sent to Alice and Bob through quantum channels, and are characterized by azimuthal angles designated as φ^a for Alice and φ^b for Bob. The analyzers of Alice and Bob determine the azimuthal angles and the analyzer orientations are randomly chosen by Alice and Bob. Both announce their analyzer orientations over a public channel. Following a series of entangled particle pair transmission measurements, two groups are formed, the first group where the analyzer orientations are the same, and the second group where the analyzer orientations are different. Then Alice and Bob announce over a public channel their measurement results for the first group only. A value for the CHSH parameter S from the generalized Bell theorem (Sect. 8.2.5) is then calculated. A value $S \approx -2\sqrt{2}$ implies that these first group particles are not disturbed and therefore the particle measurements in the second group are anticorrelated. The second group measurements are then converted into a cryptographic code or secret key. This technique is referred to as the EK91 protocol. Since the analyzer orientations of Alice and Bob are not known by an eavesdropper, it is difficult for the eavesdropper to escape detection.

Using these principles, Ekert and collaborators proposed an experimental scheme to produce a quantum encrypted message using a pair of correlated photons [5]. As shown in Fig. 9.2, a λ = 441.6 nm laser having a frequency $2\omega_0$ pumps a SPDC crystal, and a pair of spatially separated photons is emitted. The signal photon frequency ω_{signal} and idler photon frequency ω_{idler} are inputted to fiber optic cables and transmitted to Mach-Zehnder interferometers, each having a long and short path length. Tilted plates in one path of each interferometer produces signal and idler phase shifts φ_{signal} and φ_{idler}. The degree of correlation between the signal and idler photons is described by a correlation coefficient $J(\varphi_{signal}, \varphi_{idler}) = \cos(\varphi_{signal} + \varphi_{idler})$. Thus the two non-local photons are perfectly correlated when $\varphi_{signal} + \varphi_{idler} = 0$. Alice, receiving the signal photon, randomly sets her phase shift for either $\varphi_{signal} = 0$ or $\varphi_{signal} = \pi/2$, while Bob, receiving the idler photon, randomly sets his phase shift for either $\varphi_s = 0$ or $\varphi_s = \pi/2$. All uncorrelated detector registrations where $\varphi_{signal} + \varphi_{idler} \neq 0$ are then discarded, while the correlated interferometer measurements of a large subset are publicly shared. From this it can be inferred that the remaining unmeasured subset is also correlated and can form a cryptographic key. The use of interference rather than polarization would permit longer transmission fidelity in optical fibers, with expected secure transmission over a distance of 10–20 km.

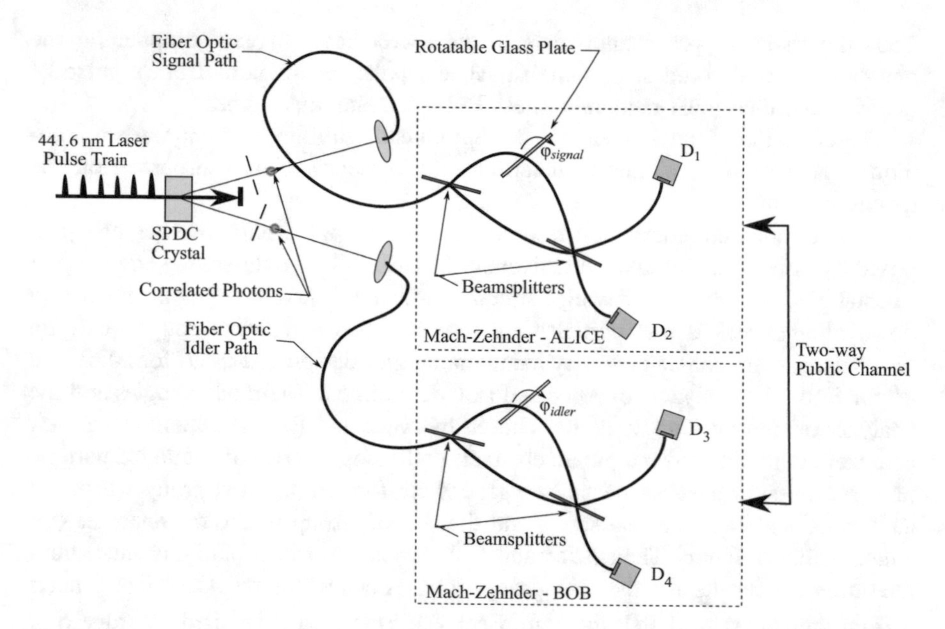

Fig. 9.2 Experimental proposal to generate a quantum encrypted message using a pair of correlated photons (This figure adapted with permission from Ref. [5]. Copyrighted by the American Physical Society)

9.2.2 Single-Photon Beam for Secure Quantum Communication

One method of eavesdropping on an encrypted quantum channel without detection would be to capture one or more of the photons in the burst of photons produced by most triggered laser sources. A captured photon could then be utilized to gain basis information without detection by Alice or Bob, since the remaining photons would preserve the information transfer between the parties. This is referred to as a *photon number splitting attack*. However, for a source producing only a single-photon at a time, any such tampering would be more easily detected.

9.2.3 Ideal Properties of a Pulsed Single-Photon Light Source

For many photonic quantum communication applications, a reliable source of pulsed single-photons is essential. Some ideal properties of a pulse train of single-photons are:

- One photon per pulse
- Stable and tunable pulse wavelength
- Narrow and tunable pulse bandwidth
- Equal temporal separation between pulses (anti-bunching)
- On-demand pulse emission
- Defined and fixed source position
- Near-room temperature operation.

Several technologies have been developed to address these needs.

9.2.4 Single-Photon Light Source Using Diamond Nanocrystals

A single-photon source was reported in 2002 that was stable, useable at room temperature, and easy to control [6]. The source material was a synthetic diamond nanocrystal having a nitrogen vacancy (NV) center created by irradiation. It was then spin coated in a polymer solvent to form a 30 nm thick layer with evenly dispersed 50 nm diamond nanocrystals on the surface. In one arrangement, the nanocrystal layer was pumped by a pulsed Nd:YAG laser with 1.2 ns pulse duration at a 10 MHz repetition rate. The pulse wavelength was 532 nm at an amplified energy of 2.5 nJ, and the emitted fluorescent radiation was collected by a confocal microscope setup. By the use of a 50/50 beamsplitter, intensity correlations were measured using two avalanche diodes and a coincidence counter. Figure 9.3 shows the raw coincidence rate count $c(\tau)$ versus time delay τ over a total acquisition time $T = 588$ s. The near-zero value of $c(\tau)$ at $\tau = 0$ indicated a low probability of two photons being in the same pulse, confirming the presence of single-photon emission. The time delay between consecutive single-photon pulses was greater than 5 ns to prevent photon bunching. Photon *antibunching* implies a lower probability of

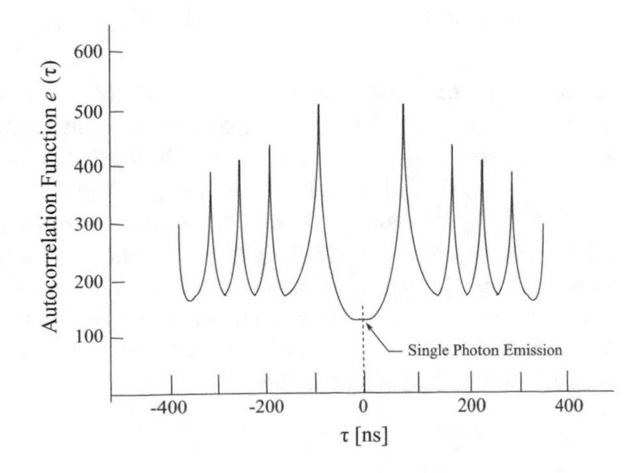

Fig. 9.3 Intensity autocorrelation function c(τ) versus time delay τ between detectors for a pulsed diamond nanocrystal NV center. Pulse repetition time is 100 ns. Absence of a peak at τ = 0 indicates emission of a single-photon

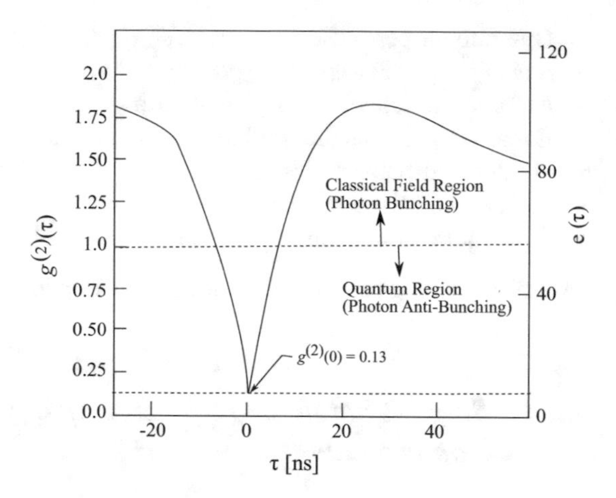

Fig. 9.4 Fitted curve for both $g^{(2)}(\tau)$ and $c(\tau)$ versus time interval τ for a single NV center in a diamond nanocrystal

simultaneously detecting two photons than the probability of detecting two photons over the time interval τ. Another useful parameter to qualify a single-photon source is the second-order autocorrelation function $g^{(2)}(\tau) = [I(t)I(t+\tau]/[I(t)^2]$, where $I(t)$ is the measured intensity at time t. When $g^{(2)}(\tau) \geq 1$, the source photons exhibit correlation and bunching. When $g^{(2)}(\tau) \leq 1$, the source photons become uncorrelated and exhibit antibunching with an equal time separation. A low value of $g^{(2)}(\tau)$ at $\tau = 0$ signifies a low probability that two photons would appear at the same time, and the photons are *anticorrelated*. For example, for the NV center of a diamond nanocrystal a value $g^{(2)}(0) = 0.13$ was measured. Figure 9.4 illustrates a fitted curve for both $g^{(2)}(\tau)$ and $c(\tau)$ versus time interval τ for a single NV center in a diamond nanocrystal.

9.2.5 Single-Photon Light Source Using a Whispering Gallery

Another type of single-photon light source utilizes an optical *whispering gallery mode resonator* (WGMR). The term "whispering gallery" was named by Lord Rayleigh around 1878 for acoustic transmission around a 34 m diameter circular wall in St. Paul's Cathedral in London. It was known that a person's whisper at one point of the wall could be heard by listeners at other positions around the wall, while shouting to a listener directly across the wall was barely audible. Rayleigh explained this through the formation of a layer of acoustic wave modes of high and low pressure that clung to the wall and were guided by the curvature of the wall. The resultant inverse decay of the sound wave intensity with distance around the wall was much less than the inverse-square decay by shouting directly across the wall.

An analogous effect occurs for light waves. In one type of single-photon source, a nano-size cylindrical disk of lithium niobate crystal ($LiNbO_3$) with highly-polished walls is used as a WGMR [7]. The experimental arrangement is illustrated in Fig. 9.5. Input photons from a continuous Nd:YAG laser at wavelength 532 nm were pumped into the WGMR using an external diamond coupling prism, through the evanescent wave layer of total internal reflection. The injected laser photon beam continuously circled the outer wall of the WGMR about 100 times and increased in intensity, guided by total internal reflection. The nonlinear optical properties of the lithium niobate WGMR produced SPDC, where the output beam was split into a pair of correlated or entangled photon beams having wavelengths of 1030 and 1100 nm. These correlated idler and signal photons were extracted from the WGMR through the same diamond prism by evanescent wave coupling. Detection of either photon at a given time ensured that the other photon was in a single-photon state by *heralding*. By controlled heating of the disk nanoresonator, each output photon beam could be precisely wavelength-tuned over a range of about 50 nm. The distance between the coupling prism and the disc was adjusted such that each beam bandwidth could be varied from 7.2 to 13 MHz. All three photon beams were spatially separated by a dispersing prism. The pump photons were detected by a PIN diode D_{pump}, while the idler photon was detected by an avalanche photodiode APD_{idler}. The signal photon, after passing through a 50/50 beamsplitter, was detected by avalanche photodiodes APD_{sigmal} and APD'_{signal}. Outputs from the split signal photons were analyzed by a Hanbury-Brown-Twiss (HBT) intensity interferometer configuration with coincidence counting. Anti-bunching of single-photons was verified with $g^{(2)}(0) < 0.2$.

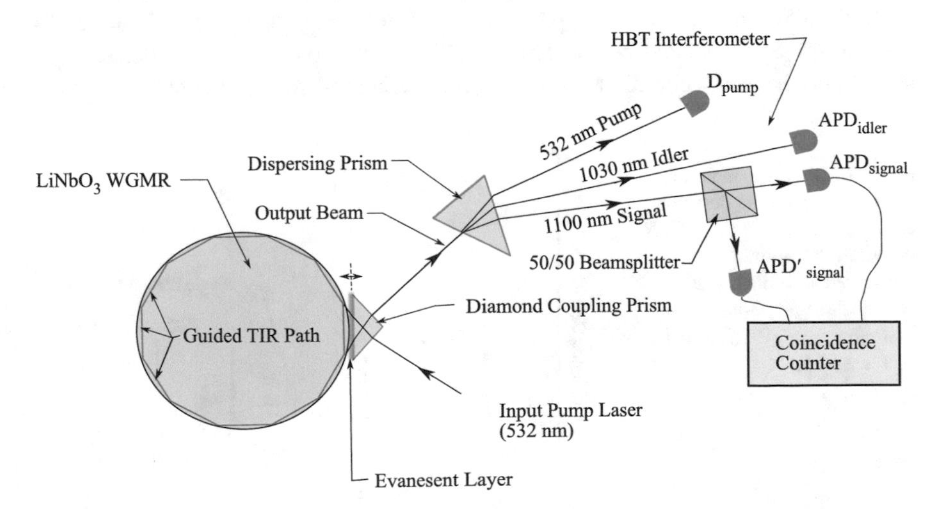

Fig. 9.5 Optical arrangement of a single-photon source using a lithium niobate whispering gallery mode resonator with SPPC to produce idler and signal photons

9.2.6 Other Room Temperature Single-Photon Light Sources

Quantum dots have been shown to function as an emitter of single photons. However, many of these sources require cryogenic cooling, decreasing their portability for quantum communication repeater use [8, 9]. A quantum dot source of single photons was demonstrated in 2014 that operated at 300 K at a defined physical position [10]. First, an array of 25 nm diameter apertures was fabricated on a sapphire substrate. The gallium-nitride quantum dots were imbedded in nanowires positioned in these apertures, defining the source positions. The quantum dots were excited by a pulsed laser, with subsequent intensity measurements using a beam-splitter and two single-photon detectors. The time delay τ between a series of recorded pulses was then recorded. Figure 9.6 shows the measurements of time delay τ versus detector coincidence counts. The suppression of simultaneous detection of photons at $\tau = 0$ confirmed the emission of single-photons, where an autocorrelation value $g^{(2)}(0) = 0.13$ was obtained.

Another room temperature single-photon source has been demonstrated using a silicon photonic chip emitting in the telecommunications band [11]. Spontaneous photon pair production was achieved by creating strong third-order nonlinearity in a silicon nanophotonic waveguide by means of a four-wave-mixing configuration, and using a pump laser beam to simultaneously generate an entangled photon pair with different frequencies. The 500 nm long silicon *coupled-resonator optical waveguide* (CROW) consisted of a series of 35 micro-ring resonators which introduced a nonlinearity coefficient about four times that of a highly nonlinear optical fiber. See Fig. 9.7. The strong 1549.6 nm pump beam, combined with a weaker 1570.5 nm probe beam, produced a 1529.5 nm signal photon beam and a 1570.5 nm idler photon beam. The heralding idler photons were detected by single-photon avalanche diodes and photon correlation measurements were made

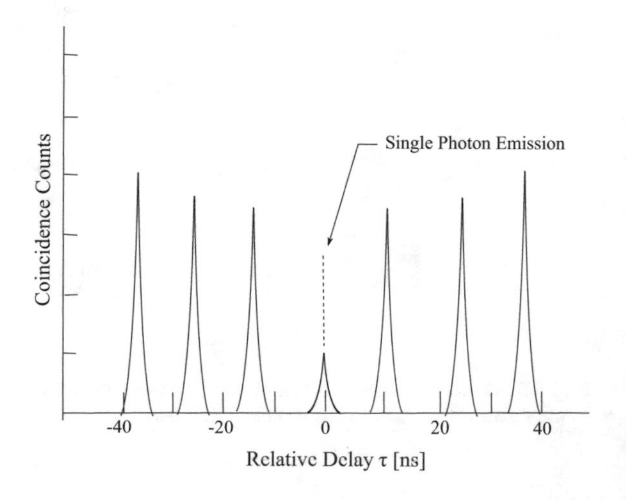

Fig. 9.6 Measurements of time delay τ versus detector coincidence counts for a GaN quantum dot emitter embedded in a nanowire, showing single-photon emission with $g^{(2)}(0) = 0.13$ at T = 300 K

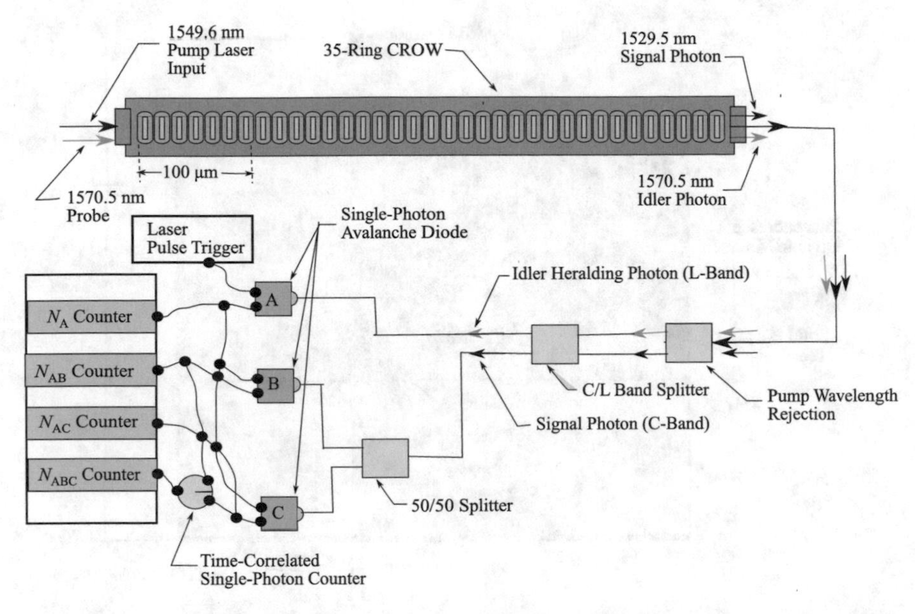

Fig. 9.7 Operating principle of a silicon coupled-resonator optical waveguide (CROW) single-photon source with a heralding photon. Size of CROW is exaggerated to show illustrative detail (Adapted from Ref. [11] with the permission of AIP Publishing)

on the 50/50 split path signal photons using a Hanbury-Brown Twiss intensity interferometer. The autocorrelation function $g^{(2)}(0)$ was defined as $g^{(2)}(0) = \frac{N_{ABC}N_A}{N_{AB}N_{AC}}$, where N_{ABC} was the number of triple coincidence at detectors A, B, and C, N_A and N_B were the number of single counts at detectors A and B, and N_{AB} and N_{AC}- were the number of double coincidence counts between the designated detectors. The low calculated value of $g^{(2)}(0) = 0.19 \pm 0.03$ showed the dominating presence of single-photons with anti-bunching. This room temperature single-photon silicon source has the advantage of using established chip fabrication techniques and easier integration with other chip photonic devices.

Another type of room temperature single-photon light source was developed at the University of Stuttgart that utilized an electrically driven organic light-emitting diode structure [12]. Figure 9.8 illustrates the principle of operation. Organic color molecules having three organic complexes grouped around a central iridium atom were created—designated as $Ir(PIQ)_3$. These phosphorescent molecules were distributed with a controlled density in a solid polymer matrix layer deposited on a conducting glass substrate. A cathode was formed on the top side by deposition of a 1 nm thick barium layer and a 100 nm thick aluminum layer. An ITO anode was formed on the glass substrate. When the device was electrically driven, phosphorescence of single molecules emitted single-photons over an active area of about 2 mm diameter. Photon correlation measurements then verified the emission of single-photons.

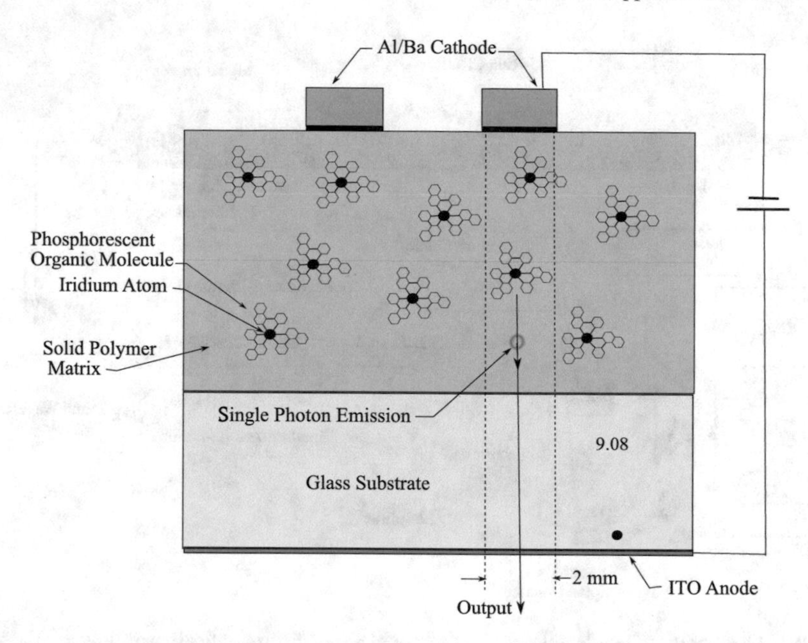

Fig. 9.8 Construction and principle of operation of an electrically driven single-photon light source using organic molecule emission. Shown is a single iridium-centered organic molecule emitting a single-photon

9.2.7 Single-Photon Detectors for Encrypted Photons

This use of weak single-photons for QKD requires single-photon detectors with very low background noise, and a fast response to the large number of incident photons required for a high secure key rate. The high background noise of conventional room temperature single-photon avalanche photodiode detectors limits QKD using single-photons to about 150 km over telecommunication optical fiber. To achieve the required low background noise required for longer distance optical fiber transmission, a cryogenically cooled ($T \approx 4$ K) *superconducting nanowire single-photon detector* (SNSPD) can be used. In 2001 a SNSPD having a response time of several picoseconds was demonstrated [13]. A 5 nm thick, 200 nm wide, 1 μm long NbN nanowire strip was deposited on a sapphire substrate. It was current-biased to slightly below the value where the wire loses it superconductivity, and connected to external voltage measurement circuitry by gold contacts. The nanowire was cooled to $T = 4.2$ K. An incident photon on the nanowire induced a "hotspot", and the nanowire temperature rose above the critical temperature $T_{crit} \approx 10$ K over a small region. For this conductive nanowire state, a voltage pulse could be measured. However, the nanowire returned to the superconducting zero-voltage state in ≈30 ps, and was ready to receive the next incident photon. Figure 9.9a–c illustrate the construction and principle of operation.

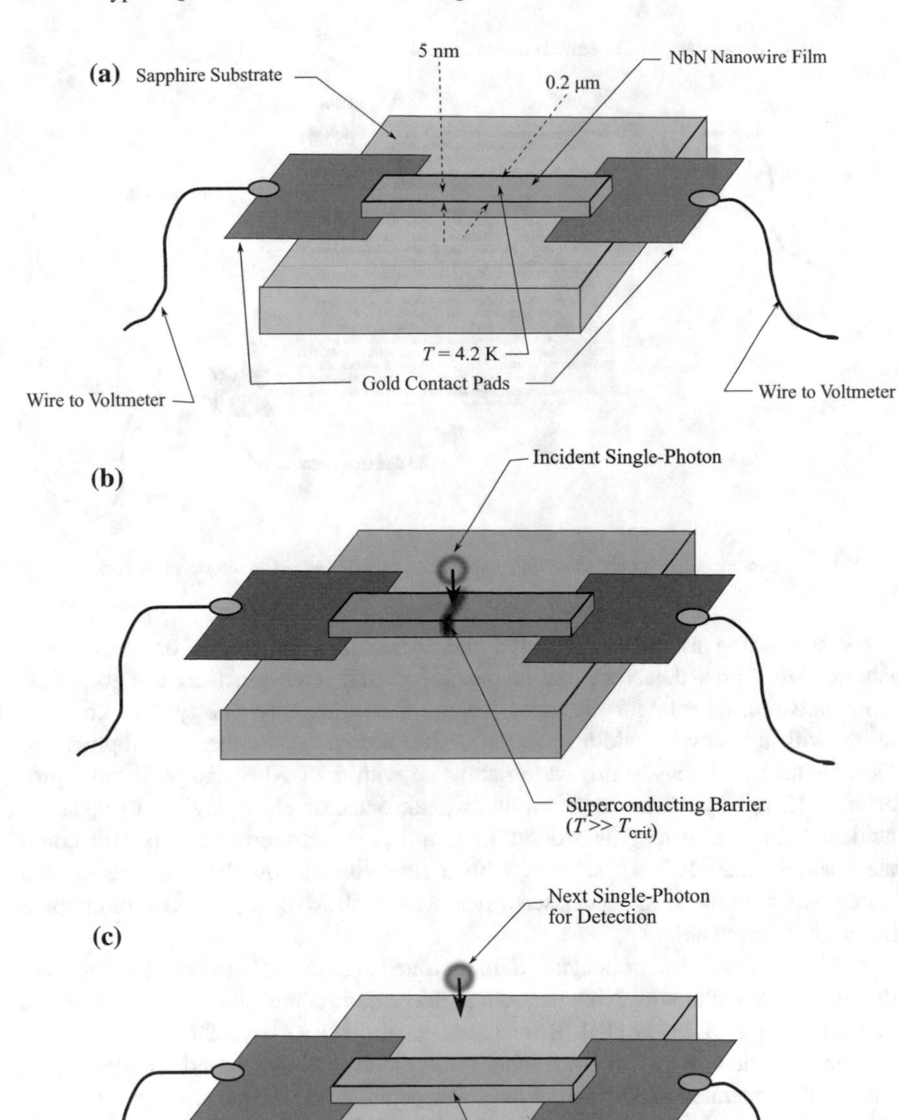

Fig. 9.9 **a** Basic construction of a superconducting nanowire single-photon detector (SNSPD). **b** Operation of a SNSPD. An incident single-photon produces a superconducting barrier region, allowing a voltage pulse to be measured. **c** Operation of a SNSPD. Nanowire returns to a continuous superconductive state, ready to receive next incident single-photon

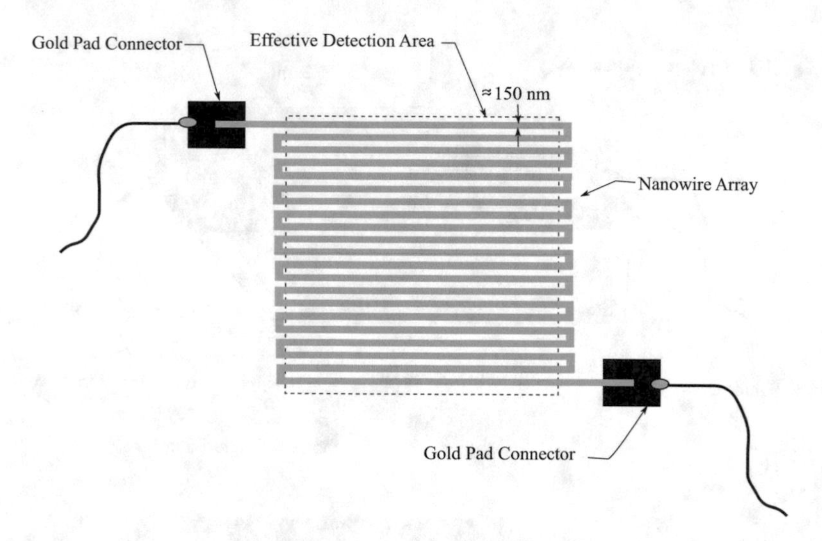

Fig. 9.10 Illustration of a superconducting nanowire meandering array single-photon detector

By connecting a series of parallel nanowires in a serpentine or meandering pattern, larger area detectors can be produced that have practical use for QKD communication over larger distances. Figure 9.10 illustrates one type of configuration, with a nanowire width in the 100–200 nm range. Further development of SNSPDs using thinner 30 nm wide nanowire widths of NbN and WSi have produced a 12 μm by 12 μm detector with peak detector efficiency up to 93% for incident photon wavelengths around 1500 nm [14]. Furthermore, the dark count rate was less than 100 per second, with a reset time below 100 ns. The photon source was a 1550 nm pulsed laser diode with a 50 MHz repetition rate coupled into a fiber optic cable.

In 2007 Takesue and colleagues demonstrated secure QKD over a 200 km long fiber using a SNSPD with NbN nanowires and a dark count rate of a few Hz and a secure rate of 12.1 bits/s [15]. Researchers at the University of Geneva in 2014 developed a new type of avalanche photodiode InGaAs based single-photon detector that operated at a non-cryogenic temperature of −110 C with an efficiency of 10% and a very low dark count of one per second [16]. Using this detector, QKD transmission over a 307 km long optical fiber was demonstrated a year later [17].

9.2.8 Secure Quantum Communication Through Free-Space

Another important application of QKD is secure quantum transmission between two stations in free-space, for example, between ground-based stations and satellites. One of the first practical demonstrations of free-space QKD used the BB84 protocol for transmission over a distance of 40 m in Munich, Germany [18]. The quantum bit error rate was 5.6%. QKD over 144 km in free-space was demonstrated in 2007, from a source at the Canary Island of La Palma to a receiver at the European Space Agency Optical Ground Station at Tenerife [19].

The first transmission of a QKD from a moving aircraft to a fixed position ground telescope was demonstrated in 2013 [20]. Compensation for aircraft-generated air turbulence and precise positioning of the telescope receiver were obtained by the use of cameras, lasers, and rapidly moving mirrors. The resultant line-of-sight deviation was less than 3 m over a 20 km distance.

Using an electrically driven InAs quantum dot single-photon infrared light source, researchers in Germany demonstrated free-space QKD transmission between attic rooms in downtown Munich over a 500 m distance [21, 22]. The 900 nm wavelength emission of the InAs single-photon light source allowed the use of a silicon avalanche photodiode detector, and an atmospheric window around 770 nm produced minimal dispersion and birefringence. The light source was cooled by a commercial cryostat to 57 K and the polarization of each emitted single photons was varied at a 125 MHz repetition rate. They then used a polarized photon BB84 protocol to obtain secure key rates up to 17 kHz at error rates of 6–9%.

9.3 Ultra-High Data Transmission Using Photons

9.3.1 Angular Momentum of a Photon

Each photon in an ordinary beam of light has energy $hf = h\omega/2\pi = \hbar\omega$ and linear momentum $hf/c = 2\pi\hbar f/c = k\hbar$, where $k = 2\pi f/c$ is the propagation direction of the planar wavefront. When a photon is right or left circular polarized with a helical path in the propagation direction, a photon *spin angular momentum* (SAM) with a value $\pm\hbar$ is acquired. The SAM is directed along the propagation direction and is a consequence of the photon spin. See Fig. 9.11a–b. However, it is also possible to produce a light beam with l intertwined helical *wavefronts*, where l is the helicity of the beam ($l = \pm 1, \pm 2, \pm 3, \pm 4, \ldots$, and $l\hbar$ is the *orbital angular momentum* (OAM) of each photon. Pure OAM is independent of polarization and photon spin. The resultant corkscrew wavefront path is referred to as *twisted light*, and can be produced by transmitting a laser beam through a special phase hologram or spatial light modulator [23]. See Fig. 9.12a–b. If the incident laser beam is circularly polarized, the OAM has both spin and orbital components. A large number of these intertwined OAM modes can travel independently over large distances, where each mode carries separate digital information. This multiplexing increases the data transmitting capacity of the beam by the number of superimposed modes. The number of encoded OAM modes theoretically has no limit, but on a practical basis, four, eight, and sixteen mode transmission has been experimentally studied. The photon wavelength is yet another parameter that can be used in conjunction with spin and orbital angular momentum for mode multiplexing.

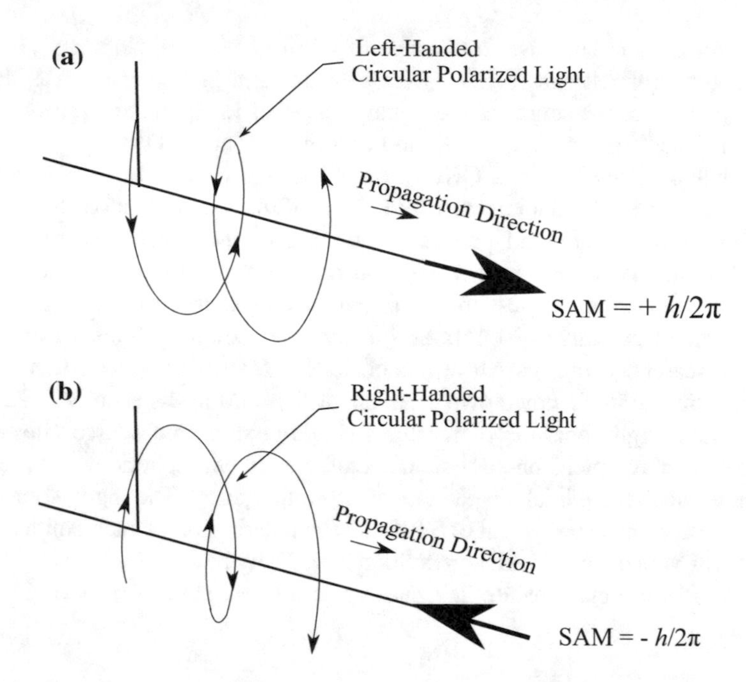

Fig. 9.11 **a** Illustration of left-handed circular polarization of a light beam (rotation is counter-clockwise when viewed from receiver to light source). Spin angular momentum SAM = $+h/2\pi$, pointing in the beam propagation direction. **b** Illustration of right-handed circular polarization of a light beam (rotation is clockwise when viewed from receiver to light source). Spin angular momentum SAM = $-h/2\pi$, pointing opposite to the beam propagation direction

9.3.2 Data Transmission Through Free-Space Using OAM Photons

For free-space optical frequency transmission experiments, it was initially found that the use of AOM modes to increase transmission capacity was sensitive to air turbulence. The resulting crosstalk between the multiplexed modes produces ambiguities at the receiver. However, using a mode intensity recognition system at the receiver, researchers in Vienna could distinguish between 16 superposed OAM modes with about 1.7% accuracy. Using this technique they transmitted grayscale images over a 3 km distance in Vienna through a turbulent atmosphere [24]. Although the rate of data transmission was not high, it was also found that the relative phase between the multiplexed modes was insensitive to air turbulence.

In 2012 an international team under Alan Willner at the University of Southern California achieved a data transmission rate of 1.37 Tb/s through free-space using four coaxial multiplexed OAM modes combined with polarization multiplexing [25]. The transmission rate was further increased to 2.56 Tb/s by the use of two concentric rings of eight polarization multiplexed OAM photon beams. Team

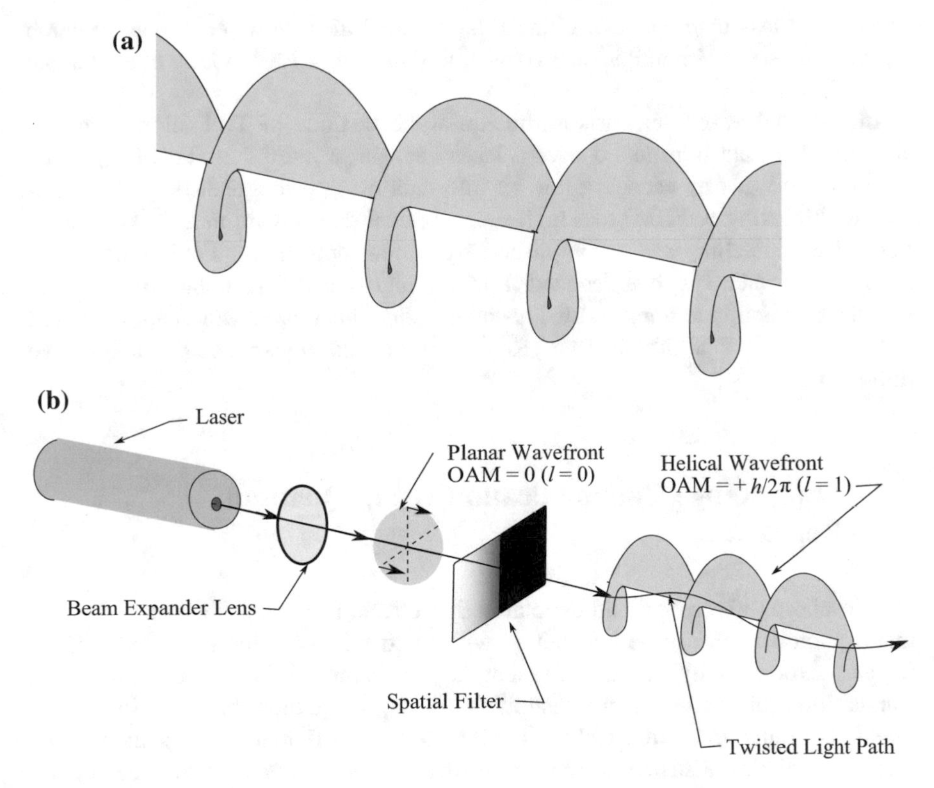

Fig. 9.12 **a** Illustration of twisted light with orbital angular momentum OAM = $lh/2\pi$. **b** Production of a photon beam having OAM using a laser and spatial modulator

researchers then demonstrated 2 Tb/s data transmission through free-space using two orthogonal OAM channels, each carrying 25 wavelength-division-multiplexed modes [26].

9.3.3 Ultra-Fast Data Transmission Through Optical Fibers Using Photons

Standard telecommunication optical fibers use other multiplexing techniques, such as wavelength and polarization to increase the data carrying capacity over large distances. Up to now, the data carrying capacity using standard broad band fibers has been limited to about 30 Mb/s. However, the use of OAM in these standard fibers has not been successful due to coupling and crosstalk between the OAM modes By using a multi-core optical fiber the rate was increased to 10 Gb/s by maintaining separation between multiplexed OAM channels [27]. In a further advancement, the fabrication of a special ring-like optical fiber has reduced channel crosstalk to less than −10 dB. Using this "vortex" fiber, an international research team has used OAM multiplexing to transmit data over a 1.1 km long fiber at a rate of 1.6 Tb/s [28].

In a related effort, scientists at the Karlsruhe Institute of Technology encoded and decoded data transmitted over a laser beam at a rate of 26 Tb/s [29]. They accomplished this by encoding low bit rate data using orthogonal frequency division multiplexing (OFDM) to a high-speed optical data stream in a 50 km optical fiber. This encoding was accomplished by a new optical fast Fourier transform routine. The data was then decoded from the high-speed data stream by a similar routine. The ability to transmit 26 Tb/s over optical fiber cable would allow the full video content of a 50 Gb Blue Ray DVD to be transmitted in about two milliseconds.

9.4 Long-Range Communication Using Quantum Repeaters

Due to photon absorption and depolarization effects in optical fibers, the distance that a photon qubit can be transmitted with good fidelity is limited. These effects increase exponentially with the fiber length. For example, with 0.2 dB/km loss in a standard telecom optical fiber transmitting a 1550 nm photon, high-quality photon fidelity is limited to about a 10 km fiber length. One method to increase the useable fiber transmission distance is by the use of *quantum repeaters*, where each node contains a memory of the quantum information from the previous node. This photon state is sequentially teleported to the next node, allowing the total transmission distance to be extended to an intercontinental range of over 10,000 km. Photon quantum repeaters make use of *entanglement swapping*, *quantum memory*, *quantum networking,* and *quantum teleportation*.

9.4.1 Quantum Memory

For quantum repeater applications, an efficient quantum memory should absorb a single-photon, store its quantum state for a sufficient time, and then reemit the photon in a well-defined quantum state for further transmission. The emitted photon could then be transferred to another quantum memory relay station. Failure of the quantum memory mode can occur when a photon is not reemitted (low-efficiency) or if of the quantum state of the emitted photon overlaps the state of the absorbed photon (low-fidelity). Early experiments in photon stoppage (Sect. 7.2) used atomic gases at cryogenic temperatures and electromagnetically induced transparency. One of the first experimental quantum memories for a weak pulse of light used a vapor ensemble of cesium atoms, where a quantum state of light was transferred into the atomic memory with 70% fidelity with a reported 4 ms quantum memory lifetime [30]. In another experiment, a solid-state quantum memory consisted of a 14 mm long lightly doped Pr^{3+}:Y_2SiO5 crystal held at 3 K. Using a 605.98 nm transition in the praseodymium ions, a single photon pulse and subsequent recall was realized by a gradient echo method. Using controlled external fields, the absorbed photon radiated an opposite canceling pulse (the echo) in the crystal, where the echo contained the quantum information of the incident pulse. This echo pulse was then reemitted from the crystal, after a determined storage time. The storage efficiency was up to 69% with a 1.3 μs storage time [31].

A favored approach for single-photon quantum memories uses an entangled photon pair, where one photon of the emitted pair is stored in the memory. When the other heralding photon is detected, this triggers the release of the photon in memory to act as another single-photon source. The memory storage time must be long enough for this entanglement process to occur, usually on the order of microseconds or milliseconds. For quantum repeater use, these deterministic single-photon sources must exhibit high-efficiency. Other factors important for quantum memories in repeater protocols are sufficient storage time for entanglement creation between distant nodes, and an efficient photon wavelength for minimum absorption in long length optical fibers.

A type of quantum memory developed at the California Institute of Technology stored a polarization-entangled pair of photons into two separated quantum memories in an ensemble of cryogenically cooled cesium atoms [32]. This proof-of-principle experiment is illustrated in Fig. 9.13. Collinear pulses of polarized single-photon pairs of 25 ns duration were emitted at a 1.7 MHz rate from transitions in a dense cold ensemble of cesium atoms. An accompanying strong coaxial control field $\Omega_c(t)$ could be switched ON and OFF at a 20 ns time interval. Right and left circular-polarized photon beams with a one millimeter separation were then formed by a beam-displacing polarizing prism and polarization control elements. Each beam entered the cooled cesium ensemble to form two spatially separated quantum memory regions. With the control field laser OFF the entangled incoming photon states were mapped and stored in the quantum memory regions as

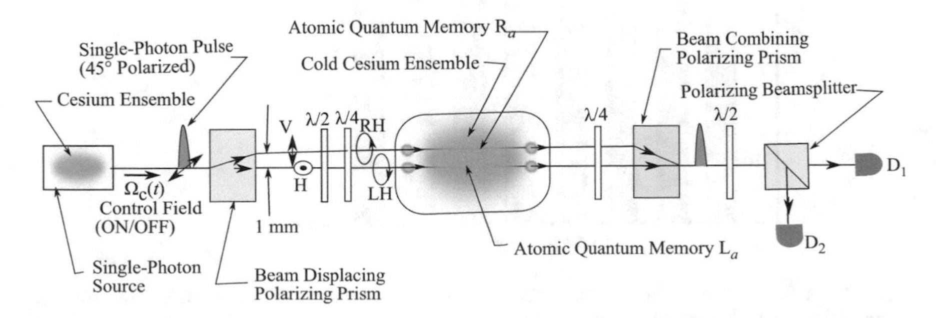

Fig. 9.13 Schematic representation of an atomic ensemble quantum memory that stores a pair of entangled input photons and retrieves the entangled pair after a determined time interval

atomic excitations in about 20 ns. The useable memory storage time τ_m was measured as 8 ± 1 μs. Heralded entanglement of these excitation states occurred between these separated quantum memories. After a determined storage time (≈1.1 μs), the control field laser was turned ON, and the cesium atomic ensemble became transparent by an electromagnetically induced transparency. The atomic excitations were then coherently remapped into entangled emitted photons with an efficiency of 17%. The fidelity of the outgoing photon entanglement was verified, and by removal of the quantum memory, the fidelity of the input photon entanglement was also verified.

Another crystal-based quantum memory was demonstrated in 2008 at the University of Geneva [33]. A YVO_4 crystal was doped with an inhomogeneous ensemble about 10^7 Nd^{3+} rare-earth ions and cooled by a pulse tube cooler to ≈3 K. The quantum state of a single-photon of 880 nm wavelength was absorbed by the 880 nm resonant transition lines of the niobium ions, illustrated in Fig. 9.14a. This storage was realized by first preparing the ions in the form of a spectral grating

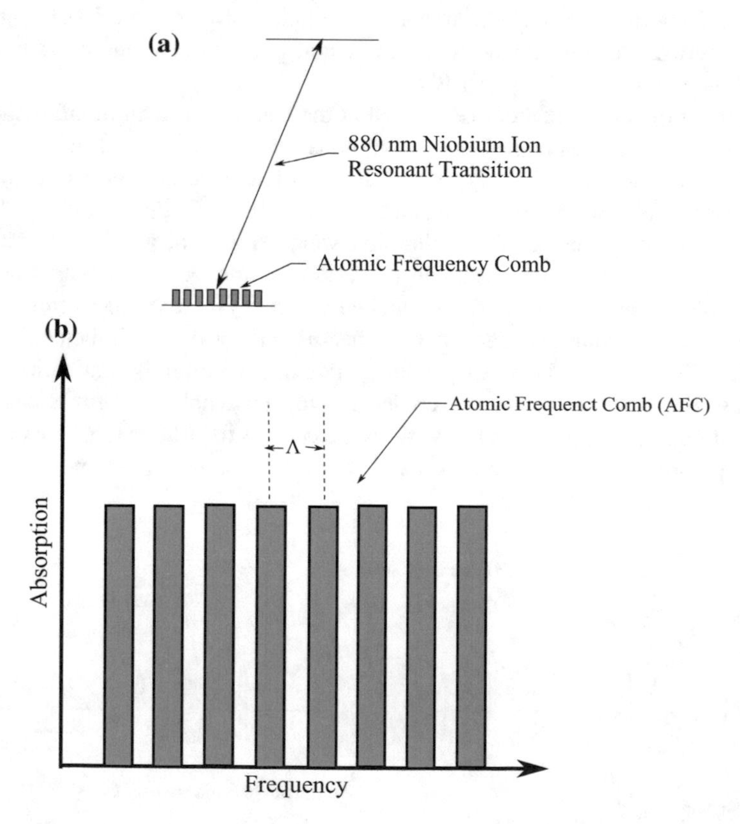

Fig. 9.14 **a** Illustration of the quantum state of a single-photon of 880 nm wavelength absorbed by an 880 nm resonant transition line of the niobium ions. **b** Detail of the atomic frequency comb (AFC) prepared in an ensemble of niobium ions, with a 4 MHz periodicity

known as an *atomic frequency comb* (AFC) with a periodicity $\Lambda = 4$ MHz. See Fig. 9.14b. This spectral grating periodicity determined the storage or delay time $\tau \approx 1/\Lambda$ of the photon, and the comb structure maintained the fidelity of the spatial mode. Reversible absorption then emitted the photon with high coherence and fidelity, which was experimentally verified. The observed storage time for a single temporal mode was 250 ns, close to the value predicted from the comb periodicity.

The use of an optical transition in a cryogenic-cooled praseodymium-doped crystal memory with an atomic frequency comb and spin-wave storage further extended the quantum memory storage time [34]. Using a control pulse, the stored line transition excitation was converted to a stored spin-wave excitation, which in combination with the comb structure, increased the storage time up to 20 μs for 450 ns light pulses.

Photon entanglement was combined with solid-state quantum memory in 2011 when entanglement was demonstrated between an external telecommunication wavelength photon and a photon stored in quantum memory [35]. Figure 9.15 illustrates the experimental arrangement. The Nd:Y_2SiO_5 doped crystal had a resonance transition wavelength of 883 nm and was maintained at a temperature of 3 K. An entangled photon pair was produced by SPDC in a PP-KTP waveguide pumped by a 532 nm laser, emitting a signal photon at 883 nm and an idler photon at a telecommunication wavelength 1338 nm. After angular separation of the two beams by a diffraction grating, the two beams were spectrally filtered. The 883 mm beam was sent through the quantum memory in a double-pass mode and detected by a silicon avalanche diode. The 1338 nm photon was sent to a superconducting single-photon detector through a 50 m long optical fiber. The memory atomic ensemble was prepared into an AFC state over a 15 ms time interval by an 883 mm

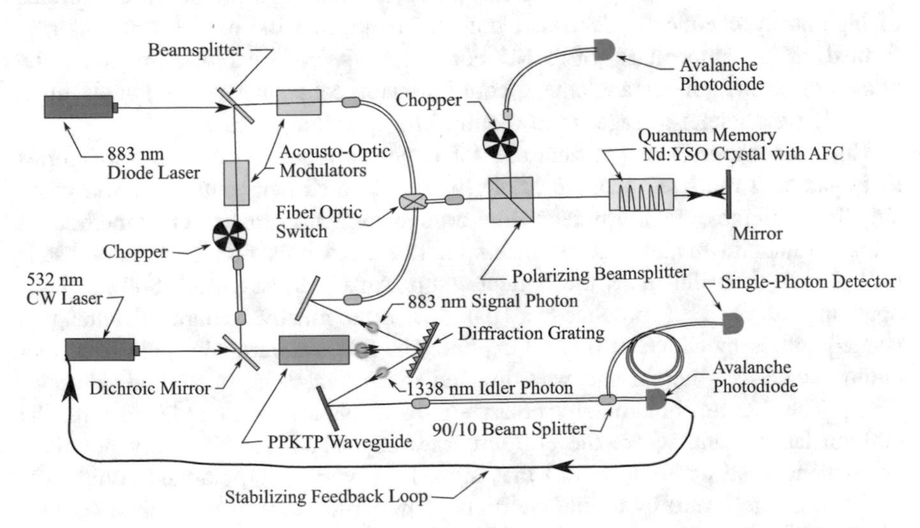

Fig. 9.15 Demonstration of quantum entanglement between an external telecommunication wavelength photon and a photon stored in quantum memory. Some additional spectral filtering components are not shown

diode laser. This atomic comb maintained the fidelity of the stored collective excitation state. The storage time could be set at an optimal 25 ns (21% efficient) or 100 ns (12% efficient) by control of the AFC period. Preservation of quantum behavior between the retrieved 883 nm memory photon and the 1338 nm telecommunication photon was verified by calculation of a second-order cross-correlation function $g_{si}^{(2)} \approx 30$ for a 25 ns storage time and 3 mW pump power. Since $g_{si}^{(2)} > 2$, the lower bound for quantum behavior, the stored photon maintained its quantum properties. Verification of the quantum entanglement between the telecommunication photon, the photon stored in memory as a collective excitation, and the released photon was determined in a separate series of interferometric measurements. Using Bell's inequality in the form of the CHSH parameter S (Sect. 8.2.5), a value of $S = 2.64 \pm 0.23$ was calculated. Since $S > 2$, the CHSH inequality was violated and non-local photon pair entanglement was preserved during memory storage, along with conservation of the entanglement.

Another type of quantum memory used a diamond crystal with a nitrogen-vacancy (NV) center as used for a single-photon source, since the emission of a single photon from a quantum memory can be considered as a type of single-photon source [36]. In this method an absorbed red photon of wavelength 637.2 nm was entangled with the stored electron spin in the NV center. The electron spin state was initially prepared by irradiation of the diamond by 532 nm green light pulses, accompanied by 2.88 GHz pulsed microwave radiation. Entanglement of the absorbed single photon with the electron spin then occurred with a fidelity of about 95%. Then another pulsed red read-out photon, synchronized with the microwave pulse, read out the spin state to emit another red photon. A strong correlation between the outgoing photon polarization and electron spin states was measured. The NV center was positioned about 50 μm below the surface of high-purity chemically deposited bulk diamond, and the experiment was performed at a temperature of 5 K. For this type of diamond memory the photon-electron spin entanglement could remain coherent over a longer time, producing the longer storage times required for quantum repeaters.

However, in terms of practical use for quantum repeaters, quantum memories that operate at room temperature or at slightly elevated temperatures would probably be preferable to those requiring cryogenic temperatures. One method to achieve quantum memory at room temperature used bulk diamond nanocrystals [37]. Figure 9.16 illustrates the optical setup. An ultra-fast pulsed 800 nm laser operating at 2 THz produced a pair of orthogonally polarized entangled single-photons by SPDC at a BBO crystal. The 895 nm vertically polarized idler photon acted as a herald and was directed to an avalanche photodiode detector APD_1. The 723 nm horizontally polarized signal photon was combined with the 800 nm laser beam, where the 800 nm laser beam had a horizontally polarized component (read or probe beam) that lagged the vertically polarized component (write or excite beam) by a time interval τ. The write beam pulse excited 40 THz bandwidth *optical phonons* (resonance optical vibrations in the diamond crystal lattice), which were stored in the diamond quantum memory for a short time, on the order of 3.5 picoseconds. The lagging read beam then triggered emission of a single

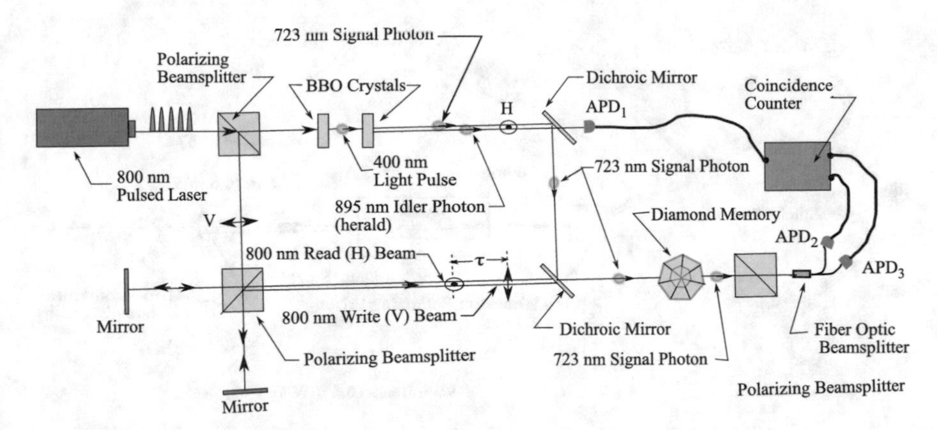

Fig. 9.16 Operation of a single-photon read/write memory using a room-temperature bulk diamond. Some polarization control components are not shown (This figure adapted with permission from Ref. [37]. Copyrighted by the American Physical Society)

photon from the optical phonons in the diamond memory. The high terahertz storage and retrieval rates of the memory avoided any decoherence influence on the quantum states. Using the APD_1, APD_2 and APD_3 detectors and a coincidence counter, calculation of the correlation coefficient $g^{(2)}(0) = 0.65 \pm 0.07$ of output photons demonstrated the quantum nature of the memory. The large 40 THz energy of the diamond optical phonons allowed low-noise operation at room temperature, and the memory was tunable for use at several photon wavelengths. However, a longer storage or coherence times would probably be required for long-distance communication repeater application.

In another approach, a vapor cell at a warm temperature was developed to function as a quantum memory [38]. Spontaneous Raman scattering from an atomic ensemble of ^{87}Rb vapor produced emission of a pair of correlated single-photons. One emitted photon had less energy than the pulsed pump beam (Stokes photon), and the other had more energy than the pump beam (anti-Stokes photon). The experimental setup is shown in Fig. 9.17. The 7.5 cm long cell was filled with ^{87}Rb vapor at an atomic density $N = 4.3 \times 10^3\ \text{cm}^{-3}$, with added pressurized Ne buffer gas to reduce the effect of fluorescence on the fidelity of the stored atomic excitations in the quantum memory. The cell was magnetically shielded and held at a temperature of 37 °C. The one microsecond duration pulsed 0.6 mW co-propagating read and write beams were orthogonally polarized using Glan-Thompson polarizers. The 20 kHz pulsed counter-propagating pump beam, tuned to a transition emission of the ^{87}Rb vapor, was switched on shortly after the read pulses and switched off shortly before the write pulses, entering the opposite side of the cell as shown in the figure. Spurious fluorescence caused by buffer gas collisions in the cell was reduced for the signal beams by detuning the read and read pulses from the ^{87}Rb resonance transition. The collinear emitted photons were separated and orthogonally polarized by a Wollaston polarizing beam splitter. After further spectral filtering, the Stokes and

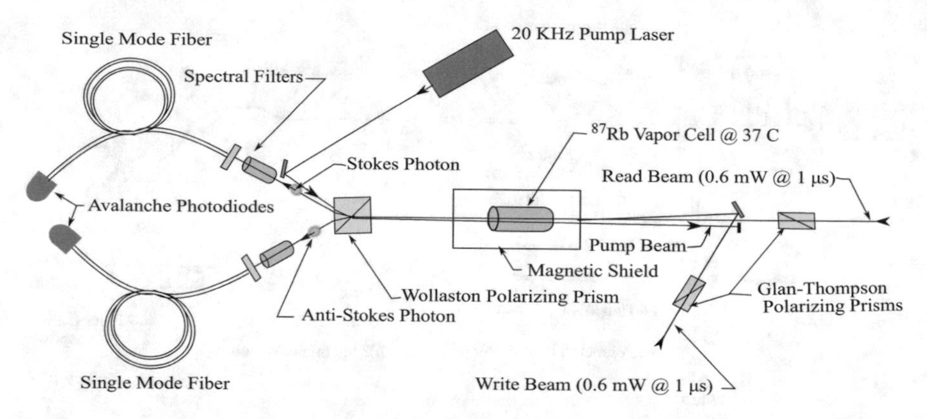

Fig. 9.17 Experiment setup for demonstration of photon storage in a warm rubidium vapor cell quantum memory (Adapted from Ref. [38] with permission from the Optical Society of America)

anti-Stokes photons were transmitted through a length of single mode optical fiber and tested for cross-correlation by photon-counting avalanche diode detectors. The calculated correlation factor gave $g^{(2)}(\tau) > 2$, where τ was the delay time between the write and read pulses, indicating a quantum memory storage time τ of about 4 μs.

Another type of warm vapor quantum memory was developed at the University of Warsaw [39]. A 25 mm diameter glass tube of 100 mm length was filled with warm (90 °C) rubidium gas and magnetically shielded. Incoming photons were stored by interfacing with off-resonant Raman transitions in the gas, and a photon storage time up to 30 ms was reported.

9.4.2 *Principles of Quantum Teleportation*

The concept of *quantum teleportation* for photon particles refers to the transfer of the quantum state of one photon to that of another photon. Teleportation is an essential operation for quantum state transfer between a series of quantum memory units to achieve long-range quantum communication. It also is an important technique to transport quantum cryptographic keys. A scheme for teleportation of an unknown quantum state $|\psi\rangle_1$ from one observer (Alice) to another observer (Bob) was first described by Bennett and an international team of theorists [40]. The methodology used quantum spin states, spin up (↑) and spin down (↓), related to the spin angular momentum of a quantum particle such as an electron. As shown in Fig. 9.18, an entangled pair of quantum particles (2 and 3) is emitted from an EPR-source, with particle 2 with state $|\Psi^-\rangle_{23}$ being sent to Alice and particle 3 with state $|\Psi^+\rangle_{23}$ being sent to Bob. Alice then performs a Bell-state measurement between the unknown state particle 1 and particle 2 to project them into four equally probable entangled states $|\Psi^-\rangle_{12}$, $|\Psi^+\rangle_{12}$, $|\psi^-\rangle_{12}$, and $|\psi^+\rangle_{12}$, where

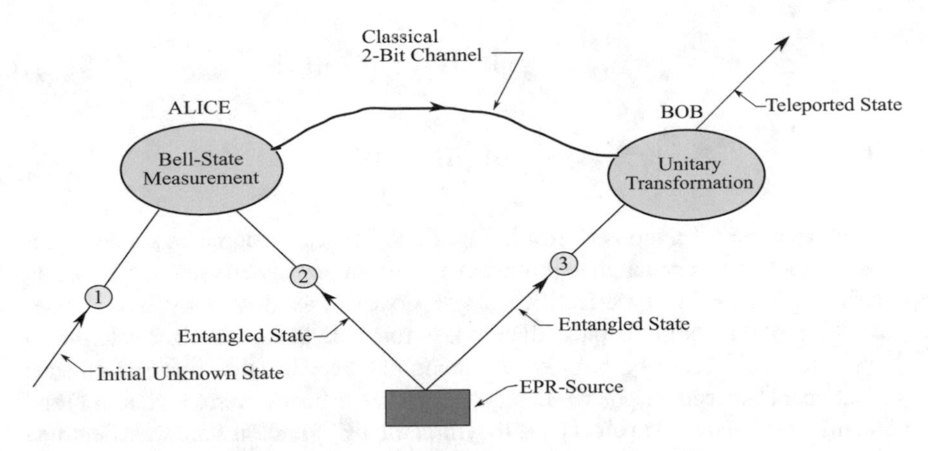

Fig. 9.18 Principal of quantum teleportation using an entangled pairs of spin quantum states

$$|\Psi^{\pm}\rangle_{12} = \frac{1}{\sqrt{2}}[|\uparrow\rangle_1|\downarrow\rangle_2 \pm |\downarrow\rangle_1|\uparrow\rangle_2], \tag{9.5}$$

$$|\psi^{\pm}\rangle_{12} = \frac{1}{\sqrt{2}}[|\uparrow\rangle_1|\uparrow\rangle_2 \pm |\downarrow\rangle_1|\downarrow\rangle_2]. \tag{9.6}$$

Since particle 1 is now entangled with particle 2, all three particles become entangled, and the single original state $|\psi\rangle_1$ disappears and cannot be sent to Bob as a clone. Alice then sends a classical outcome of her Bell-state measurement to Bob over a 2-bit classical channel. The correlation of Bob's particle state with the other two particle states then allows Bob to replicate or teleport the original state $|\psi\rangle_1$ by applying one of four possible unitary transformations, specified by the classical outcome sent by Alice. The teleported $|\psi\rangle_3$ can then be transferred for further processing.

A useful set of rotational operators for Bob to analyze the Bell states created by Alice are the Pauli complex matrices X, Y, Z, and I. These are defined by $X = \begin{pmatrix} 0 & 1 \\ 1 & 0 \end{pmatrix}$, $Y = \begin{pmatrix} 0 & -i \\ i & 0 \end{pmatrix}$, $Z = \begin{pmatrix} 1 & 0 \\ 0 & -1 \end{pmatrix}$, along with the identity matrix $I = \begin{pmatrix} 1 & 0 \\ 0 & 1 \end{pmatrix} = X^2 + Y^2 + Z^2$. When Bob knows which Pauli matrix to apply to Alice's orthonormal Bell states, the original state $|\psi\rangle_1$ is extracted and transformed into Bob's teleported state $|\psi\rangle_3$. Application of the identity matrix leaves the original Bell-state unchanged.

The first experimental accomplishment of quantum teleportation was performed in 1998 at the University of Innsbruck, Austria [41]. For this experiment, horizontal $|H\rangle$ and vertical $|V\rangle$ polarized photon states replaced the spin states in Eqs. (9.5) and (9.6), where

$$|\Psi^{\pm}\rangle_{12} = \frac{1}{\sqrt{2}}[|H\rangle_1|V\rangle_2 \pm |V\rangle_1|H\rangle_2], \quad (9.7)$$

$$|\psi^{\pm}\rangle_{12} = \frac{1}{\sqrt{2}}[|H\rangle_1|H\rangle_2 \pm |V\rangle_1|V\rangle_2]. \quad (9.8)$$

The experimental setup is shown in Fig. 9.19. The EPR-source was a nonlinear beta-barium-borate crystal that produced a pair of entangled and orthogonally polarized photons by Type-II SPDC. The crystal was driven by a UV laser ($\lambda = 400$ nm) with a 200 fs pulse duration. A rotatable linear polarizer determined the polarization direction of the unknown quantum state. Alice then made a Bell-state measurement between the polarization quantum state to be transferred (photon 1) and photon 2 of the entangled pair. The anti-symmetric $|\Psi^-\rangle$ photon state was identified by coincidence counts at the two detectors. In total, two of the four Bell states could be identified, yielding a 50% success probability for the teleportation. The other entangled state $|\Psi^+\rangle_{23}$ of photon 3 was sent directly to Bob for polarization analysis. The polarization direction of the photon to be teleported was then sent from Alice to Bob over a classical two-bit channel. When the detector coincidence of photons 1 and 2 was adjusted to be close to simultaneous, the linear polarization of the state transported to Bob was maximized at about 70%. Bob then extracted the polarization state $|\psi\rangle_1$ of photon 1 from the entangled polarization state of particle 3 using the known polarization direction (unitary transformation). The polarization state $|\psi\rangle_1$ was essentially teleported to the polarization state $|\psi\rangle_3$ of photon 3.

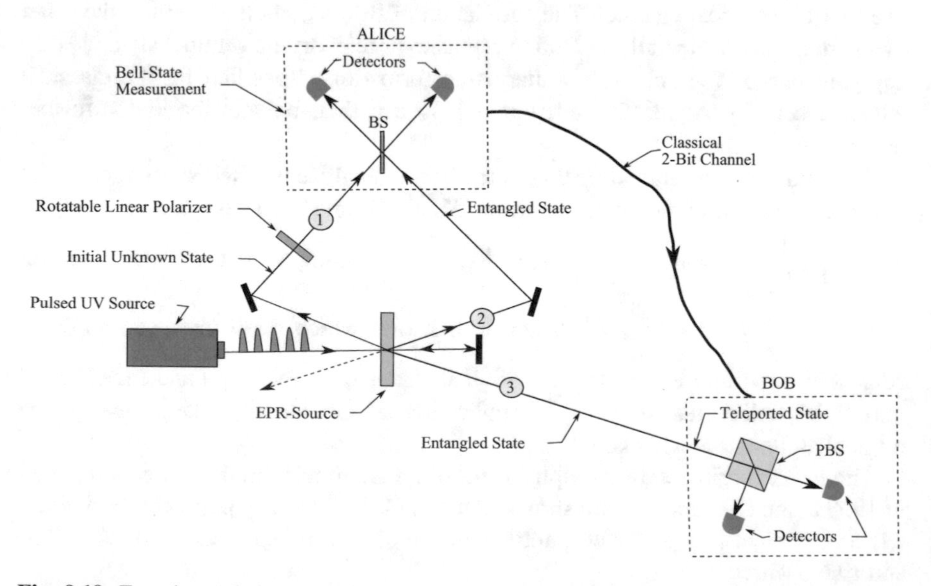

Fig. 9.19 Experimental demonstration of quantum teleportation using entangled photons with polarization quantum states

Subsequent quantum teleportation research has focused on increasing the discrimination of entangled states in Bell-state measurements. A group of collaborators in Italy and Great Britain developed a Bell-state measurement scheme that simultaneously distinguished between all four Bell states, allowing 100% teleportation success [42]. This was accomplished by using two photons instead of three photons for the teleportation process. Figure 9.20 shows the basic experimental arrangement. A pair of polarization entangled photons was produced by Type-II SPDC from a BBO crystal. A pair of beam displacing calcite prisms created an additional *path entanglement* between the separated paths now sharing a single photon. One path entangled photon in each leg was sent to Alice, the other to Bob. Using polarization rotators, a "Preparer" brought the photons sent to Alice into a state unknown to Alice and having the form $|\psi\rangle_1 = \alpha|H\rangle + \beta|V\rangle$, where $|\alpha|^2 + |\beta|^2 = 1$. The other entangled photons were sent directly to Bob for analysis. For example, the Preparer brought photon 1 into one of three possible linear polarized quantum states at rotation angles of 0°, 120°, or −120°. Using polarization analyzers, path length adjustment and detector counts, Alice was able to identify all four Bell states. Since photon 1 and photon 2 were entangled, Bob possessed the outcomes of the Bell-state measurements of Alice. Using this information, Bob could then transform photon 2 into the unknown state of photon 1. The quantum nature of the teleportation was verified by measuring the average success probability S and comparing it with the classical teleportation success probability S_c requirement: $S_c \leq 3/4$. It was found that $S = 0.831 \pm 0.009$, violating the classical S_c requirement by nine standard deviations.

In 2013 Lee and Jeong in Korea proposed an optical method of quantum teleportation that used hybrid qubits, combining two-photon pairs with entangled

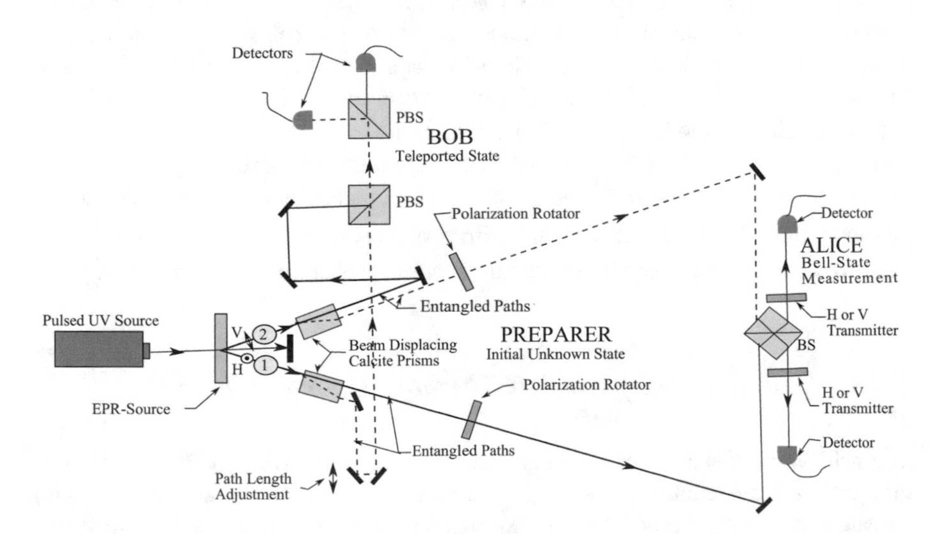

Fig. 9.20 Experimental setup to achieve 100% teleportation success using two photons instead of the usual three

coherent qubit states [43, 44]. This technique would allow completely deterministic Bell-state measurements to be made in the teleportation process, with a high success probability. The two-photon entangled pairs have the Bell states given by Eqs. 9.7 and 9.8. The coherent state qubit Bell states are given by

$$|\Psi^{\pm}\rangle = N^{\pm}[|\alpha\rangle|-\alpha\rangle \pm |-\alpha\rangle|\alpha\rangle] \tag{9.9}$$

$$|\psi^{\pm}\rangle_{12} = N^{\pm}[|\alpha\rangle|\alpha\rangle \pm |-\alpha\rangle|-\alpha\rangle], \tag{9.10}$$

where $|\alpha\rangle$ and $|-\alpha\rangle$ are coherent states with amplitudes $\pm\alpha$, and $N^{\pm}$ is a normalization factor. A deterministic Bell-state measurement can be made using coherent states, with a success probability $P_s = 1 - \exp(-2\alpha^2)$. However, the Bell-state measurement failure rate increases as α increases, and the non-orthogonality of $|\alpha\rangle$ and $|-\alpha\rangle$ produces unitary transformation problems for the receiver Bob. The nondeterministic two-photon Bell-state measurement only identifies two of the four Bell states with a success probability of 50%. The hybrid entanglement approach avoids these problems by combining single photon and coherent state qubits, where the hybrid qubits $|0_L\rangle$ and $|1_L\rangle$ are defined as

$$|0_L\rangle = |\frac{1}{\sqrt{2}}[|H\rangle|V\rangle + |V\rangle|H\rangle]|\alpha\rangle \tag{9.11}$$

$$|1_L\rangle = |\frac{1}{\sqrt{2}}[|H\rangle|V\rangle - |V\rangle|H\rangle]|-\alpha\rangle. \tag{9.12}$$

Near-deterministic quantum teleportation of an unknown hybrid qubit input state $|\psi\rangle = a|0_L\rangle + b|1_L\rangle$ can be realized when Alice makes two Bell-state measurements, one for the coherent state mode and another for the photon mode. Both modes are entangled with a hybrid channel $|\Psi_C\rangle$ having the form $|0_L\rangle_{Alice}|0_L\rangle_{Bob} + |1_L\rangle_{Alice}|1_L\rangle_{Bob}$. Bob can perform the Pauli X-matrix operation on the single-photon mode by a polarization rotator, and by a phase shift on the coherent state mode. The Pauli Z-matrix operation can be effected by a phase shift on the single-photon mode. These deterministic operations of Bob can restore the unknown hybrid qubit input state $|\Phi\rangle$ with a success probability $P_s = 1 - \frac{\exp(-2\alpha^2)}{2}$. This would result in a 99% teleportation success probability when $\alpha = 1.4$.

9.4.3 Quantum Teleportation Through Optical Fibers

The achievable distance for quantum teleportation from Alice to Bob at telecommunication wavelengths through a continuous optical fiber without quantum repeaters is usually limited by the exponential drop in light intensity and degradation of polarized photon quantum states with increasing fiber length. One method for quantum teleportation through a continuous fiber uses a type of quantum entanglement between photon pairs separated in time, referred to as *energy-time entanglement* or *time-bin entanglement* [45]. This type of photon entanglement remains robust for longer fiber transmission lengths, and the states are discriminated by their arrival times at suitable detectors. The time-bin qubits $|short\rangle$ and $|long\rangle$ are created by passing femtosecond laser pulses through an unbalanced pump interferometer. A reference phase difference φ is then introduced between the qubits, producing a photon output state of the form $|\Psi\rangle_{photon} = \frac{1}{\sqrt{2}}[|1,0\rangle - e^{i\varphi}|0,1\rangle]$, where the time separation $\Delta\tau = 1.2$ ns of the time-bin qubits significantly exceeds the photon coherence time [46]. A team of scientists in Switzerland and Denmark have demonstrated the teleportation of 1550 nm photons over a 2 km long telecommunication fiber between two laboratories 55 m apart [47]. The experimental setup is illustrated in Fig. 9.21. A Ti:Sapphire 710 nm laser light source produces 150 fs pulses at a 76 MHz rate. A polarization beamsplitter and half-wave plate send the reflected vertically polarized photons to a lithium triborate (LBO) nonlinear crystal. Polarization-entangled 1550 and 1310 nm photons are emitted from the crystal by SPDC, and coupled into an optical fiber. The transmitted horizontally polarized photons are sent through an unbalanced pump (Michelson) interferometer and then to a second LBO crystal which emits another pair of entangled 1550 and 1310 nm photons. Using a wavelength dependent angular-separation device, such as a prism or wavelength division multiplexer, the 1330 nm photons from the first LBO crystal are sent to Alice, while the 1550 nm

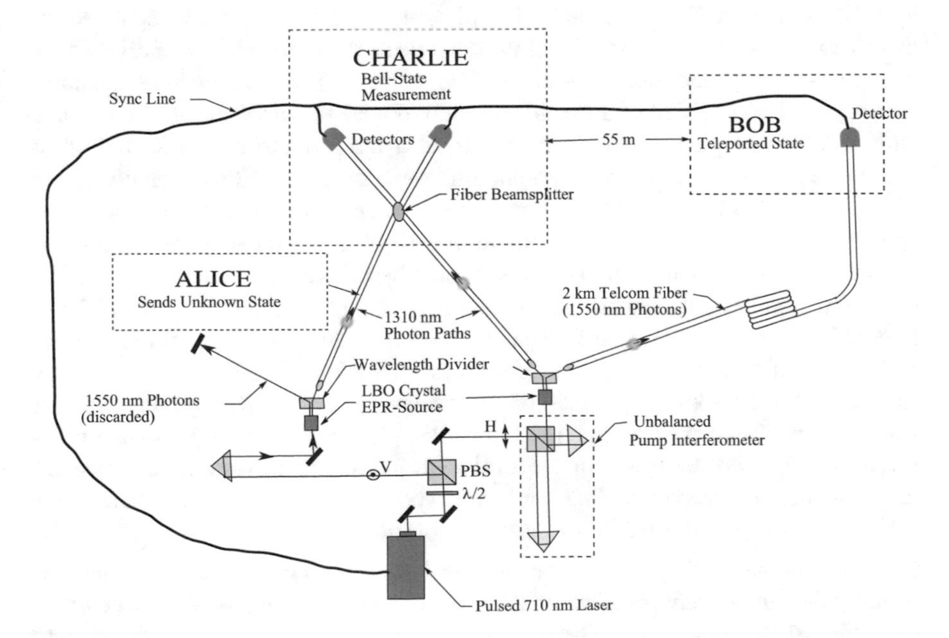

Fig. 9.21 Demonstration of teleportation of 1550 nm photons over a 2 km long telecommunication fiber between two laboratories 55 m apart

photons are discarded. Using an unbalanced fiber interferometer, Alice then creates the 1330 nm qubits into a state having the form $|\psi\rangle_{Alice} = \frac{1}{\sqrt{2}}\left[a_0|1,0\rangle_{Alice} + a_1 e^{i\alpha}|0,1\rangle_{Alice}\right]$, where $a_1 = \sqrt{1 - a_0^2}$, and the phase α is defined relative to the reference phase. This is the unknown state to be transported to Bob. The entangled photons from the second LBO crystal are also separated into two directions and coupled into optical fibers. The 1550 nm photons are sent to Bob over a 2 km length of optical fiber.

Another agent "Charlie" is introduced to receive the 1310 nm unknown state from Alice and perform a Bell-state measurement between this state and the 1310 nm photons from the second LBO crystal. Out of four possible results, Charlie selects only the entangled state $|\psi^-\rangle = \frac{1}{\sqrt{2}}\left[|1,0\rangle_{Alice}|0,1\rangle_{Charlie} - |0,1\rangle_{Alice}|1,0\rangle_{Charlie}\right]$ and sends this result to Bob via a classical channel. Bob's photon is then in the state $|\psi\rangle_{Bob} = \frac{1}{\sqrt{2}}\left[a_1 e^{i\alpha}|1,0\rangle_{Bob} - a_0|0,1\rangle_{Bob}\right]$. By applying the unitary Pauli *Y* matrix to this state, a bit flip and a phase flip are realized, and the unknown 1550 nm state of Alice is successfully transported to Bob over a 2 km length of optical fiber with calculated average teleportation fidelity of 81.2%.

Long-distance teleportation through optical fibers to a solid-state quantum memory has been realized using polarization entanglement between flying telecommunication 1338 nm idler photons and 883 nm signal photons stored in a rare earth ion doped crystal quantum memory [48]. Two beams of polarization entangled pairs of signal and idler photons were created by SPDC using a nonlinear waveguide and a neighboring nonlinear waveguide, both simultaneously pumped by a 532 nm laser. The signal and idler photons in each beam were separated by dichroic mirrors. The 883 nm signal photons were then recombined and sent to a quantum memory and stored as an excitation state. The 14 mm long quantum memory device consisted of a two neodymium doped yttrium orthosilicate crystals with a half-wave plate interface, operating at a temperature of 2.5 K. Using an atomic frequency comb, the 883 nm photons were stored for 50 ns at an efficiency of 5%. The 1338 nm idler photons were recombined and transmitted to a beam-splitter for a Bell-state measurement about 10 m from the quantum memory. Another 1338 nm polarized photon was created by difference-frequency generation in a third nonlinear waveguide simultaneously pumped by 532 nm and 883 nm laser light. This 1338 nm *input state* photon was indistinguishable from the original 1338 nm idler photon, and was the qubit state to be teleported. This input state was sent through a 12.4 km long coiled telecom fiber to a 50/50 beamsplitter, where it was combined with the original 1338 nm idler photon for a joint Bell-state measurement. The two detectors for the Bell-state measurement were high-efficiency tungsten-silicide superconducting nanowire types operating at 2.5 K, where coincidence detection indicated a successful projection of the $|-\rangle = \frac{1}{\sqrt{2}}[|H\rangle - |V\rangle]$ teleportation state. The emitted or retrieved signal photon from the quantum memory then underwent polarization analysis to determine the purity and fidelity of the retrieved teleported state. The measured average fidelity of the teleported state for the $|-\rangle$ input state was 89%. In another configuration the original 1338 nm idler photon state was sent through an additional 12.4 m long coiled fiber before the Bell-state measurement, producing teleportation of the state $|+\rangle = \frac{1}{\sqrt{2}}[|\mathrm{H}\rangle + |\mathrm{V}\rangle]$ over a combined 24.8 km distance.

9.4.4 Quantum Teleportation Through Free-Space

One of the ongoing areas of research is quantum teleportation in free-space. This teleportation has applications in long-range surface and satellite-to-ground communication. In 2010 a team of scientists in China demonstrated surface photon teleportation over a 16 km distance between Badaling and Huailai on mainland China [49]. A blue laser created a photon pair from a nonlinear BBO crystal, one spatial entangled and the other polarization entangled. Using a path entanglement scheme as described in Ref. [42], a full Bell-state measurement was performed, and average teleportation fidelity of 89% was achieved. The 16 km teleportation distance would be enough to penetrate the earth's atmosphere, but is not long enough to be received by any earth-orbiting satellites.

Since this experiment there have been a series of demonstrations of free-space quantum teleportation over increasing distances. In 2012 another team of Chinese scientists transported a photon quantum state from Alice to Bob, each situated on opposite shores of Qinghai Lake in China, with a free-space teleportation distance of 101.8 km [50]. Alice and Bob were not directly visible to each other and Fig. 9.22 illustrates the geographic distances. From a third location on the island Haixin Hill near the center of the lake, Charlie created an entangled pair photon state $|\psi^+\rangle$, which was the state to be teleported. The entangled photons were sent to both Alice and Bob, and received by accurately positioned tracking telescopes. Charlie was situated 51.2 km from Alice and 52.2 km from Bob, producing a 3 μs delay between the measurements of Alice and Bob. Thus the nonlocality condition was satisfied. The states to be teleported consisted of horizontal (H), vertical (V), right circular (R), and left circular (L) polarization states $|\pm\rangle = \frac{1}{\sqrt{2}}[|\mathrm{H}\rangle \pm |\mathrm{V}\rangle]$, $|\mathrm{R}\rangle = \frac{1}{\sqrt{2}}[|\mathrm{H}\rangle + i|\mathrm{V}\rangle]$, and $|\mathrm{L}\rangle = \frac{1}{\sqrt{2}}[|\mathrm{H}\rangle - i|\mathrm{V}\rangle]$. The receiving stations of Alice

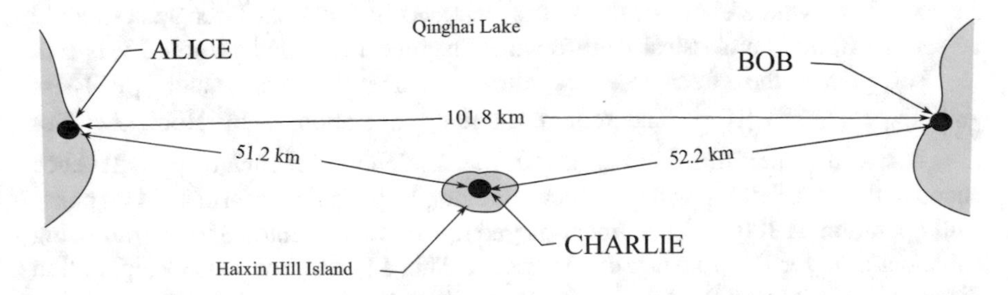

Fig. 9.22 Free-space teleportation over a 101.8 km distance between opposite shores of Qinghai Lake in China

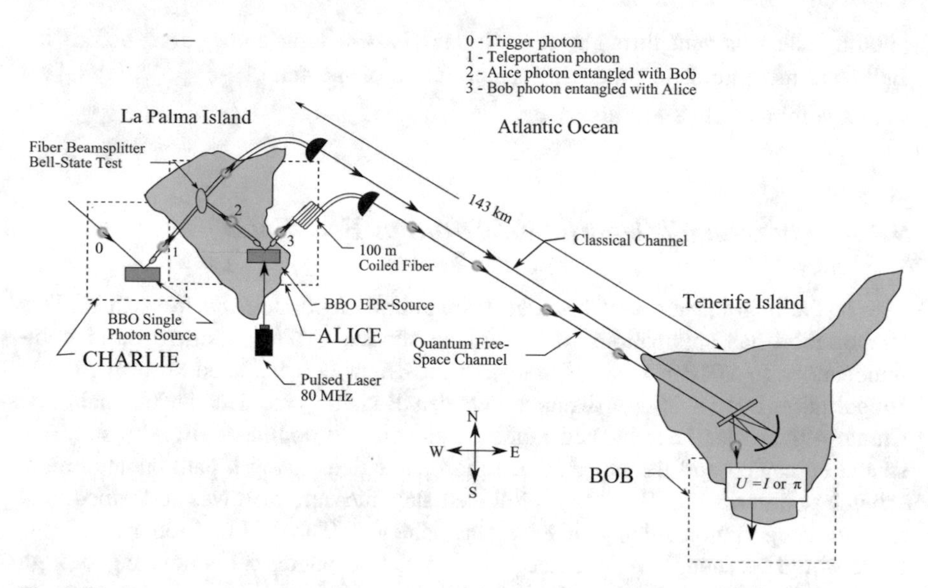

Fig. 9.23 Free-space teleportation over a distance of 143 km between La Palma and Tenerife in the Canary Islands

and Bob analyzed the incoming photons using a set of rapid (20 μs interval) and randomly selected polarization settings. Counting the detector coincidences over a specified measurement period, a CHSH parameter $S = 2.51 \pm 0.21$ was obtained, violating Bell's inequality by 2.4 deviations, with an average teleportation fidelity of 80%.

Another free-space teleportation experiment was reported in 2102, demonstrating teleportation of photon states over a distance of 143 km between La Palma and Tenerife in the Canary Islands [51]. The experiment is illustrated in Fig. 9.23. An input photon 0 generated by Charlie acted as a trigger for a BBO crystal to emit an unknown teleportation photon 1 in the state $|\psi\rangle_1 = \alpha|\mathrm{H}\rangle + \beta|\mathrm{V}\rangle$, where $|\alpha|^2 + |\beta|^2 = 1$. Thus photon 0 heralded the presence of photon 1. Another BBO crystal, driven by an 80 MHz pulsed laser, emitted a pair of entangled photons 2 and 3. Alice performed a Bell-state measurement on photons 1 and 2 to produce two states $|\Psi^{\pm}\rangle_{12}$, where either of the states, encoded in 1064 nm laser pulses, could be sent to Bob via a classical feed-forward channel. Entangled photon 3 was sent to Bob over the free-space quantum channel as the state $|\Psi^{-}\rangle_{23} = \frac{1}{\sqrt{2}}\left[|\mathrm{H}\rangle_2|\mathrm{V}\rangle_3 - |\mathrm{V}\rangle_2|\mathrm{H}\rangle_3\right]$. The required unitary operation U by Bob, $U = I$ or $U = \pi$, was then performed to extract $|\psi\rangle_3$ as a teleported replica of $|\psi\rangle_1$. If Alice encoded the state $|\Psi^{-}\rangle_{12}$ on the classical channel, Bob would perform a $U = \pi$ phase shift operation. A 100 m coiled fiber delayed the arrival of photon 3 to Bob allowing sufficient time for Bob to enact the operation. This 143 km free-space teleportation distance was close to the orbiting height of low-earth orbit navigation satellites, typically about 100–500 km above the earth's surface.

9.4.5 *Quantum Repeaters for Long-Range Optical Fiber Transmission*

Considering the exponential absorption loss through a standard telecom fiber at 1550 nm wavelength, about 95% of a photon stream would be transmitted through a 1 km long fiber, and a 100 km long fiber would have only about 2% transmission. Quantum repeaters are the preferred method to maintain high transmission over a great distance using optical fibers. One of the first quantum repeater protocols was described by Briegel and collaborators in 1998 [52]. This is a type of quantum network where entanglement swapping between a series of fiber optic links connected by paired quantum memories (repeater nodes), creates entanglement between two distant points **A** (Alice) and **B** (Bob). This would allow quantum teleportation of photon qubit states between these points. The separation of these repeater nodes depends on the useable absorption length of the fiber.

Figure 9.24 illustrates an in-line scheme for a long-distance optical fiber quantum repeater network with Bell-state measurement entanglement between quantum memories in each repeater node. For a total transmission length L, a distance between repeater nodes L_0, the number of node segments $N = L_0/L$. For example, with $L_0 = 10$ km and $N = 128$, a continental distance $L = 1290$ km could be

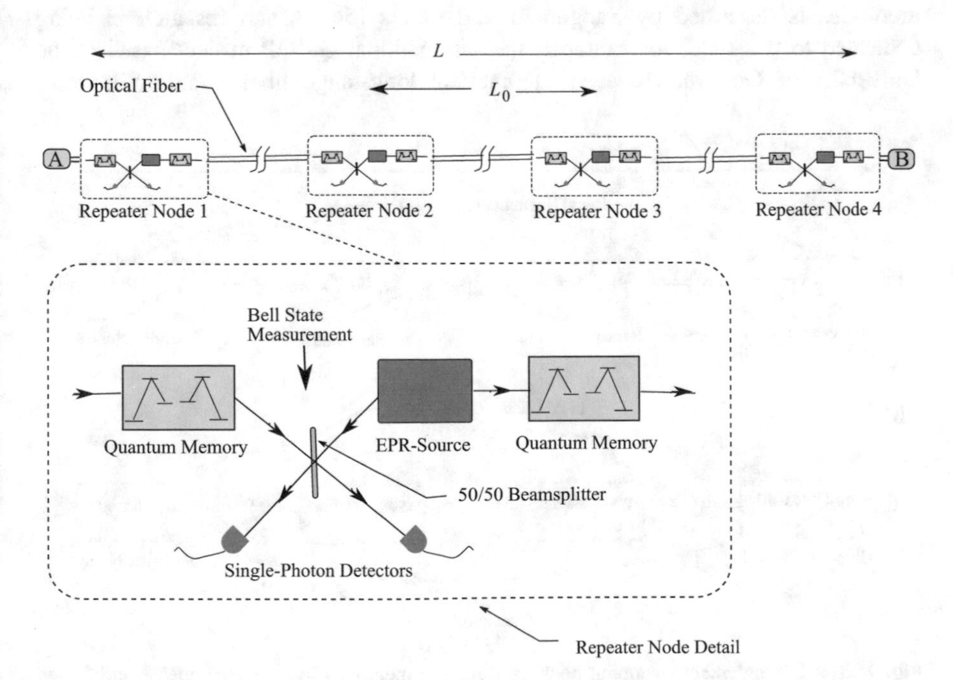

Fig. 9.24 Long-distance optical fiber quantum repeater network with Bell-state measurement entanglement between quantum memories in each repeater node. Only four repeater nodes are shown in this illustration

traversed, and with $L_0 = 10$ km and $N = 1024$, an intercontinental communication distance $L = 10240$ km could be possible.

The most useful type of quantum memory for these repeater protocols are those using atomic ensembles. Quantum memories of this type were described in Sect. 9.4.1, and first applied to repeater quantum communication in 2001 by an international team from Austria, China, and the United States [53]. This technique is referred to as the *DLCZ protocol* (after Duan, Lukin, Cirac, and Zoller). Figure 9.25a, b illustrates the basic process of entanglement swapping between adjacent quantum memories that can provide direct teleportation of a photon quantum state between points **A** and **B** over a very large distance. To illustrate the process, a network consisting of four repeater nodes is shown.

Using the entanglement between the quantum memories in each repeater node, the process of entanglement swapping continues until entanglement occurs between the quantum memories in the first and last repeater nodes. Using a teleportation scheme requiring an additional classical channel, the quantum state $|\psi\rangle_A$ at position **A** is transferred to $|\psi\rangle_B$ at long-range position **B** with good fidelity.

The technical requirements for the implementation of a practical fiber repeater network are difficult. For example, the quantum memories must maintain the coherence of the stored photon quantum state and store the state long enough for entanglement swapping to occur with the next adjacent node and over the entire network. A review of current quantum repeater techniques using atomic ensemble memories is described by Sangouard and others [54]. Much research is being dedicated to this field, for example, the international QuReP project based at the University of Geneva. However, a practical long-range fiber quantum repeater

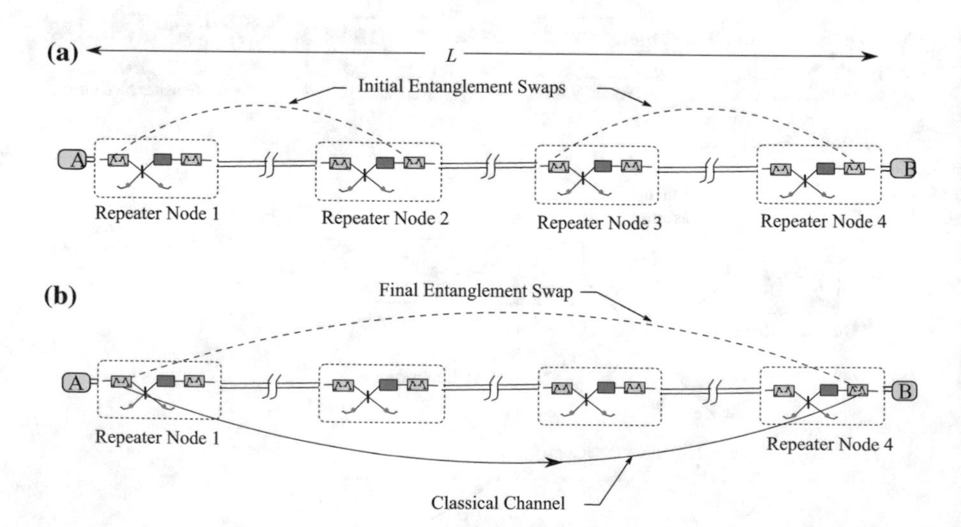

Fig. 9.25 **a** Entanglement swapping between quantum memories in repeater nodes 1 and 2, and between repeater nodes 3 and 4. **b** Final entanglement swapping between quantum memories in repeater nodes 1 and 4. Then quantum state $|\Phi\rangle_A$ at A is teleported to quantum state $|\Phi\rangle_B$ at B with the use of a classical channel

network has yet to be demonstrated. The use of quantum repeaters for quantum cryptography, such as QKD, is a complementary activity in quantum communication research [55].

9.4.6 *Photonic Quantum Networks*

The use of quantum repeater nodes for long-range one-way communication through optical fibers is a type of photonic quantum network for sending at one end and receiving at the other end. More generally, the repeater node should be reversible between the atomic and photonic states to allow two-way information transfer. One of the proposed methods of realizing an effective reversible quantum node uses an emitting atom contained in a high-Q optical cavity [56]. An external laser pulse shifts the atomic state in this cavity (node **A**) to the cavity optical state, producing a photon state. This technique is often called "cavity QED". The photon leaks out of the cavity wall and enters a second cavity (node **B**), where its quantum state is converted to the atomic state of another contained atom. Another laser shifts the atomic state to an optical quantum state which leaks through the cavity wall for transfer to another cavity node. Usually photon polarization and atomic ground-state spin are the encoded quantum states. This coherent mapping between the atom and photon quantum states can produce efficient and reversible transfer between multiple nodes with high fidelity. Two-way entanglement and photonic information exchange between distant nodes might then be possible.

In 2012 an experiment was performed at the Max Planck Institute for Quantum Optics to demonstrate a simple quantum network using cavity light-matter interactions [57]. A single neutral rubidium (^{87}Rb) atom was trapped within an optical reflective cavity and used to store and retrieve the quantum state polarization of a single photon. Then the cavity atomic state was successfully transferred from one node to another by transmitting a single photon through a fiber optic channel. Lastly, it was verified that a choice of entangled single atom quantum states were shared between the nodes, with high fidelity for a time period up to 100 μs. Figure 9.26 illustrates the experimental setup. The two nodes were connected in independent laboratories by a 60 m fiber optic channel. In node **A**, initial Zeeman states of the ^{87}Rb atom were prepared, and a polarized control laser generated a photon from these atom states. This transmitted photon was subsequently stored in node **B**, with a reversal of the storage process for photon emission. The resultant quantum network was both symmetric and reversible.

Several other variations of this cavity approach have been realized by independent investigators using atomic ensembles, single ions, and the cavity containment of two ions or two neutral atoms. In the case using two atoms it was found that the photon production was not significantly increased, but there was a speed increase in the conversion of the atomic quantum state to the photon quantum state in the cavity.

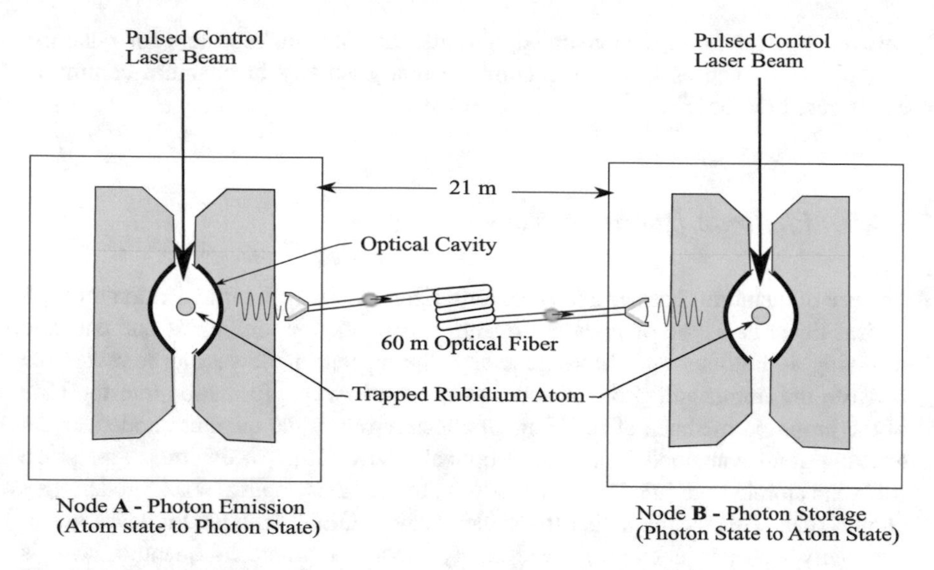

Fig. 9.26 Simple quantum network using cavity light-matter interactions. A single-photon fiber optic connection is made between two cavities over a 60 m distance to transfer atomic states

9.4.7 A Quantum Internet

A photonic quantum network could be constructed from a two- or three-dimensional architecture of quantum storage and processing nodes connected to each other by fiber optic channels. To function as a *quantum internet*, the photonic quantum information flow between all quantum nodes would need to be rapid, switchable, storable, secured by encryption, and reversible [58]. In addition, the network should be expandable or scalable, with no degradation of information transfer. Figure 9.27 illustrates an envisioned two-dimensional quantum network that could function as a quantum internet. Practical functioning networks of this type would require further technical breakthroughs. A recent development has produced heralded single-photon entanglement between separated crystals containing doped rare-earth ions [59]. Use of solid-state repeater nodes to convert collective excitation to photon modes might facilitate the development of a workable quantum internet. However, Los Alamos Scientific Lab has reported a type of "quantum internet" that has been in operation for several years [60]. This encrypted quantum communication network uses a "hub and spoke" architecture, as shown in Fig. 9.28. A central quantum communications "hub" receiver (Trent) is connected by 50 m long fiber optic "spoke" channels to three client transmitters (Alice, Bob, and Charlie). The information flow through the channels used polarization qubits, and each client uses random number generators and computers to establish QKD with Trent using the BB84 protocol. The receiver Trent performs the polarization analysis using four super-cooled single photon detectors. Once Trent has the quantum key, he acts as a "trusted authority" and sends information securely to

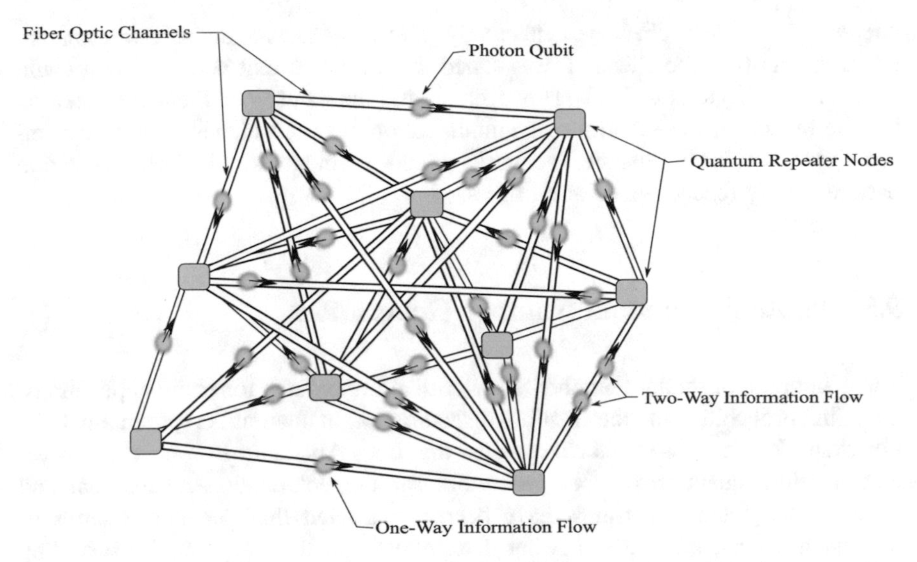

Fig. 9.27 Idealized quantum network functioning as a quantum internet. Only nearest neighbor connections are shown

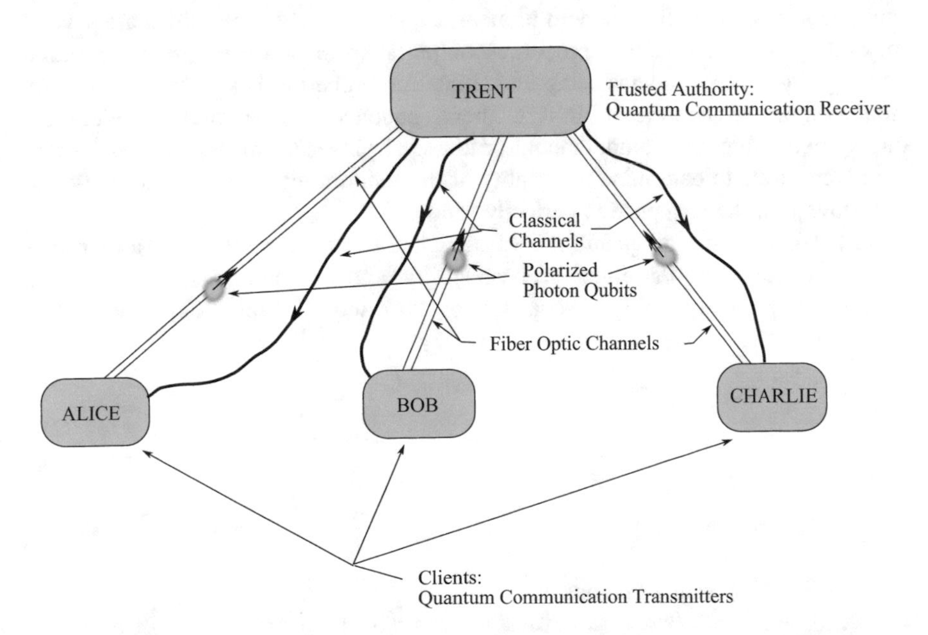

Fig. 9.28 Encrypted quantum communication network using a "hub and spoke" architecture. In this type of "quantum internet", a central quantum communications "hub" receiver (Trent) is connected by 50 m long fiber optic "spoke" channels to three client transmitters (Alice, Bob, and Charlie)

clients by classical channels, effectively allowing secure indirect information exchange between the clients. It was stated that in the current version, Trent could accommodate up to 100 clients. However, should the security of Trent in the central hub be breached, the security of communication between clients would be compromised. An ideal quantum network would maintain secure communication between every repeater node at all times.

9.5 Photonic Random Number Generation

The generation of random numbers is of critical importance for many applications, including probability and the creation of secure keys in quantum cryptography [61]. For example, the BB84 protocol requires that both Alice and Bob have their own local random quantum number generators for the polarization modes sent and received. Computer algorithms have been constructed that generate a series of random numbers, but are somewhat deterministic in that they require a starting number, or "seed", to initiate the computation and are usually *biased*, where the number of "0" and "1" bits are not perfectly balanced. Secure quantum cryptography requires a bias-free random number sequence. Fortunately, there are several methods using the quantum properties of particles to produce perfectly random number sequences. A device using these quantum properties is known as a *quantum random number generator* (QRNG). These sequences can produce a randomness quality exceeding any computational or classical physical process. Although there are several tests to determine the quality of the random bit sequence, these tests do not *prove* that the sequence is perfectly random.

In a 1994 review of quantum cryptography, Rarity, Owens, and Tapster suggested the use of a 50/50 quantum beamsplitter and a stream of single-photons to generate a perfectly random sequence of "0" and "1" bits [62]. Figure 9.29

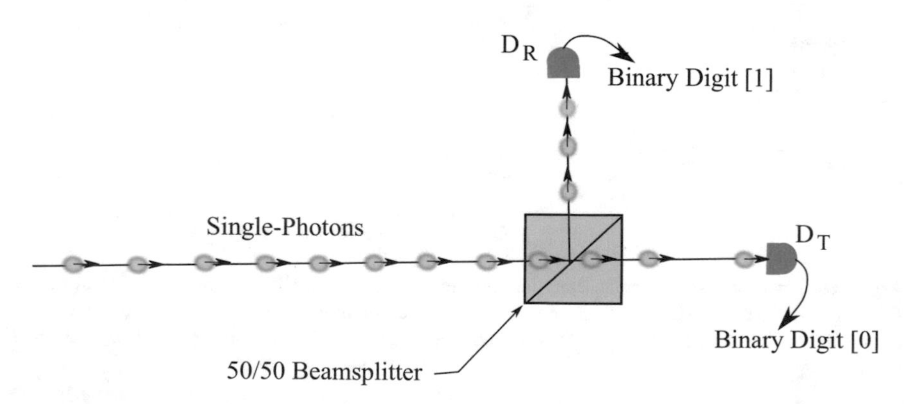

Fig. 9.29 Arrangement of a 50/50 quantum beamsplitter and a stream of single-photons to generate a perfectly random sequence of "0" and "1" bits

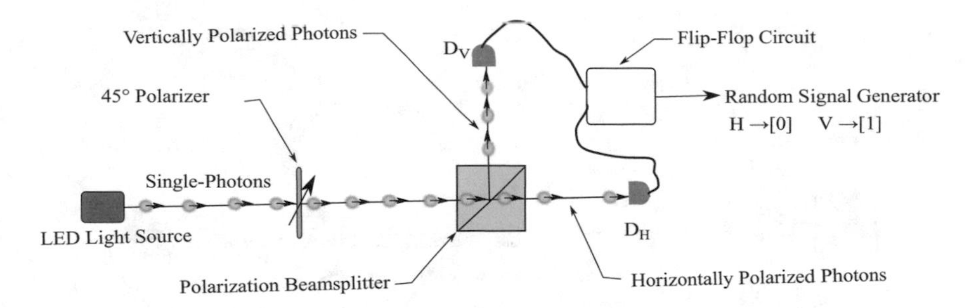

Fig. 9.30 Experimental setup for random number generation using a 50/50 polarizing beamsplitter and 45° incident polarized light

illustrates the concept. A single-photon light source sends a series of single-photons to a perfect 50/50 beamsplitter and the transmitted and reflected photon are registered by two single-photon detectors. A sequence of bits is then generated by assigning a "0" bit to photons counted by the D_T detector and a "1" bit to photons counted by the D_R detector. Since the probability of transmission or reflection is exactly 50%, this quantum process has no predetermination effect or bias and the bit sequence is perfectly random.

This technique was experimentally realized at the Institute for Experimental Physics in Vienna in 2000 by the construction of several beamsplitting systems [63]. The experiment was run in two setup modes. In one mode unpolarized incoming single-photons were incident on a 50/50 beamsplitter (BS). These single-photons had a 50% probability of being transmitted or reflected. In another mode incoming single-photons, linearly polarized at 45°, had a 50% chance of being transmitted with horizontal (H) polarization or reflected with vertical polarization (V) at a 50/50 polarizing beamsplitter (PBS). Figure 9.30 shows the experimental setup using the polarizing beamsplitter. Single-photons were emitted from a red LED source having a very short coherence length and slight rotation of the linear polarizer allowed setting an exact 50/50 split between the horizontal and vertical polarizations at the beamsplitter. Photomultiplier detectors D_T and D_R registered the detection in the two output paths. A binary random signal was then generated using electronic flip-flop circuitry that switched between high and low states, dependent on which detector registered the last detection. A continuous sequence of thousands of bits was periodically sampled at a 1 MHz sampling frequency and then cyclically sent as a string of random numbers to a personal computer.

About the same time, another QRNG was demonstrated at the University of Geneva, based on random arrival times of photons at one single-photon detector [64]. Figure 9.31 illustrates the experimental arrangement. An 830 nm LED was pulsed at 1 MHz to emit photons incident on a 2 m long small core single-mode fiber cable. These photons were inputted to two adjacent larger core multimode fibers, with an approximate 50% probability of entering one fiber (path A) or the other (path B). Path B was lengthened to provide a 60 ns second delay at the

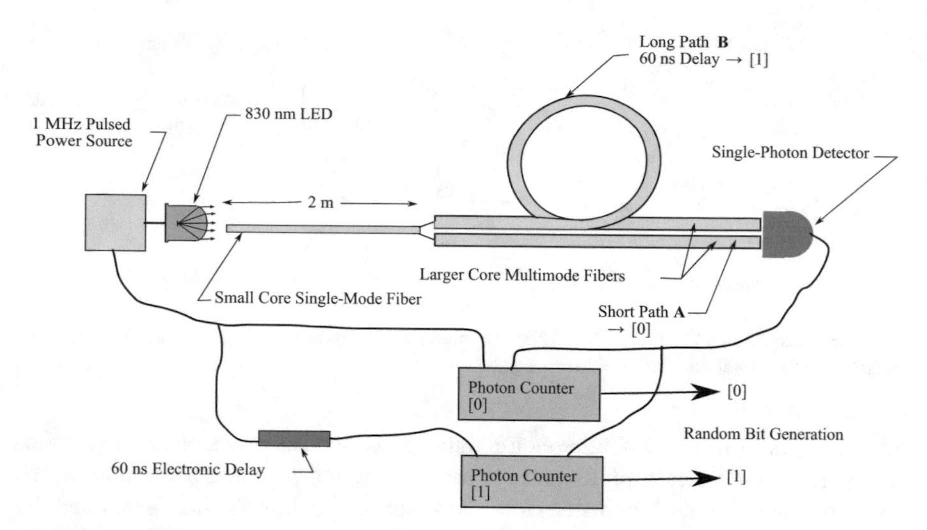

Fig. 9.31 A QRNG using a single detector based on random arrival times of photons at the detector (Adapted from Ref. [64], courtesy of Taylor & Francis)

coupled output to a single silicon avalanche photodiode detector. The bit "0" was assigned to path A, and the bit "1" was assigned to path B. Photons transmitted by the multimode fiber emerged as a weak beam having a single-mode wavelength, and were indistinguishable. A series of random digits could then be generated and distinguished by the photon arrival times at the detector. Since a perfect 50% probability of a photon from the single-mode fiber entering either of the multimode fibers does not occur, the bit distribution was unbiased using several available mathematical computations. This type of QRNG was hardly affected by detector sensitivity variations or external perturbations.

Another quantum random number generator based on the arrival times of photons was developed at the Chinese Academy of Sciences, Beijing [65]. This technique used the random time interval distribution of photons emitted by a strongly attenuated pulsed laser. The use of a single detector contributed to the system robustness for mechanical and temperature disturbances.

A third type of QRNG produced random numbers by sequential measurements of the position coordinates of photons arriving at an imaging detector [66]. The random detector positions were a consequence of the quantum randomness inherent in the photon emission process. For an operational prototype, a random number generation of 8 Mb/s was reported.

A fiber optic beamsplitting QRNG was demonstrated in 2009 at the Pohang University of Science and Technology in Korea using a pair of entangled photons and four detectors [67]. This methodology used both quantum state projection and superposed quantum states, and more easily produced pure quantum states. Figure 9.32 illustrates the experimental setup. Entangled 816 mm photon pairs were produced by SPDC in a Type-II BBO crystal pumped by a 408 nm diode laser operating at 40 mW. The entangled photon paths were coupled to optical fibers and

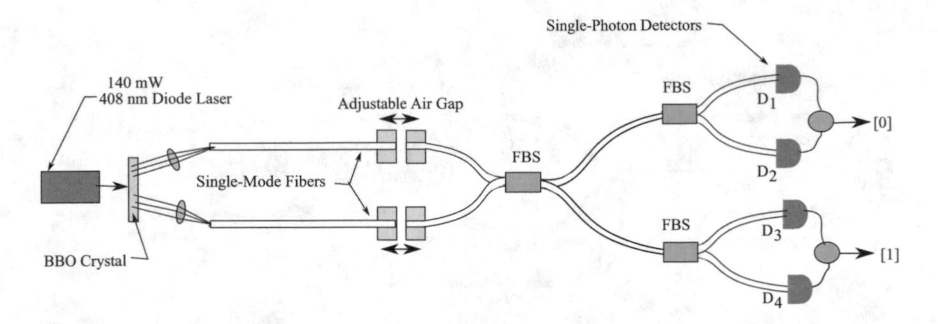

Fig. 9.32 A fiber optic beamsplitting QRNG using a pair of entangled photons and four detectors. Using both quantum state projection and superposed quantum states, pure quantum states are more easily produced (Adapted from Ref. [67] with permission from the Optical Society of America)

fed into a 50/50 fiber optic beamsplitter FBS. An air-gap delay of several micrometers was varied in each path to enable simultaneous arrival of each photon at the FBS such that the photons were indistinguishable. The Hong-Ou-Mandel effect predicted that both photons would exit the FBS in one output path or the other, with the choice being completely random. Each output path was further split by fiber optic beamsplitters and coupled to single-photon detectors D_1, D_2, D_3, and D_4. Coincidence counts between detectors D_1 and D_3 verified the simultaneous arrival of the entangled photons at the FBS. Random coincidence counts between D_1 and D_2 were then assigned a bit value [0], and random coincidence counts between D_3 and D_4 were then assigned a bit value [1]. A reference timing frequency allowed true [0] and [1] bits to be identified and recorded, while error bits could be discarded.

Another basis for random number generation utilizes the random quantum electromagnetic field variations in the quantum vacuum, often referred to as *quantum noise*. Researchers at the Max Planck Institute for the Science of Light devised an experiment to produce a QRNG from the lowest energy quantum vacuum state [68]. Laser light was sent through a beamsplitter in a vacuum where the lowest energy vacuum state interacted with the beams to produce a quantum noise component in each beam. The beams were incident on two detectors, where the difference in the electronic signals contained only electrical and quantum noise. Using a calculation to separate out the electrical noise, the remaining quantum noise was converted into a string of random bits at a rate of 6.5 Mb/s.

Another QRNG using the quantum vacuum was demonstrated in 2011 in Spain using a pulsed laser diode light source and an unbalanced fiber optic Mach-Zehnder interferometer [69]. Figure 9.33 illustrates the experimental arrangement. A 100 MHz pulsed laser diode at 850 nm was periodically brought below and above the threshold level. This produced spontaneous emission of coherent photons with random phases and subsequent amplification of these random phases, where the photons originating from the amplified quantum vacuum in the laser diode cavity had random phases Φ^{1}, Φ^{11}, Φ^{111},.... The photons entered a polarization

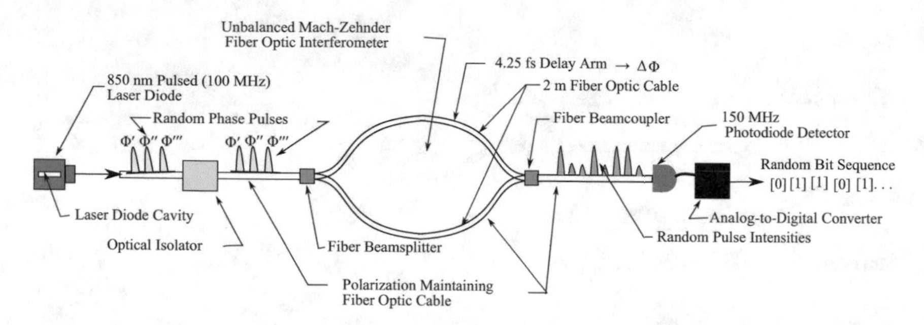

Fig. 9.33 A QRNG using the quantum vacuum. Photons originating from the amplified quantum vacuum in the laser diode cavity have random phases which are converted to a sequence of random bits

maintaining fiber with an optical isolator to suppress back reflections to the cavity. The 2 m long fiber optic arms of the Mach-Zehnder interferometer were unbalanced by introducing a 4.25 fs time delay in one of the arms through a small temperature induced refractive index change, producing a small phase change $\Delta\Phi$. The random phases were then converted by the interferometer to random energy pulses, recorded by a 100 MHz photodiode, and converted into random bits. A stream of random bits was generated at a rate of 1.1 Gb/s.

Another QRNG system using quantum vacuum states was developed by the Australian National University Quantum Optics Group [70]. Random bits were generated by measuring the vacuum field in the radio-frequency sidebands of a 1550 nm single-mode laser. Quantum random bit generation at a rate of 2 Gb/s was demonstrated and verified by standard randomization tests.

After these proof-of-principle experiments confirmed the application of quantum noise as a viable QRNG, efforts continued to increase the QRNG rate to the higher gigabit ranges required for highest security quantum encryption. In 2012 scientists at the University of Toronto used the quantum phase fluctuations of a low-intensity laser operating near the threshold level, where random phase variations of spontaneous photon emission predominated over the constant phase photons of stimulated emission [71]. The phase fluctuating photons were produced by a 1550 nm CW diode laser and transmitted through a polarization maintaining fiber to a temperature controlled planar lightwave circuit and an unbalanced fiber optic Mach-Zehnder interferometer system having a 500 ps delay in one arm. The resulting phase fluctuations were converted to intensity variations and recorded by 5 GHz InGaAs photodetector. A 6 Gb/s rate random bit stream was then produced by sampling and digitizing using an 8-bit analog-to-digital converter.

In 2014 the QRNG rate was further increased to 43 Gb/s by increasing both the laser diode bandwidth and the laser modulation frequency [72]. Shortly thereafter, the QRNG rate was again increased to a record 68 Gb/s by researchers at the Hefei National University for Physical Sciences in China [73]. They operated a laser near the threshold level to produce random photon phase fluctuations that were converted to intensity changes using an interferometer with active feedback control and a standard high-speed photodetector. These intensity variations were then digitized, and effects of classical noise were removed to produce a series of quantum generated random bits at a 68 Gb/s rate.

9.6 Quantum Computing Using Photons

9.6.1 Quantum Gates Using Linear Optics

As the development of classical electronic computers progressed, it was realized that many computational processes might be achieved faster and more efficiently by the use of optical quantum computers. The photon is a preferred information carrier in quantum computation because of its speed and definable states. In 1997, at the California Institute of Technology, several types of optical gates were proposed that used a single photon and passive linear optical components—the types of beamsplitters and phase shifters described in this chapter [74].

Although development of a complete quantum computer was not attempted, several types of one-photon and two-photon quantum logic gates were described. We use the conventional quantum bit notation $|0\rangle$ and $|1\rangle$, although for horizontally and vertically polarized photons $|0\rangle \rightarrow |\mathrm{H}\rangle$ and $|1\rangle \rightarrow |\mathrm{V}\rangle|$. One type of single qubit single-photon logic gate is the *Hadamard gate*. The *Hadamard transform H* for a single qubit is given by

$$H = \frac{1}{\sqrt{2}}\begin{pmatrix} 1 & 1 \\ 1 & -1 \end{pmatrix}, \tag{9.13}$$

This transform can be applied to a photon qubit having input states $|0\rangle$ and $|1\rangle$, producing output states such that

$$\underset{\text{Output}}{\frac{1}{\sqrt{2}}\begin{pmatrix} |0\rangle + |1\rangle \\ |0\rangle - |1\rangle \end{pmatrix}} = \frac{1}{\sqrt{2}}\begin{pmatrix} 1 & 1 \\ 1 & -1 \end{pmatrix}\underset{\text{Input}}{\begin{pmatrix} |0\rangle \\ |1\rangle \end{pmatrix}} \tag{9.14}$$

The state $|0\rangle$ can be considered as the vacuum state. A Hadamard gate can be implemented optically using a 50/50 beamsplitter and two $-\pi/2$ phase shifters, as shown in Fig. 9.34. There is a $\pi/2$ phase shift of the reflected light relative to the transmitted light at the beamsplitter. The Hadamard gate produces the following computations on a photon qubit, yielding $|0\rangle \rightarrow \frac{1}{\sqrt{2}}[|0\rangle + |1\rangle]$ and $|1\rangle \rightarrow \frac{1}{\sqrt{2}}[|0\rangle - |1\rangle]$. The output states would be determined by projective measurements.

In 2001 Knill, Laflamme and Milburn at Los Alamos National Laboratory showed that small quantum computational networks could be built using

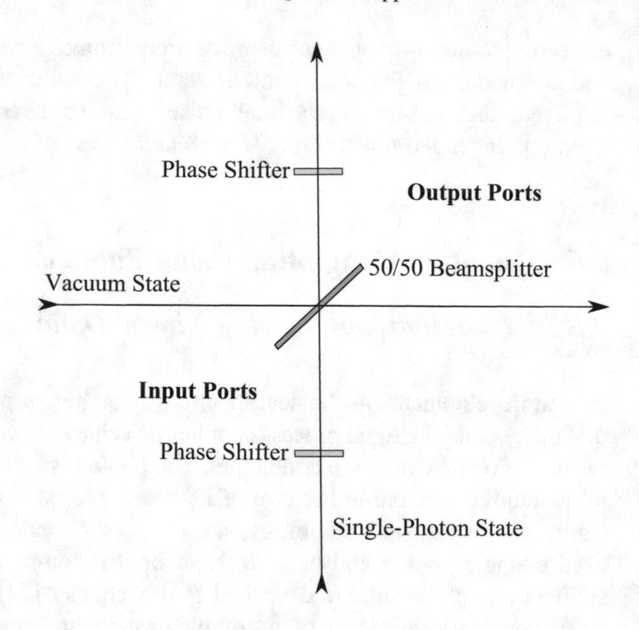

Fig. 9.34 Optical implementation of a Hadamard gate by using a 50/50 beamsplitter and two $-\pi/2$ phase shifters

single-photon light sources, single-photons, linear optical components, and single-photon detectors. This approach is often called *KLM theory*, or *linear optics quantum computation* (LOQC) [75]. Two examples of two-photon gates are the controlled NOT gate (CNOT gate) and the controlled-phase CSIGN gate (also called the CZ gate or CPHASE gate). The CNOT unitary matrix applied to a two-qubit input basis gives

$$\underbrace{\begin{pmatrix} |00\rangle \\ |01\rangle \\ |11\rangle \\ |10\rangle \end{pmatrix}}_{\text{Output}} = \underbrace{\begin{pmatrix} 1 & 0 & 0 & 0 \\ 0 & 1 & 0 & 0 \\ 0 & 0 & 0 & 1 \\ 0 & 0 & 1 & 0 \end{pmatrix}}_{\text{CNOT unitary matrix}} \underbrace{\begin{pmatrix} |00\rangle \\ |01\rangle \\ |10\rangle \\ |11\rangle \end{pmatrix}}_{\text{Input}}. \tag{9.15}$$

The first photon qubit of each two-qubit input is called the *control* qubit and the second cubit is called the *target* qubit. When the input control qubits are set at $|0\rangle$, the target output qubits are unchanged. When the input control qubits are set to $|1\rangle$, the corresponding output qubits are flipped, yielding the two-qubit computations $|00\rangle \to |00\rangle$, $|01\rangle \to |01\rangle$, $|11\rangle \to |10\rangle$, and $|10\rangle \to |11\rangle$. Also, the CNOT gate is *reversible*, where the input can be reconstructed or retrieved from the output. The CNOT gate, using this reversibility property, can also entangle and disentangle two-qubit states. An optical replication of the CNOT gate can be formed by two intersecting non-interacting single-photon paths, where two of the four two-qubit input modes contain "which-path" information and the other two modes contain $|\mathrm{H}\rangle$

and $|V\rangle$ polarization information. The CNOT operation occurs through a controlled polarization rotation in one of the output paths.

The controlled-phase CSIGN gate uses a two-qubit input basis to give

$$\underbrace{\begin{pmatrix} |00\rangle \\ |01\rangle \\ |10\rangle \\ -|11\rangle \end{pmatrix}}_{\text{Output}} = \underbrace{\begin{pmatrix} 1 & 0 & 0 & 0 \\ 0 & 1 & 0 & 0 \\ 0 & 0 & 1 & 0 \\ 0 & 0 & 0 & -1 \end{pmatrix}}_{\text{CSIGN unitary matrix}} \underbrace{\begin{pmatrix} |00\rangle \\ |01\rangle \\ |10\rangle \\ |11\rangle \end{pmatrix}}_{\text{Input}}, \tag{9.16}$$

generating the two-qubit computations $|00\rangle \rightarrow |00\rangle$, $|01\rangle \rightarrow |01\rangle$, $|10\rangle \rightarrow |10\rangle$, and $|11\rangle \rightarrow -|11\rangle$. Two-photon CSIGN gates would require a phase shift between two photons that normally don't interact. One method to obtain such a phase shift utilizes a nonlinear crystal to produce a cross-Kerr effect. However, it is possible to construct nonlinear sign (NS) gates that use only linear optical components, beamsplitters and two single-photon detectors for projective measurements. Especially important is the Hong-Ou-Mandel effect on the beamsplitters, where two spatially-separated input photons are "bunched" together along a common output path. By combining two beamsplitters with two NS gates, a controlled-phase CSIGN gate can be constructed. Various configurations of NS gates have been reviewed by Kok et al. [76].

The CNOT, CSIGN, and NS gates are all non-deterministic when used with projective measurements, having a probability of success $P \leq 1$. For example, the NS gate has a maximum success probability $P = 0.5$. To increase the success probability in a quantum computation to a near-deterministic level, one can scale the system by increasing the number of gates. However, this would soon lead to an impractically large number of gates. Another approach is to create deterministic entangled states within the gates and then apply quantum teleportation of modes and quantum error correction to increase the success probability. For example, using a number of teleportation steps a near-deterministic CSIGN computation could be obtained.

In 2011 a team of researchers in Japan and the United Kingdom demonstrated an experimental CNOT gate using polarization-encoded photon qubits [77]. Three types of polarization beamsplitter were used in a quantum optical circuit, each having different transmission and reflection for horizontal and vertically polarized light. Two pairs of input entangled photons were produced by Type-I SPDC by passing a 390 nm laser pulse twice through a beta-barium borate crystal. The first pair functioned as control (C) and target (T) qubits, and the second pair (A1, A2) as auxiliary photons. The CNOT gate was tested by polarization analysis of coincidence measurements at four output detectors DC, DT, DA1, and DA2.

Another type of quantum gate is the three-qubit input *Toffoli gate*, devised by Tommaso Toffoli in 1980, and often referred to as a controlled CNOT gate (CCNOT gate) [78]. The three-qubit input is mapped to a three-qubit output, given by

$$\begin{pmatrix} |000\rangle \\ |001\rangle \\ |010\rangle \\ |011\rangle \\ |100\rangle \\ |101\rangle \\ |111\rangle \\ |110\rangle \end{pmatrix} = \begin{pmatrix} 1 & 0 & 0 & 0 & 0 & 0 & 0 & 0 \\ 0 & 1 & 0 & 0 & 0 & 0 & 0 & 0 \\ 0 & 0 & 1 & 0 & 0 & 0 & 0 & 0 \\ 0 & 0 & 0 & 1 & 0 & 0 & 0 & 0 \\ 0 & 0 & 0 & 0 & 1 & 0 & 0 & 0 \\ 0 & 0 & 0 & 0 & 0 & 1 & 0 & 0 \\ 0 & 0 & 0 & 0 & 0 & 0 & 0 & 1 \\ 0 & 0 & 0 & 0 & 0 & 0 & 1 & 0 \end{pmatrix} \begin{pmatrix} |000\rangle \\ |001\rangle \\ |010\rangle \\ |011\rangle \\ |100\rangle \\ |101\rangle \\ |110\rangle \\ |111\rangle \end{pmatrix}. \quad (9.17)$$

Output ⟵ Toffoli (CCNOT) unitary matrix ⟶ Input

When the first two control qubits of each three-cubit input are set to $|1\rangle|1\rangle$, the third target output qubits are flipped, such that $|110\rangle \rightarrow |111\rangle$ and $|111\rangle \rightarrow |110\rangle$. Otherwise, the three-cubit outputs remain unchanged. The quantum Toffoli gate is universal (performs any Boolean function computation) *and* reversible, and can replace a sequence of two-bit gates, such as six CNOT gates. The three-qubit Tiffoli gate is an important element in any general quantum computer architecture where arbitrary quantum networks are required. Also, the Toffoli gate reversibility allows, in principle, construction of computers that consume little or no energy, since there is no loss of information that is dissipated as heat. Toffoli gates have been experimentally realized using trapped ion computation and also with photons and linear optics [79, 80]. It has been shown that the Hadamard and Tiffoli gates constitute a set of universal gates, whereby any other gate required for a quantum computation can be derived from these gates with extreme accuracy.

9.6.2 Boson Sampling

Several types of specialized photonic quantum computers have been described that provide a solution to the *boson sampling problem*, where a group of photons (bosons) are inputted into a network of waveguides and beamsplitters, and the output distribution of these photons is determined. Figure 9.35 illustrates a hypothetical boson sampler network, consisting of 14 fiber optic input/output modes m and 37 fiber optic lossless 50/50 beamsplitter nodes. A series of single-photon light sources launches n indistinguishable photons into specified input ports. A series of single-photon detectors sample the outputs from the boson sampling network. Through a series of repeated quantum calculations for a specific input, the probability of specific photon distributions occurring at the outputs can be sampled. The sampler network can be considered as a unitary matrix U performing a linear transformation that produces the output. In this quantum calculation, two indistinguishable photons arriving simultaneously at a 50/50 beamsplitter node from

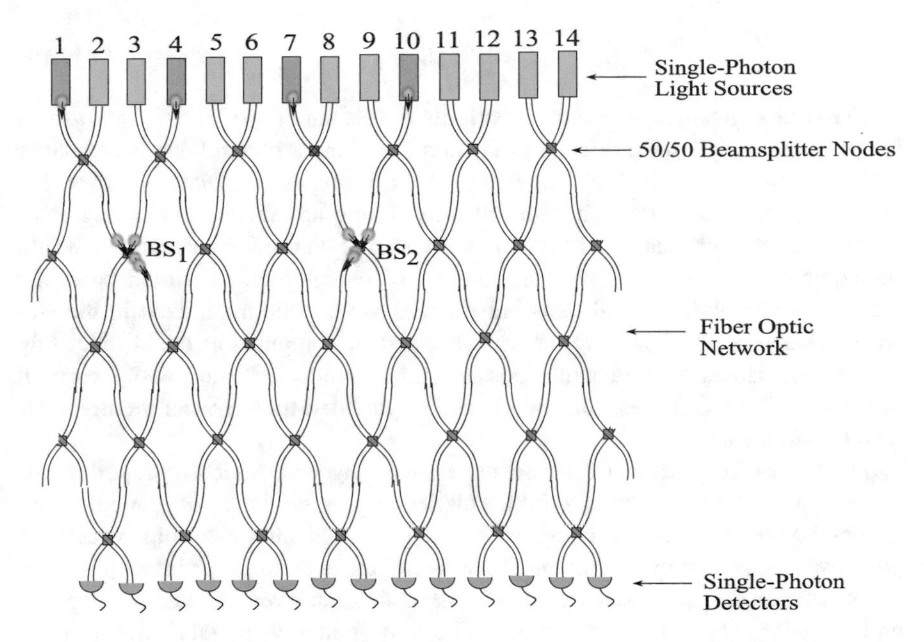

Fig. 9.35 Architecture for a quantum boson sampler

opposite directions exit the node in a common direction due to the Hong-Ou-Mandel effect. As shown in Fig. 9.35, this direction can be to the right (BS_1) or to the left (BS_2) with a 50% probability. For a general beamsplitter, the probability P that two incident indistinguishable photons will exit in separate paths or a single path is often calculated using a quantity called the *permanent* or *Per*. The permanent describes the wave function of indistinguishable photons, and the path probability P is given by

$$P = \left| Per\begin{pmatrix} T & iR \\ iR & T \end{pmatrix} \right|^2 = \left| T^2 - R^2 \right|^2, \tag{9.18}$$

where $\begin{pmatrix} T & iR \\ iR & T \end{pmatrix}$ is the beamsplitter matrix, and T and R are the transmission and reflection percentages at the beamsplitter surface. For a 50/50 beamsplitter the probability P that the photons will exit in two paths is zero, affirming the Hong-Ou-Mandel effect. More generally, for a boson sampling network described by unitary matrix U with m possible input and output modes ($m \geq n$), the probability P that indistinguishable single-photons inputted into n input modes are outputted into different modes is specified by the permanent *Per* of the unitary $n \times n$ sub-matrix U'

$$P = [Per(U')]^2 \tag{9.19}$$

Classical computer probability distribution calculations of the U' permanent become increasingly more difficult as the number of input photons and beamsplitter nodes increases. A classical computer could probably calculate U' over the range$10 \leq n \leq 50$, although with difficulty. Using an efficient computing algorithm it has been estimated that when $n = 20$, about 800 million calculations would be required [81]. The corresponding number of modes required would be of the order $m \approx n^2 = 400$, if "collisions" of photons were to be minimized in the network. This is at the boundary of what a classical computation could efficiently accomplish. However, a quantum boson sampler computer could easily perform this calculation using linear optical elements, provided the physical requirements could be implemented.

In the year 2013 several research teams, working independently, reported different types of experimental boson samplers using an input of three to four single-photons. One device consisted of an integrated photonic chip, where the photons were guided by silicon-on-silicon waveguides and beamsplitters [82]. The boson sampler had six possible spatial input modes, ten cross-coupled beamsplitter nodes, and six spatial output modes. The waveguides were fabricated using a focused CW UV laser on a photosensitive layer to produce a local increase in the refractive index. By precise translation of the substrate, the waveguide paths were formed to produce the network shown in Fig. 9.36. Two laser pumped parametric

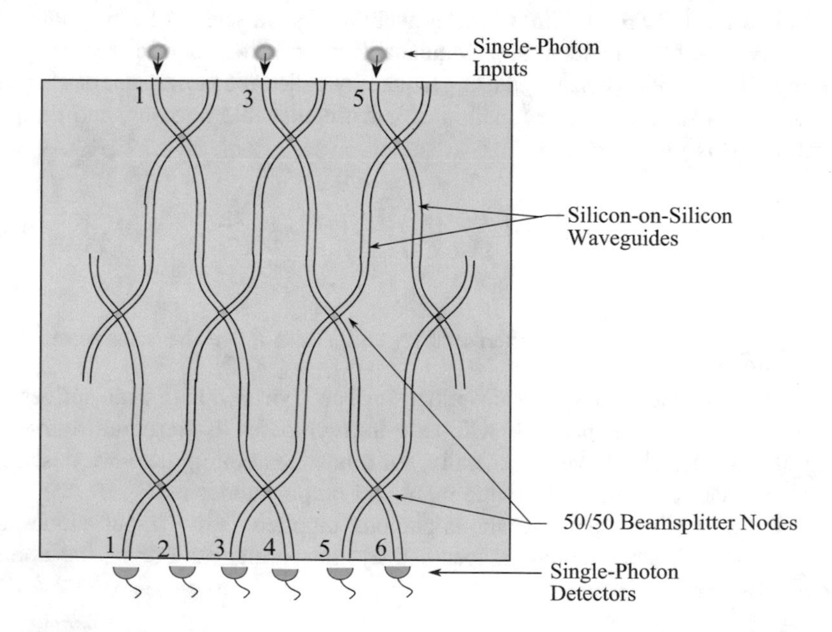

Fig. 9.36 Schematic of a boson sampler as an integrated photonic circuit. There are 6 accessible spatial inputs and 10 beamsplitters

down-conversion sources produced four nearly-indistinguishable photons in 100 fs pulses. In one experiment, three photons were coupled into input ports 2, 3, and 5, and the fourth photon was sent to a heralding detector. Single-photon avalanche photodiodes registered the output from each of the six output modes. The output probabilities of this quantum calculation were found to favorably agree with a classical computation for this limited number of photons. This gave confidence that the quantum calculation would be valid for higher numbers of photons and spatial modes, where classical computations become increasingly difficult.

Another three-photon boson sampler used a fiber optic unitary transformation matrix U with three input spatial modes and six possible output modes [83]. See Fig. 9.37. The three input single-photons were selected from two pairs of nearly indistinguishable photons emitted from double-pass SPDC in a BBO crystal driven by a Ti:S femtosecond laser. The fourth photon was sent to a trigger detector. The three polarized input photons were integrated by a three-mode fiber beamsplitter, called a *tritter*, and exited as three single-photons in three spatial modes. Possible Hong-Ou-Mandel two-photon exit paths disappeared due to an effect called *bosonic coalescence*. In a boson sampler the probability of a certain output from U should be proportional to the wave functions of indistinguishable photons in sub-matrices U' of the unitary matrix U. This was verified by a counting circuit from output

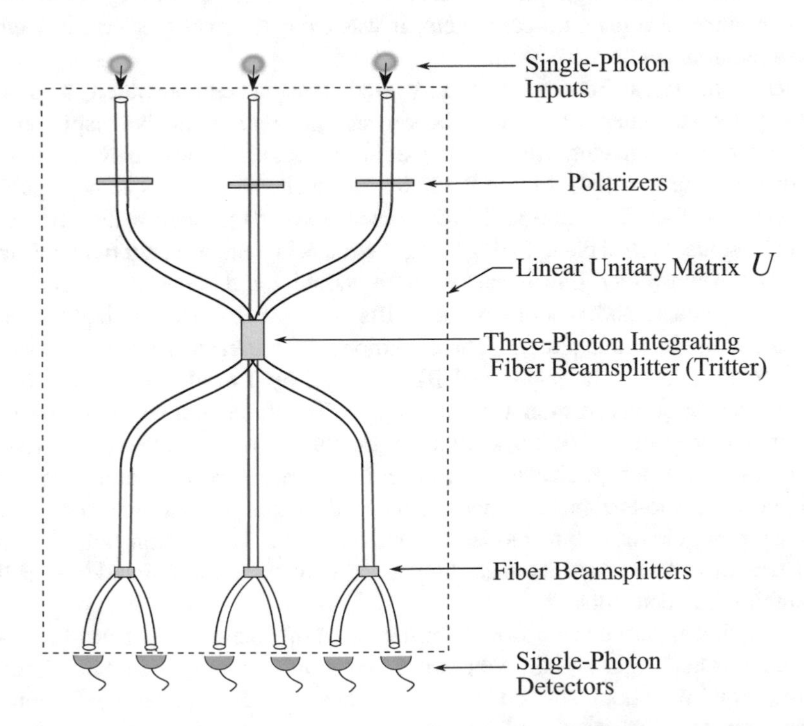

Fig. 9.37 A photon boson sampler using a fiber optic unitary transformation matrix U with three input spatial modes and six possible output modes. Three polarized input photons are integrated by a three-mode fiber beamsplitter, and exit as three single-photons in three spatial modes

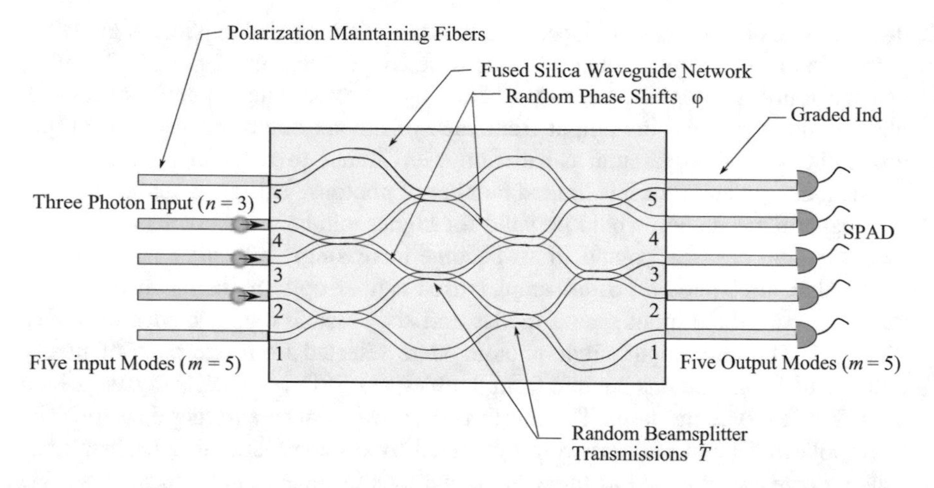

Fig. 9.38 A 5 × 5 input/output mode boson sampler waveguide network with eight beamsplitters having *random* transmission values *T*, and eleven *random* network phase shifts φ. Three indistinguishable polarized photons are inputted

photo-avalanche diode single-photon detectors. When scaled to the large number of photons required for quantum computers, it was expected to outperform any classical computation.

A team of researchers in Germany and Austria demonstrated a 5 × 5 input/output mode integrated circuit boson sampler with eight beamsplitters or directional couplers having *random* transmission values *T*, and eleven *random* network phase shifts φ [84]. Figure 9.38 shows the circuit layout. The waveguides were formed by focusing a 200 nJ 150 fs pulsed laser 370 μm below the surface of a 10 cm long fused silica plate, and moving the focus in a precise pattern at constant speed. The beamsplitter transmissions were varied by the degree of coupling between the modes, and random phase shifts were created by varying the path length between the beamsplitters. Three photons were created from two pairs of photons emitted from a laser-pumped BBO crystal by Type-II SPDC, and made indistinguishable in polarization and wavelength. The three photons were sent into three input modes of the boson sampler circuit using optical fibers. The fourth photon was sent to a trigger detector to confirm coincidence measurements. Five single-photon photo-avalanche diode (SPAD) detectors were connected to the boson sampling circuit output modes by optical fibers. The photon output distribution was then determined by recording a series of threefold coincidence at the five output mode detectors.

Another integrated chip boson sampler used a focused laser to fabricate waveguides inside a glass plate, where the waveguides could be formed in three dimensions and were not restricted to a single plane [85]. The optical interferometer network chip had five input and output modes, and ten beamsplitting nodes as shown in Fig. 9.39. Evanescent coupling by closely spaced waveguides at the

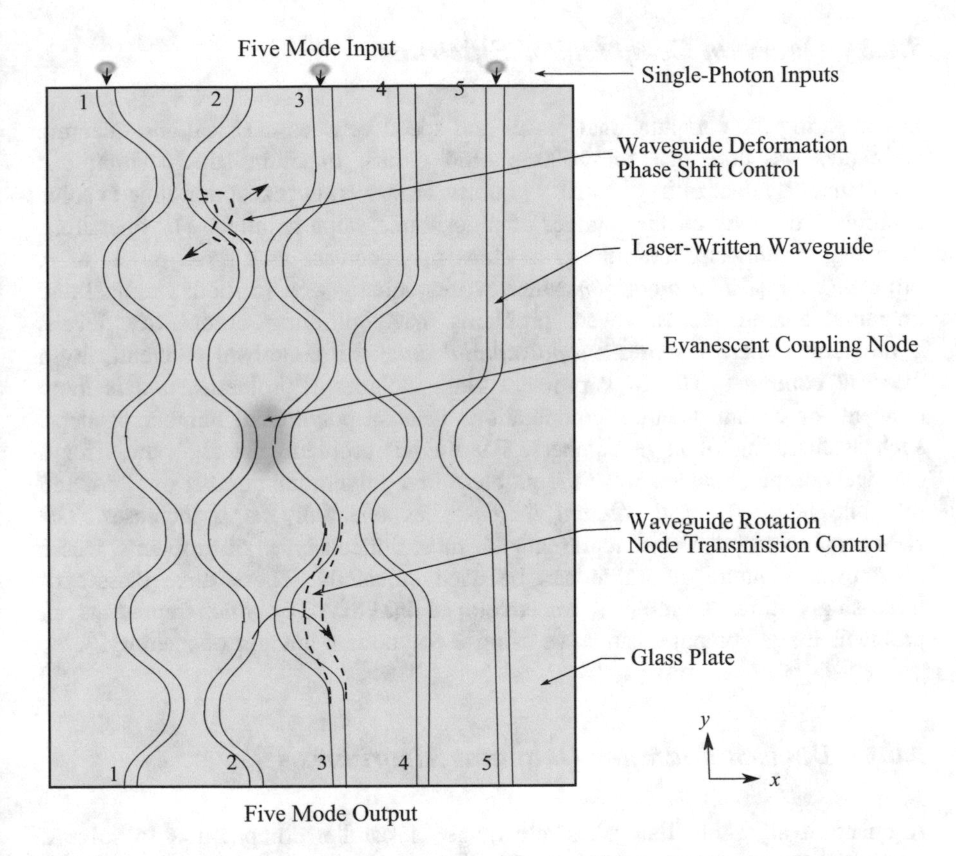

Fig. 9.39 A 5 × 5 input/output mode boson sampler where the waveguides are not restricted to a single plane. The transmission T of each of the ten beamsplitter modes was randomized by twisting the waveguide out of the x-y plane. Random phase shifts φ were produced by deforming the waveguide bend before each beamsplitting node region

beamsplitter nodes determined the variable transmission. The transmission T of each beamsplitter was randomized by twisting the waveguide out of the x-y plane to change the waveguide separation at the coupler node. Random phase shifts φ were produced by deforming the waveguide bend before each beamsplitting node region. Three synchronized non-interacting single-photons were inputted into modes 1, 3, and 5, and the probability distributions of the outputs where three photons exited the chip were determined and agreed well with a classical computation. Since these probabilities were proportional to calculated permanents of the 3 × 3 sub-matrices of the 5 × 5 mode unitary matrix U, the quantum mechanical performance for this three photon boson sampler was verified. By scaling up of the number of modes and input photons, classical computer calculations of the permanents of all sub-matrices would be increasingly difficult compared to the quantum mechanical boson sampling methodology.

9.6.3 Quantum Computation Efficiency

In comparing the computational power and speed between classical and quantum computers we note that for 1000 photon qubits, quantum superposition can simultaneously control 2^{1000} ($\approx 10^{300}$) qubits. However, the execution time to solve a problem depends on the number of algorithmic steps required. For operations involving n numbers, the number of steps may increase as a fixed power of n, producing an *efficient algorithm* which works equally well for both classical and quantum computers. However, problems have differing complexity levels. P problems, where P represents *polynomial time*, can be solved efficiently by a classical computer. The BQP problem class includes P problems, and is more efficient for certain quantum computations using a polynomial number of steps, such as factoring of large numbers. For an NP problem, the algorithms for a classical computer cannot solve the problem in a polynomial number of steps, and the time required for a solution increases exponentially as n increases. The NP-complete problem class represents the most difficult types of problems, where no known quantum algorithm can be used efficiently. These three classes of increasingly difficult problems are enveloped in PSPACE, which represents all problems that a computer can solve using a polynomial amount of memory.

9.6.4 Deutsch and Shor Quantum Algorithms

A quantum algorithm that takes advantage of parallel computation to solve a problem was developed in 1985 by David Deutsch, and is known as *Deutsch's algorithm* [86]. Two input qubits to a "black box", such as $|0\rangle$ and $|1\rangle$, are mapped to output qubits from the box by a set of four binomial functions $f(x)$. In the case where $f(0) = f(1)$ the algorithm outputs the same qubits and $f(x)$ is *constant*. If $f(0) \neq f(1)$, then different output qubits occur for different input qubits, and $f(x)$ is *balanced*. Deutsch's algorithm then determines whether $f(x)$ is constant or balanced. Although this is not a very useful calculation, the algorithm's value lies in demonstrating that by using four superposed states, only one query to the box is required to make this determination. A classical algorithm making this determination would require two queries to resolve the ambiguity of a single query, and therefore takes twice as much time.

The algorithm is implemented for the two input qubits in the black box by simultaneous processing all possible values of $f(x)$ using four types of quantum unitary gates or *oracles*. These gates are Hadamard and CNOT types. A measured output qubit $|0\rangle$ shows that $f(x)$ is constant and a measured output qubit $|1\rangle$ shows that $f(x)$ is balanced. This type of quantum computation, using a series of gates, is usually referred to as the *circuit model* or *network model*, and the computation is usually reversible. Appendix 9A shows some circuit model calculations for Deutsch's algorithm.

Deutsch's algorithm was experimentally realized in 2007 for *one-way computation* using four entangled photons in a *cluster state*, where all possible functions interacting with a two-qubit input are implemented using the four-qubit cluster state [87] This approach differs somewhat from the circuit model in that entangled states are used for single-qubit measurements. However, these measurements dismantle the state, making it a one-way process. Figure 9.40 shows the experimental arrangement. A BBO crystal, driven by a pulsed UV laser, produces two pairs of polarization-entangled photons by SPDC. They have the following Hadamard gate forms: $(|HH\rangle + |VV\rangle)/\sqrt{2}$ and $(|HH\rangle - |VV\rangle)/\sqrt{2}$. Two main polarizing beam-splitters form a spatially separated four-photon cluster state

$$|\Psi_c\rangle = (|HHHH\rangle + |HHVV\rangle + |VVHH\rangle - |VVVV\rangle)/2. \tag{9.27}$$

This cluster state is properly prepared when coincidence occurs at all four detectors. To execute the algorithm a series of half-wave and quarter-wave plates determine the measurement basis over a wide-range of photon polarization states (linear polarization at various angles, and right-and left-handed circular polarization). These polarization states represent various $f(x)$ modes. Four auxiliary beamsplitters and four single-photon detectors perform single qubit projective measurements. In the actual computation one input qubit $|x\rangle = |+\rangle$ is encoded on photon 1, while the other input qubit $|y\rangle = |-\rangle$ becomes encoded on photon 3. When a photon qubit is measured, then it is disentangled and removed from the cluster state. This moves the computation in a one-way direction and makes the computation irreversible. Interaction between qubits in the form of CNOT and CPHASE gates produces the measured outputs of the four qubits, where output qubits 1 and 3 show the results of the interaction with $f(x)$. It was found that a

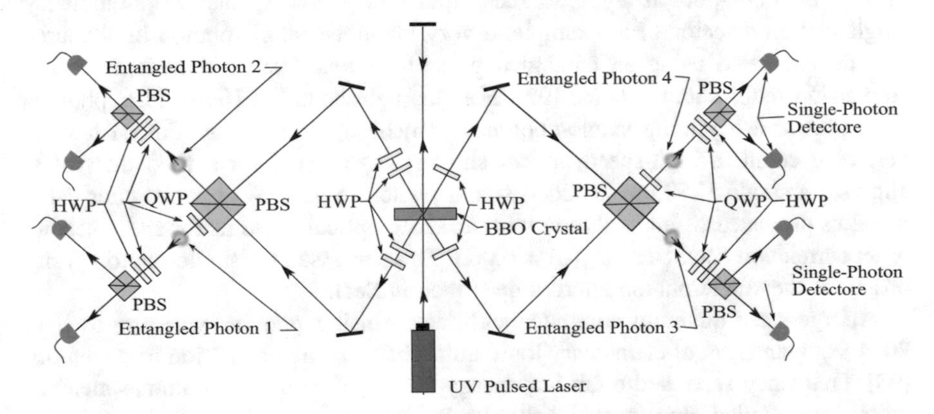

Fig. 9.40 Experimental realization of Deutsch's algorithm using a cluster state of four entangled photon qubits representing all possible $f(x)$ functions acting on a qubit pair. Single-photon detectors measure a polarized photon qubit, thereby removing it from the cluster state. The process is not reversible, making it a one-way quantum computation (This figure adapted with permission from Ref. [87]. Copyrighted by the American Physical Society)

success rate of up to 90% was obtainable by a single run of the algorithm using only qubit 1 as an output measurement.

Deutsch's algorithm was generalized in 1992 as the *Deutsch-Josza algorithm*, which provided a deterministic outcome with an n-bit input [88]. Only one query is required. Now the real advantage of the quantum algorithm appears, since a classical computation would require up to $2^{n-1} + 1$ queries to achieve a deterministic outcome. Algorithms of this type might solve a specific set of problems in exponentially less time than a classical computer, and initiated interest in the field of quantum computing.

Perhaps motivated by the Deutsch algorithm, in 1994 Peter Shor at AT&T Bell Labs developed a powerful quantum algorithm, widely-known as *Shor's algorithm*. This algorithm reduced the exponential time for a specific NP problem to polynomial time [89]. For example, the Shor algorithm is useful for finding the prime factors (prime numbers that divide an integer exactly) of a large number comprised of n digits. Several independent teams of investigators have performed proof-of-principle photonic experiments of a simplified form of Shor's algorithm for factoring an $n = 15$ integer [90, 91]. Both experiments showed the quantum nature of the experiment by the use of multiple photon entanglement.

9.6.5 A Universal Quantum Computer Architecture

A complete and general quantum computer using a large number of controlled photons has yet to be demonstrated. Rather, several types of architectures have been proposed and experimental research has focused on the critical components that would make such a computer operational. These components include logic gates, ultra-pure single-photon light sources, quantum memory, and high-efficiency single-photon detectors. For example, a very low-noise single-photon light source was demonstrated using an entangled pair of photons from a parametric-down-conversion multi-photon source [92]. For this light source, 810 nm idler photons were sent to a heralding single-photon avalanche diode detector. For each valid heralding count, a high-speed optical shutter was opened for a time interval τ, allowing a single 1550 nm heralded signal photon to pass. An open shutter value $\tau = 2$ ns produced a single-photon with a 0.25% optical noise factor and a second order correlation coefficient $g^{(2)}(0) = 0.005$. This surpassed the single-photon purity of any of the single-photon sources described in Sect. 9.2.

A large-scale quantum computer architecture using photons was proposed in 2014 using an array of elementary logic units (ELUs) with crystal ion trap memory [93]. Frequency separated qubit ion energy levels $|\uparrow\rangle$ and $|\downarrow\rangle$ in atomic memory units are coupled and provide the basis for the operation of deterministic phonon-based quantum gates between the ions. This operation could be extended for all qubits throughout the processor by qubit entanglement between ELUs. Figure 9.41 shows the general layout, which is referred to as a *modular universal scalable ion trap quantum-computer* (MUSIQC). For this illustration, the ELU

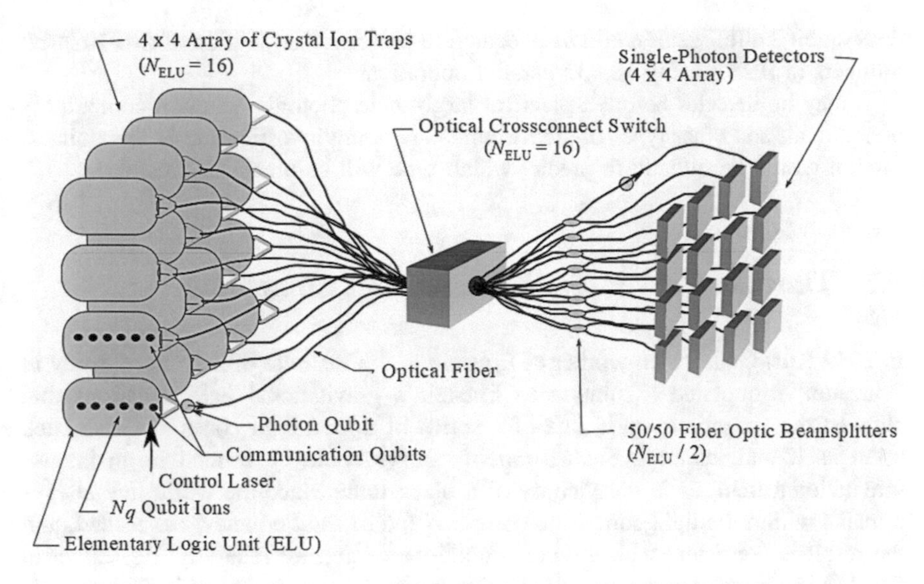

Fig. 9.41 Proposed universal quantum computer architecture (MUSIQC) using elementary logic unit (ELU) ion traps, fiber optic photon output channels, optical cross-connect switch, fiber optic 50/50 beamsplitters, and single-photon detectors. Illustration shows 4 × 4 array of ELUs (cross-section), 4 × 4 optical switch, 8 beamsplitters, and 4 × 4 array of single-photon detectors for Bell-state measurements (This figure adapted with permission from Ref. [93]. Copyrighted by the American Physical Society)

registers are arranged in a 4 × 4 array, where the ELU number $N_{ELU} = 16$. The number of controllable ion qubits N_q per ELU register is estimated to be $10 \leq N_q \leq 100$, producing $N = N_{ELU}\ N_q$ useable qubits for the entire system. A laser pulse causes one or two of the communication qubits within any ELU to emit a correlated photon by energy level transitions. Photons emitted from the ELU outputs are coupled to single-mode optical fibers and sent to a microelectromechanical system (MEMS) cross-connect switch, allowing any ELU output channel to be connected to any other output channel. Pairs of ELU channels are connected to $N_{ELU}/2$ fiber optic 50/50 beamsplitters, followed by a pair of single-photon detectors to form Bell-state analyzers. Thus N_{ELU} Bell-state tests can be performed in parallel. When the detectors indicate coincidence or interference, the correlated communication qubits between the corresponding ELUs are entangled. The deterministic quantum operations of the ion qubits within an ECU can then be combined with the probabilistic entanglement of the ion qubits between the ELUs to form a scalable hybrid device for quantum computation. The mean connection time depends on the repetition rate of the ion transition process and the success probability of the photon entanglement, and is expected to be in the millisecond range. It is anticipated that the MUSIQC architecture could be scaled to thousands of ELUs with up to 10^6 qubits, all operating in parallel with an acceptable fault tolerance of the photonic gates. For example, one-qubit X gates, two-qubit CNOT gates, and three-qubit Toffoli gates could be executed to perform the addition of two 1024-bit numbers in 0.28 s using 18,432 parallel operations.

It may be decades before a practical large-scale photonic quantum computer is operational, and other types of architectures are being investigated. At this point in time, it would be difficult to predict which type will be practically realized.

9.7 Timelike Curves, Time Travel, and Computing

In 1949 Kurt Gödel, a co-worker of Einstein at the Institute of Advanced Study in Princeton, formulated a solution to Einstein's gravitational field equations that allowed the spacetime world line of a series of tipped light cones to curve back upon itself and intersect. Such a *timelike curve* could be formed in an intense gravitational field, as in the vicinity of a black hole. Since the world line always remains within the light cones, the cosmic speed of light c is never exceeded, and the result is consistent with the field equations of general relativity. As shown in Fig. 9.42a a traveler in the present at point **A** could move backward in time to point **B** in the past. For example, the traveler could move from a time and place today to a time and place yesterday. This type of timelike curve is called an *open timelike curve* (OTC). In the case where the timelike curve forms a continuous loop, as shown in Fig. 9.42b, it would be possible for the traveler to meet himself in the past, even forming a clone. This is referred to as a *closed timelike curve* (CTC). Timelike curves have never been observed, but have initiated many theoretical and experimental investigations.

Several conclusions drawn from the analysis of timelike curves run contrary to current views in classical physics and quantum mechanics. Using timelike curves, several types of causality seem to be violated [94]. For the OTC in Fig. 9.42a the *consistency paradox* arises, allowing travel to the past to affect the present status of an event without the traveler actually returning to this event. For a traveler moving along a CTC in Fig. 9.42b, a paradoxical *causal loop* occurs when the traveler performs an action in the past and then returns to the starting point. As the journey begins again, this action would then appear to self-existing with no apparent cause from an earlier event. Thus this causal loop might be used as an argument for the "self-creation" of the Universe

The much-discussed *grandfather paradox* asserts that if I could travel to the past and kill my grandfather (or grandmother), or prevent them from ever meeting, then I would not be born. But not being born would then prevent me from traveling to the past—hence the logical paradox. This leads to inconsistencies in theories developed using the properties of timelike curves. One way to resolve this paradox would be the use of probabilistic quantum particles (e.g. photons) rather than deterministic ordinary travelers.

In 1991 theoretical physicist David Deutsch at Oxford University created quantum computational networks that simulated the chronological discontinuities that would occur in a continuous CTC spacetime loop [95]. These spatial networks produced

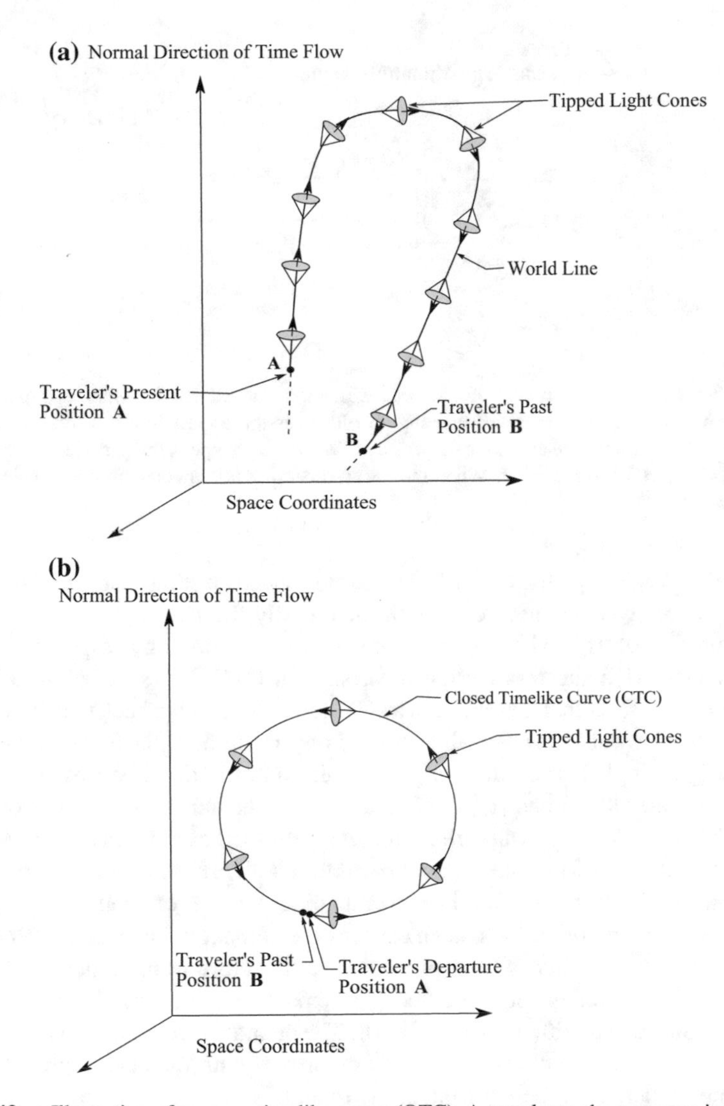

Fig. 9.42 **a** Illustration of an open timelike curve (OTC). A traveler to the past may influence the present, but the traveler in the present and past never meet. This produces the grandfather paradox. **b** Illustration of a closed timelike curve (CTC). In this case the traveler in the past may meet with himself in the present. This presents the logical paradox of how the travel could ever begin, and offers the possibility of producing a clone of the traveler

chronology-violating behavior by the use of quantum logic gates (G) and time-delay links ($-T$). Deutsch analyzed four types of logical paradoxes and showed that quantum mechanical computations removed the paradoxes. In this CTC model it is required that any output state is of the same form as the input state, or that these states satisfy a causal *consistency condition*. This usually requires nonlinear evolution of a quantum state within the network. In one type of paradox simulation, a single-particle can

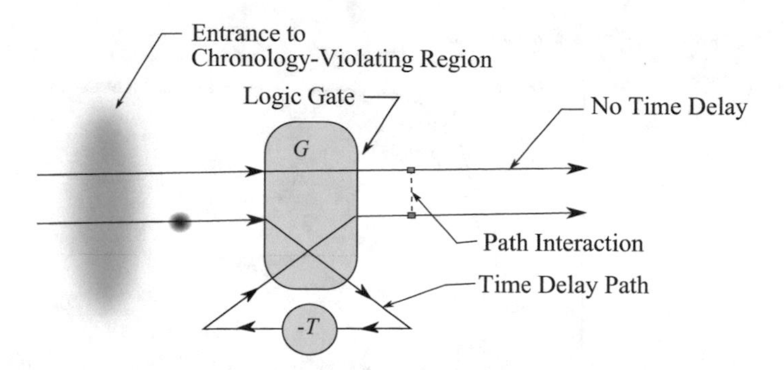

Fig. 9.43 Schematic of an optical circuit that simulates the choice between two paths in a chronology-violating region. Particles meeting an older version appear in a mixed state of being both present at an earlier time and not being present. When this happens the particle switches from one possible trajectory to the other. When the older version is not encountered, it remains on its current trajectory

traverse two possible paths: one path is chronology-respecting, and the other CTC path is chronology-violating. See Fig. 9.43. Initially the two input qubits are in a mixed state $\Psi\ (|0\rangle, |1\rangle)$ where $|0\rangle$ indicates that the chronology-respecting path is unoccupied, and $|1\rangle$ indicates a particle traversing the CTC. This simulation addresses the statement of critics that if a carrier in a CTC loop could move back to a time before it started out, it would never be able to start out in the first place. The input state representing a single-particle in one path or the other is described by a subspace between the states $|0\rangle|1\rangle$ and $|1\rangle|0\rangle$. Now the two paths interact in the network, and when a CTC particle reaches an earlier time, it meets itself as older in a mixed state of presence and absence. Thus the younger particle is both *prevented* and *not prevented* to travel further back in time. If it becomes an older version of itself, it appears in a mixed state of being *both* present at an earlier time and not being present. When this happens the particle switches from one possible trajectory to the other. In the case where the older version is not encountered, it remains on its current trajectory. Then the output state must be either $|0\rangle|1\rangle$ or $|1\rangle|0\rangle$. The quantum probability of exactly one particle appearing in the output is zero, as compared with the classic prediction of exactly one particle always appearing at the output.

Quantum theory for a chronology-respecting system does not allow the cloning of a quantum state, with subsequent verification that it has been cloned. However, Deutsch theorized that in a chronology-violating CTC, cloned quantum states could be achieved. This affects the Heisenberg uncertainty principle, which states that measurements on a pair of conjugate variables (usually position and momentum) on a particle cannot be simultaneously measured without a minimum error. However, cloning would permit separate highly accurate measurements to be made on conjugate quantum state variables: one variable measurement on the original state and the other variable measurement on the cloned state. The widely-accepted uncertainty principle for state measurements in chronologically-respecting systems is therefore violated.

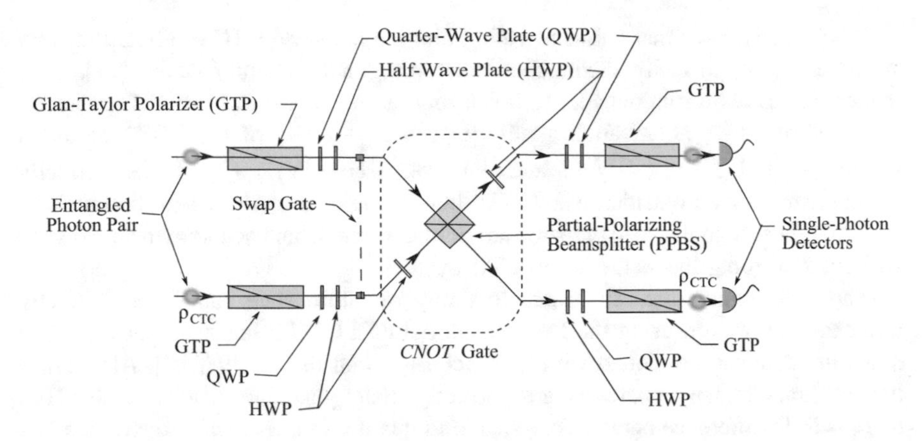

Fig. 9.44 Illustration of an experimental network to perfectly distinguish non-orthogonal quantum states using an entangled pair of polarized photons, a partial-polarizing beamsplitter, and two single-photon detectors for projective measurements of the states

Experimental realization of a qubit interacting with a former version of itself using the Deutsch CTC model was reported in 2014 by a team at the University of Queensland, Australia [96]. A pair of polarized single-photons acted as a single-photon in a CTC, with one photon performing as a past form of the other. The experiment was carried out with single-photons with a unitary interaction producing nonlinear evolution of an input state Ψ within the CTC network. Figure 9.44 illustrates an experimental network to perfectly distinguish non-orthogonal quantum states. An entangled pair of single-photons was produced by spontaneous parametric down-conversion in a nonlinear crystal. These single-photons were incident on a pair of Glan-Taylor polarizing prisms, rotatable quarter- and half-wave plates, such that arbitrary polarization input states Ψ could be produced. The upper leg represents a chronologically-respecting qubit ρ and the lower leg represents a qubit ρ_{CTC} moving in a closed timelike curve. The upper and lower beams are incident on a partial-polarizing quantum beamsplitter with a reflectivity for horizontally polarized light $R_H = 0$, and a reflectivity for vertically polarized light $R_V = 2/3$. After traversing another set of quarter and half-wave plates and Glan-Taylor polarizers, projective measurements of the two beams are made with avalanche photodiode single-photon detectors.

This network can provide perfect discrimination between non-orthogonal quantum states. For example, input states $|\Psi\rangle_0$ and $|\Psi\rangle_1$ were prepared such that $|\Psi\rangle_0 = |H\rangle$ and $|\Psi\rangle_1 = \cos(\phi/2)|H\rangle + \sin(\phi/2)|V\rangle$, where ϕ is the polarization axis rotation angle. The swap gate can be realized by switching inputs. The control gate then rotates a target qubit by angle π about an axis in the *x*-*z* plane defined by angle θxz, dependent on the state of the control qubit. By setting $\theta xz = (\phi - \pi)/2$, the state $|\Psi\rangle_1$ rotates to $|V\rangle$ and becomes orthogonal to $|\Psi\rangle_0$. Then the projective measurement provides proof of perfect discrimination of the two input states.

It has been shown that qubits traveling along a localized CTC can be utilized by an observer to perfectly distinguish non-orthogonal quantum states [97]. This makes it possible for an outsider to break any prepare-and-measure QKD protocol. For example, four polarization states $|\mathrm{H}\rangle, |\mathrm{V}\rangle, |+\rangle, |-\rangle$ of the BB84 protocol, where $|-\rangle \equiv (|\mathrm{H}\rangle - |\mathrm{V}\rangle)/\sqrt{2}$ and $|+\rangle \equiv (|\mathrm{H}\rangle + |\mathrm{V}\rangle)/\sqrt{2}$, can be perfectly distinguished by an outsider. The QKD then becomes insecure when the outsider, knowing these signal states, prepares an identical state from each known signal state without disturbing the states in the CTC system.

The non-cloning theorem states that any quantum state cannot be perfectly replicated. However, using the Deutsch model of a CTC it has been shown that a quantum state can be cloned with high accuracy and fidelity [98, 99]. Also, since two arbitrary density operators are perfectly distinguishable in a Deutsch CTC, there is little difference between classical and quantum mechanical properties in the Deutsch model.

Perfect cloning in a CTC implies violation of the Heisenberg uncertainty principle, since a variable could be measured to high accuracy for one state, and the conjoined variable could be measured to high accuracy in the cloned state. However, a team of researchers in Australia has theoretically shown that even for an open timelike curve, that the uncertainty principle can be violated [100]. They showed that the nonlinear mathematics in the Deutsch CTC model can also be applied to an OTC, if a traveling OTC qubit is entangled with an external qubit. Using an optical circuit containing an OTC, two quantum states were squeezed in orthogonal directions, and after several traversal cycles, the orthogonal components of the pair could be measured to any degree of accuracy.

It was recognized that by sending computer calculation information into the past using CTCs, that difficult computational problems could be solved more efficiently and faster. Todd Brun at the Institute for Advanced Study at Princeton University showed how this could be accomplished by factoring large numbers using a classical CTC-computer [101]. The use of CTCs in quantum computing is more speculative and controversial. In addition to arguments against the existence of CTCs, quantum effects in CTCs might prevent the algorithms from achieving deterministic outcomes.

It's not difficult to show that classical computers using CTCs can be simulated in PSPACE [102]. However, it's harder to simulate a CTC quantum computer in PSPACE using the Deutsch model, with the added requirement that a CTC must support a polynomial number of bits. These simulations showed that computers using CTCs would have no more computational power than a conventional computer.

Many scientists question the purpose and legitimacy of analysis and scientific research on timelike curves, since OTCs and CTCs have never been observed or proved to exist. Time will tell whether scientific breakthroughs will show the usefulness of timelike curves to solve practical problems, or to change the nature of current quantum and gravitational theories.

References

1. C.H. Bennett, S.J. Wiesner, Communication via one- and two-particle operators on Einstein-Podolsky-Rosen states. Phys. Rev. Lett. **69**, 2881–2884 (1992). doi:10.1103/PhysRevLett.69.2881
2. K. Mattle et al., Dense coding in experimental quantum communication. Phys. Rev. Lett. **76** (25), 4656–4659 (1995). doi:10.1103/PhysRevLett.76.4656
3. C.H. Bennett, G. Brassard, Quantum cryptography: public key distribution and coin tossing, in *Proceedings of IEEE International Conference on Computers, Systems, and Signal Processing*, Bangalore (1984), pp. 175–179
4. A.K. Ekert, Quantum cryptography based on Bell's theorem. *Phys. Rev. Lett.* **67**(6), 661–663 (1991). doi:10.1103/PhysRevLett.67.661
5. A.K. Ekert et al., Practical quantum cryptography based on two-photon interferometry. Phys. Rev. Lett. **69**(9), 1293–1295 (1992). doi:10.1103/PhysRevLett.69.1293
6. A. Beveratos et al., Room temperature stable single-photon source. Eur. Phys. J. D **18**, 191–196 (2002). doi:10.1140/epjd/e20020023
7. M. Förtsch et al., A versatile source of single photons for quantum information processing. Nat. Commun. **4**, 1818–1822 (2013). doi:10.1038/ncomms2838
8. C. Becher et al., A quantum dot single-photon source. Physica E **13**, 412–417 (2002). doi:10.1016/S1386-9477(02)00156-X
9. V.D. Verma et al., Photon antibunching from a single lithographically defined InGaAs/GaAs quantum dot. Opt. Express **19**(5), 4183–4187 (2011). doi:10.1384/OE.19.004182
10. M.J. Holmes et al., Room temperature triggered single photon emission from a III-nitride site-controlled nanowire quantum dot. Nano Lett. **14**(2), 982–986 (2014). doi:10.1021/nl404400d
11. M. Davanço et. al., Telecommunications-band heralded single photons from a silicon nanophotonic chip. Appl. Phys. Lett. **100**, 261104-1-5 (2012). doi:10.1063/1.4711253
12. M. Nothaft et al., Electrically driven photon antibunching from a single molecule at room temperature. Nat. Commun. **3**(628), (2012). doi:10.1038/ncomms1637
13. G.N. Gol'tsman et al., Picosecond superconducting single-photon detector. Appl. Phys. Lett. **79**(6), 705–707 (2001). doi:10.1063/1.1388868
14. F. Marsili et al., Detecting single infrared photons with 93% system efficiency. Nat. Photonics **7**, 210–214 (2013). doi:10.1038/nphoton.2013.13
15. H. Takesue et al., Quantum key distribution over a 40-dB channel loss using superconducting single-photon detectors. Nat. Photonics **1**, 343–348 (2007). doi:10.1038/nphoton.2007.75
16. B. Korzh et al., Free-running InGaAs single photon detector with 1 dark count per second at 10% efficiency. Appl. Phys. Lett. **104**, 081108-1-4 (2014). doi:10.1063/1.4866582
17. B. Korzh et al., Provably secure and practical quantum key distribution over 307 km of optical fiber. Nat. Photonics **1**, 343–348 (2015). doi:10.1038/nphoton.2014.327
18. R.P. Tapster et al., Developments towards practical free-space quantum cryptography. Proc. SPIE **5815**, 176–179 (2005). doi:10.1117/12.605539
19. R. Ursin et al., Entanglement-based quantum communication over 144 km. Nat. Phys. **3**, 481–486 (2007). doi:10.1038/nphys629
20. S. Nauerth et al., Air-to-ground quantum communication. Nat. Photonics, **7**, 382–386 (2013). doi:10.1038/nphoton.2013.46
21. T. Heindel et al., Quantum key distribution using quantum dot single-photon emitting diodes in the red and near infrared spectral range. New J. Phys. **14**, 083001-1-12 (2012). doi:10.1088/1367-2630/14/8/083001
22. M. Rau et al., Free-space quantum key distribution over 500 meters using electrically driven quantum dot single-photon sources—a proof of principal experiment. New J. Phys. **14**, 043003-1-10 (2014). doi:10.1088/1367-2630/16/4/043003

23. M. Padgett, L. Allen, Light with a twist in its tail. Contemp. Phys. **41**(5), 275–285 (2000)
24. M. Krenn et al., Communication with spatially modulated light through turbulent air across Vienna. New J. Phys. **16**, 113028-1-10 (2014). doi:10.1088/1367-2630/16/11/113028
25. J. Wang et al., Terabit free-space data transmission employing orbital angular momentum multiplexing. Nat. Photonics **6**, 488–496 (2012). doi:10.1038/nphoton.2012.138
26. I.M. Fazal et al., 2 Tbit/s free-space data transmission on two orthogonal orbital-angular-momentum beams each carrying 25 WDM channels. Opt. Lett. **37**(22), 4753–4755 (2012). doi:10.1364/OL.37.004753
27. Y. Awaji et al., World first mode/spatial division multiplexing in multi-core fiber using Laguerre-Gaussian mode, in *37th European Conference and Exposition on Optical Communication*, paper We.10.P1.55 (2011). doi:10.1364/ECOC.2011.We.10.P1.55
28. N. Bozinovic et al., Terabit-scale orbital angular momentum mode division multiplexing in fibers. Science **340**(6140), 1545–1548 (2013). doi:10.1126/science.1237861
29. D. Hillerkuss et al., 26 Tbit s^{-1} line-rate super-channel transmission utilizing all-optical fast Fourier transform processing. Nat. Photonics **5**, 364–371 (2011). doi:10.1038/nphoton.2011.74
30. B. Julsgaard et al., Experimental demonstration of quantum memory for light. Nature **432**, 482–486 (2004). doi:10.1038/nature03064
31. M.P. Hedges et al., Efficient quantum memory for light. Nature **465**, 1052–1056 (2010). doi:10.1038/nature09081
32. K.S. Choi et al., Mapping photonic entanglement into and out of a quantum memory. Nature **452**, 67–71 (2008). doi:10.1038/nature06670
33. H. de Riedmatten et al., A solid-state light-matter interface at the single-photon level. Nature **456**, 773–777 (2008). doi:10.1038/nature07607
34. M. Afzelius et al., Demonstration of atomic frequency comb memory with spin-wave storage. Phys. Rev. Lett. **104**, 04503-01-4 (2010). doi:10.1103/PhysRevLett.104.040503
35. C. Clausen et al., Quantum storage of photonic entanglement in a crystal. Nature **469**, 508–511 (2011). doi:10.1038/nature09662
36. H. Kosaka, N. Niikura, Entangled absorption of a single photon with a single spin in diamond. Phys. Rev. Lett. **114**(5), 053603-1-5 (2015). doi:10.1103/PhysRevLett.114.053603
37. D.G. England et al., Storage and retrieval of THz-bandwidth single photons using a room-temperature diamond quantum memory. Phys. Rev. Lett. **114**(5), 053602-1-6 (2015). doi:10.1103/PhysRevLett.114.053602
38. M. Bashkansky, F.K. Fatemi, I. Vurgaftman, Quantum memory in warm rubidium vapor with buffer gas. Opt. Lett. **37**(2), 142–144 (2012). doi:10.1364/OL.37.000142
39. M. Dąbrowski, R. Chrapkiewicz, W. Wasilewski, Hamiltonian design in readout from room-temperature Raman atomic memory. Opt. Express **22**(21), 26076-1-15 (2014). doi:10.1364/OE.22.026076
40. C.H. Bennett et al., Teleporting an unknown quantum state via dual classical and Einstein-Podolsky-Rosen channels. Phys. Rev. Lett. **70**, 1895–1899 (1993). doi:10.1103/PhysRevLett.70.1895
41. D. Bouwmeester et al., Experimental quantum teleportation. Nature **390**, 575–579 (1997). doi:10.1038/37539
42. D. Boschi et al., Experimental realization of teleporting an unknown quantum state via dual classical and Einstein-Podolsky-Rosen channels. Phys. Rev. Lett. **80**(6), 1121–1131 (1998). doi:10.1103/PhysRevLett.80.1121
43. S.-W. Lee, H. Jeong, Near-deterministic quantum teleportation and resource-efficient quantum computation using linear optics and hybrid qubits. Phys. Rev. A **87**, 022326-1-10 (2013). doi:10.1103/PhysRevA.97.022326

44. S.-W. Lee, H. Jeong, Bell-state measurement and quantum teleportation using linear optics: two-photon pairs, entangled coherent states, and hybrid entanglement, in *Proceedings of First International Workshop on Entangled Coherent States and its Application to Quantum Information Science*, Tokyo, Japan (2013), pp. 41–46
45. J. Brendel et al., Pulsed energy-time entangled twin-photon source for quantum communication. Phys. Rev. Lett. **82**(12), 2594-1-8 (1999). doi:10.1103/PhysRevLett.82.2594
46. I. Marcikic et al., Time-bin entangled qubits for quantum communication created by femtosecond pulses. Phys. Rev. A **66**(6), 062308-1-6 (2002). doi:10.1038/PhysRevA.66.062308
47. I. Marcikic et al., Long-distance teleportation of qubits at telecommunication wavelengths. Nature **421**, 509–513 (2003). doi:10.1038/nature01376
48. F. Bussières et al., Quantum teleportation from a telcom-wavelength photon to a solid-state quantum memory. Nat. Photonics **8**, 775–778 (2014). doi:10.1038/nphoton.2014.215
49. J. Xian-Min et al., Experimental free-space quantum teleportation. Nat. Photonics **4**, 376–381 (2010). doi:10.1038/nphoton.2010.87
50. J. Yin et al., Quantum teleportation and entanglement distribution over 100-kilometer free-space channels. Nature **488**, 185–188 (2012). doi:10.1038/nature11332
51. X.S. Ma et al., Quantum teleportation over 143 kilometers using active feed-forward. Nature **489**, 269–273 (2012). doi:10.1038/nature11472
52. H.J. Briegel et al., Quantum repeaters: the role of imperfect local operation in quantum communication. Phys. Rev. Lett. **81**(26), 5932-1-8 (1998). doi:10.1103/PhysRevLett.81.5932
53. L.M. Duan et al., Long-distance quantum communication with atomic assembles and linear optics. Nature **414**, 413–418 (2001). doi:10.1038/35106500
54. N. Sangouard et al., Quantum repeaters based upon atomic assembles and linear optics. Rev. Mod. Phys. **83**(1), 33–80 (2011). doi:10.1103/RevModPhys.83.33
55. N. Gisin, R. Thew, Quantum communication. Nat. Photonics **1**, 165–171 (2007). doi:10.1038/nphoton.2007.22
56. J.I. Cirac et al., Quantum state transfer and entanglement distribution among distant nodes in a quantum network. Phys. Rev. Lett. **78**, 3221–3227 (1997). doi:10.1103/PhysRevLett.78.3221
57. S. Ritter et al., An elementary quantum network of single atoms in optical cavities. Nature **484**, 195–200 (2012). doi:10.1038/nature11023
58. H.J. Kimble, The quantum internet. Nature **453**, 1023–1030 (2008). doi:10.1038/nature07127
59. I. Usami et al., Heralded quantum entanglement between two crystals. Nat. Photonics **6**, 254–237 (2012). doi:10.1038/nphoton.2012.34
60. R.J. Hughes et al., Network-centric quantum communication with application to critical infrastructure protection, in *Report LA-UR-13-22718*, Los Alamos National Laboratory, Los Alamos, New Mexico (2013)
61. M. Stipčević, Quantum random number generators and their use in cryptography, in *Proceedings of SPIE 8375, Advanced Photon Counting Techniques VI*, pp. 837504–837519, May 2012. doi:10.1117/12.919920
62. J.G. Rarity, P.C.M. Owens, P.R. Tapster, Quantum random-number generation and key sharing. J. Mod. Opt. **41**(12), 2435–2444 (1994). doi:10.1080/09500349414552281
63. T. Jennewein et al., A fast and compact quantum random number generator. Rev. Sci. Instrum. **71**(4), 1675–1680 (2000). doi:10.1063/1.1150518
64. A. Stefanov et al., Optical quantum random number generator. J. Mod. Opt. **47**(4), 595–598 (2000). doi:10.1080/0950034000823380
65. H.-Q. Ma, Y. Xie, L.-A. Wu, Random number generator based on the time of arrival of single photons. Appl. Opt. **44**(36), 7760–7763 (2005). doi:10.1364/AO.44.007760

66. Q. Yan, Multi-bit quantum random number generation by measuring positions of arrival photons. Rev. Sci. Instrum. **85**(10), 103116 (2014). doi:10.1063/1.4897485
67. O. Kwon, Y.-W. Cho, Y.H. Kim, Quantum random number generator using photon-number path entanglement. Appl. Opt. **48**, 1774–1777 (2009). doi:10.1364/AO.48.001774
68. C. Gabriel et al., A generator for unique quantum numbers based on vacuum states. Nat. Photonics **4**, 711–715 (2010). doi:10.1038/nphoton.2010.197
69. M. Jofre et al., True random numbers from amplified quantum vacuum. Opt. Express **19**(21), 20665-1-4 (2011). doi:10.1364/OE.19.020665
70. T. Symul, S.M. Assad, P.K. Lam, Real time demonstration of high bitrate quantum number generation with coherent laser light. Appl. Phys. Lett. **98**(23), 231103-1-4 (2011). doi:10.1063/1.3597793
71. F. Xu et al., Ultrafast quantum random generation based on quantum phase fluctuations. Opt. Express **20**(11), 12366-1-12 (2012). doi:10.1364/OE.20.012366
72. C. Abellán et al., Ultra-fast quantum randomness generation by accelerated phase diffusion in a pulsed laser diode. Opt. Express **22**(2), 1645–1654 (2014). doi:10.1364/OE.22.001645
73. Y.-O. Nie, 68 Gbps quantum random number generation by measuring laser phase fluctuations. Rev. Sci. Instrum. **86**, 063405-1-14 (2015). doi:10.1063/1.1922417
74. N.J. Cerf, C. Adami, P.G. Kwait, Optical simulation of quantum logic. Phys. Rev. A **57**(3), R1477–R1480 (1998). doi:10.1103/PhysRevA.57.R1477
75. E. Knill, R. Laflamme, G.J. Milburn, A scheme for efficient quantum computation with linear optics. Nature **409**, 46–52 (2001). doi:10.1038/35051009
76. P. Kok et al., Linear optical quantum computing with photonic qubits. Rev. Mod. Phys. **79**, 135–174 (2007). doi:10.1103/RevModPhys.79.135
77. R. Okamoto et al., Realization of a Knill-Laflamme-Milburn controlled-NOT photonic quantum circuit combining effective optical nonlinearities. Proc. Natl. Acad. Sci. U.S.A. **108** (25), 10067–10071 (2011). doi:10.1073/pnas.1018839108
78. T. Toffoli, Reversible computing. Technical Report MIT/LCS/TM-151 (1980)
79. T. Monz et al., Realization of the quantum Toffoli gate with trapped ions. Phys. Rev. Lett. **102**(4), 040501-1-11 (2009). doi:10.1103/PhysRevLett.102.041501
80. B.P. Lanyon et al., Simplifying quantum logic using higher-dimensional Hilbert spaces. Nat. Phys. **5**, 134–140 (2009). doi:10.1038/nphys1150
81. S. Aaronson, A. Arkhipov, The computational complexity of linear optics, in *Proceedings of the 43rd Annual ACM Symposium on Theory of Computing (STOC '11)* (2011), pp. 333–342. doi:10.1145/1993696.1993682
82. J.B. Spring et al., Boson sampling on a photonic chip. *Science* **339**. doi:10.1126/science.1231692
83. M.A. Broome et al., Photonic boson sampling in a tunable circuit. Science **339**, 6621–6629 (2013). doi:10.1126/science.1231440
84. M. Tillmann et al., Experimental boson sampling. Nat. Photonics **7**, 540–544 (2013). doi:10.1038/nphoton.2013.102
85. A. Crespi et al., Integrated multimode interferometers with arbitrary designs for photonic boson sampling. Nat. Photonics **7**, 545–549 (2013). doi:10.1038/nphoton.2013.112
86. D. Deutsch, Quantum theory, the Church-Turing principle and the universal quantum computer. Proc. R. Soc. A **400**, 97–117 (1985). doi:10.1098/rspa.1985.0070
87. M.S. Tame et al., Experimental realization of Deutsch's algorithm in a one-way quantum computer. Phys. Rev. Lett. **98**, 140501-1-4 (2007). doi:10.1103/PhysRevLett.98.140501
88. D. Deutsch, R. Jonza, Rapid solution of problems by quantum computation. Proc. R. Soc. A **439**, 553–558 (1992). doi:10.1098/rspa.1992.0167
89. P.W. Shor, Polynomial-time algorithms for prime factorization and discrete logarithms on a quantum computer. SIAM J. Sci. Stat. Comput. **26**, 1484–1509 (1997)

Note: This is an expanded version of an article originally appearing in Proceedings of the 35th Annual Symposium of Computer Science, Santa Fe, 1994

90. C.-Y. Lu et al., Demonstration of Shor's quantum factoring algorithm using photonic qubits. Phys. Rev. Lett. **99**(25), 250504-1-5 (2007). doi:10.1103/PhysRevLett.99.250504
91. B.P. Layton et al., Experimental demonstration of Shor's algorithm with quantum entanglement. Phys. Rev. Lett. **99**, 250505-1-5 (2007). doi:10.1103/PhysRevLett.99.250505
92. G. Brida et al., An extremely low-noise heralded single-photon source: a breakthrough for quantum technologies. Appl. Phys. Lett. **101**, 221112-1-11 (2012). doi:10.1063/1.4768288
93. C. Monroe et al., Large-scale quantum computer architecture with atomic memory and photonic interconnects. Phys. Rev. A **89**, 022317-1-16 (2014). doi:10.1103/PhysRevA.89.022317
94. F.S.N. Lobo, Closed Timelike Curves and Causality Violation, Chap. 6 in *Classical and Quantum Gravity: Theory, Analysis, and Applications,* Nova Scientific Publishers (2012)
95. D. Deutsch, Quantum mechanics and closed timelike lines. Phys. Rev. D **44**, 3197–3217 (1991). doi:10.1103/PhysRevD.44.3197
96. M. Ringbauer et al., Experimental simulation of closed timelike curves. Nat. Communications **5**, 4145–4152 (2014). doi:10.1038/ncomms5145
97. T.A Brun, J. Harrington, M.M. Wilde, Localized closed timelike curves can perfectly distinguish quantum states. Phys. Rev. Lett. **102**, 210402-1-4 (2009). doi:10.1103/PhysRevLett.102.210402
98. D. Ahn et al., Quantum-state cloning in the presence of a closed timelike curve. Phys. Rev. A **88**(2), 022332-1-5 (2013). doi:10.1103/PhysRevA.88.022332
99. T.A. Brun, M.M. Wilde, A. Winter, Quantum state cloning using Deutschian closed timelike curves. Phys. Rev. Lett. **111**(19), 19040-1-6 (2013). doi:10.1103/PhysRevLett.111.190401
100. J.L. Pienaar, T.C. Ralph, C.R. Myers, Open timelike curves violate Heisenberg's uncertainty principle. Phys. Rev. Lett. **110**, 060501-1-5 (2013). doi:10.1103/PhysRevLett.110.060501
101. T.A. Brun, Computers with closed timelike curves can solve hard problems. Found. Phys. Lett. **16**, 245–253 (2002)
102. S. Aaronson, J. Watrous, Closed timelike curves make quantum and classical computing equivalent. Proc. R. Soc. A **465**(2102), 631–647 (2008). doi:10.1098/rspa.2008.0350

Chapter 10
Light in Free-Space and the Cosmos

10.1 The Speed of Light in the Cosmos

10.1.1 The Cosmic Speed of Light and Planck Length

In the special theory of relativity Einstein stated that the cosmic speed of light c in the vacuum of the universe has a constant and absolute value of 299,792,458 m/s. This constancy is required so the laws of physics in one reference frame are the same in another reference frame, defined by the Lorentz transformations given in Sect. 3.4. This is called *Lorentz invariance*. Thus any evidence of a violation of Lorentz invariance could question the constant speed of light.

Also, Einstein stated that the structure of space-time was smooth and continuous. However, recent theories suggest that at the smallest dimensions of space and time, where quantum fluctuations take place, space time has a dynamic non-smooth or lumpy structure. These are caused by quantum gravity (QG) effects in the region of the smallest known length, the so-called *Planck length* l_{Planck},

$$l_{Planck} = \sqrt{hG/2\pi c^3} \approx 1.62 \times 10^{-35}\text{m}, \tag{10.1}$$

where h is Planck's constant, G is the gravitational constant, and c is the speed of light.

This graininess of space-time, sometimes called *quantum foam*, could affect the velocity of photons that propagate through it. Some theoretical models for QG predict an energy dependence of the speed of light by quantum gravity dispersion, implying a violation of Lorentz invariance.

D.F. Vanderwerf, *The Story of Light Science*,
DOI 10.1007/978-3-319-64316-8_10

10.1.2 Gamma-Ray Bursts and the Cosmic Speed of Light

The MAGIC (Major Atmospheric Gamma-ray Imaging Cherenkov) telescope at La Palma, Spain has been used to study the arrival times of energetic photons from a gamma-ray burst (GRB) millions of light years away [1]. It could measure gamma-ray energies over a range between 30 GeV and 30 TeV. On July 9, 2005 the MAGIC telescope was pointed at a flare of the active galactic nucleus Markarian 501, about 500 million light years away. The energy range of this flare was 120 GeV to over 1.2 TeV [2]. It was expected that all gamma-ray photons were emitted simultaneously from the flare, and surprisingly, there were arrival time delays between photons of different energies.

It was found that high-energy photons arrived about four minutes later than the low-energy photons. This implied that the propagation speed was dependent on the photon energies, and that the cosmic speed of light was not constant when traveling through the quantum foam at the Planck length scale. This would challenge Einstein's second postulate and indicate a violation of Lorentz invariance.

There have been several other follow-up gamma-ray galactic burst measurements to test these time delays. On July 28, 2006 a high-energy flare was observed from the blazer PKS 2155-304 using the High Energy Stereoscopic System (HESS) four-unit array of gamma-ray telescopes in Namibia, Africa. This gamma-ray outburst occurred about four times more distant than the Markarian 501 flare, and measured delays from about zero to 30 s produced no definite conclusions. The NASA Fermi Gamma-ray Space Telescope (FGST) was launched in June 2008. Several gamma-ray bursts have already been observed using the FGST, notably GRB080916C, a massive burst about 12 billion light years away in the constellation Carina, and observed on September 15, 2008 [3]. The arrival time difference between low-energy photons and high-energy photons was about 16 s. Whether subsequent GRB measurements from these telescopes will change one of the most widely-accepted laws of relativistic physics is yet to be determined.

10.1.3 The Speed of Light and the Fine Structure Constant

The *fine structure constant* α is a fundamental physical constant and was introduced by Arnold Sommerfeld in 1916. It describes the coupling strength between electrically charged particles. It can be simply defined as the ratio of the velocity of an electron in the first circular Bohr orbit to the velocity of light. It can also be defined in terms of other fundamental constants, for example,

$$\alpha \equiv \frac{e^2}{2\varepsilon_0 hc} = \frac{e^2 \mu_0 c}{2h}, \tag{10.2}$$

where e is the elementary charge of an electron, h is Planck's constant, ε_0 is the vacuum permittivity, μ_0 is the vacuum permeability, and c is the cosmic speed of light. The fine structure constant is dimensionless and has a currently accepted value $\alpha = 7.2973525698 \times 10^{-3}$.

The first evidence that the value of the fine structure constant has changed over cosmological distance and time was shown by John Webb and associates at the University of New South Wales in 1999 [4]. In 2001, from measurements of many quasars using the Keck telescope and the Very Large Telescope, they determined the value of α by comparing absorption spectra at red shifts between 0.5 and 3.5 [5]. The results indicated a slight increase in the value of α over 12–15 billion years as

$$\frac{\Delta\alpha}{\alpha} = \frac{\alpha_{further} - \alpha_{closer}}{\alpha_{closer}} = -0.72 \pm 0.18 \times 10^{-5}. \tag{10.3}$$

Some scientists have argued that this change in the fine structure constant α would most probably be due to changes in the speed of light c in Eq. (10.2). This change in α could be due to a spacetime quantum foam effect over cosmological distances.

In addressing the relationship between α and c, Antonio Alfonso-Faus in Spain introduced a new system of electromagnetic units that preserved local Lorentz invariance and local position invariance [6]. In this system $\varepsilon_0 = \mu_0$, and since $\varepsilon_0\mu_0 = 1/c^2$, then $\varepsilon_0 = \mu_0 = 1/c$. Recasting Maxwell's equations leads to a modified form of Coulomb's law $F = c\frac{qq'}{r^2}$, and a new expression for the fine structure constant $\alpha = \frac{e^2}{4\pi\hbar}$. where $\hbar$ is the reduced Planck's constant $h/2\pi$. This expression does not have c as a factor. Thus, if the fine structure constant has undergone a cosmological change, it is not due to a change in the speed of light c.

10.1.4 Isotropy of the Speed of Light in Space

Possible anisotropy of the speed of light in space has been investigated with increased accuracy in recent years. Modern variations of the Michelson-Morley and Kennedy-Thorndike experiments still test for interferometric fringe shifts in orthogonal directions, with the apparatus oriented in various directions in space. They do not absolutely establish that the speed of light is isotropic in space, but that the anisotropy limit $\Delta c/c$ is so small that a dependence of light speed c on direction cannot be inferred. From a relativity standpoint, anisotropy of the speed of light would lead to a violation of Lorentz invariance. Therefore some experiments concentrate on testing the validity of Lorentz invariance by measuring the anisotropy of the speed of light.

A 1990 version of the Kennedy-Thorndike experiment searched for variations between a laser locked to a reference line and a laser locked at the resonance frequency of a stable cavity [7]. No variations were found at a level of 2×10^{-13}, which was about 300 times better than the original 1932 Kennedy-Thorndike experiment.

Two recent Michelson-Morley type experiments have been conducted that measure the anisotropy $\Delta c/c$ to very precise levels using cryogenic cooled orthogonal optical resonators [8, 9]. The values of $\Delta c/c$ were at the 10^{-15} and 10^{-16} levels, respectively. An improved method in 2009 further reduced the anisotropy limit $\Delta c/c$ to the 10^{-17} level. using two rotating optical resonator cavities in a block of fused silica. In 2015 a team of scientists from Germany and Australia reported measurements of Lorentz symmetry at the 10^{-18} level using two sapphire resonant cavities [10]. The experiment is illustrated in Fig. 10.1. Two sapphire discs of 50 mm diameter and 30 mm thickness were held in a mount with orthogonal orientation of their crystal axes. The discs were cooled to 4 K in a helium-filled container and continuously rotated at a period of 100 s on an air bearing turntable with tilt control. Each sapphire disk was driven to a whispering gallery mode at a 12.97 GHz resonance frequency. Two looped oscillators were locked to the disc resonant frequencies such that measurements of a beat frequency difference during rotation would indicate any light speed anisotrophy as the direction was changed. Using data taken over a year, it was found that any frequency change $\Delta\nu/\nu$ due to a violation of Lorentz symmetry of the photon was less than $9.2 \pm 10.7 \times 10^{-19}$. Thus no significant violation of Lorentz symmetry was revealed.

Controversies still exist concerning Michelson-Morley type experiments that detect no significant speed of light anisotropy. For example, Gezari of the NASA

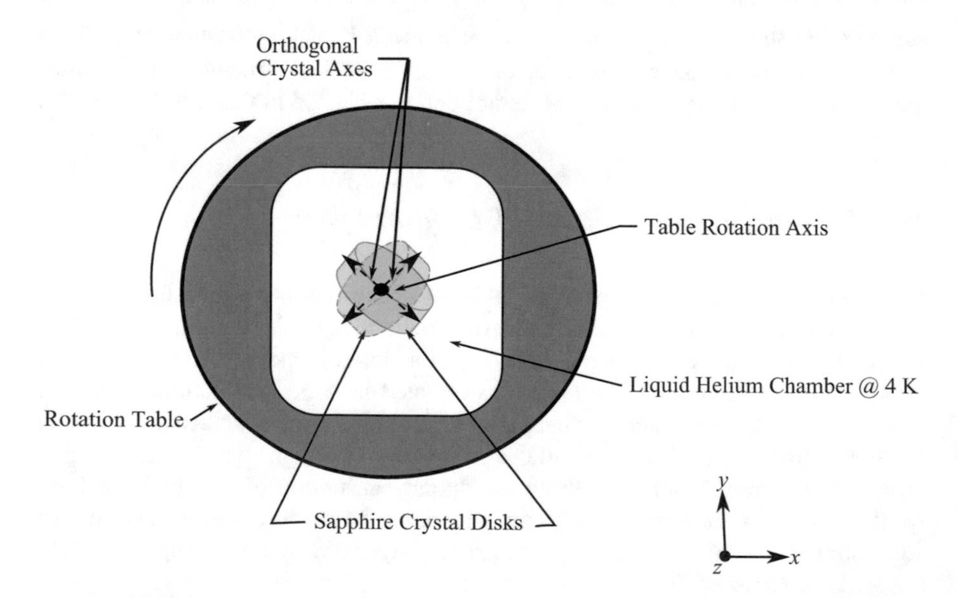

Fig. 10.1 Ultra-precise measurement of light speed anisotropy in vacuum using dual sapphire disks with orthogonal orientation of their crystal axes. Sapphire disks are held in a mount attached to a rotating platform, and are separated by a distance along the z-axis (shown in phantom overlay). Mount and output electronics are not shown

Goddard Space Flight Center has used lunar ranging to bounce light signals from Earth to the Moon and back using a retroreflector placed on the Moon [11]. The calculated speed of light was about 200 m/s higher than the accepted value of c = 299,792,458 m/s. He attributed this to the line-of-sight rotational motion of the Earth ($\approx \pm$ 300 m/s). He then concluded that the speed of light depends on the motion of the observer, that Lorentz invariance is violated, and that light speed is anisotropic. In an independent report, Cahill analyzed data from the Apache Point Lunar Laser-ranging Lasing Operation, which bounced Earth generated photons from a retroreflector on the Moon with a timing resolution of 0.1 ns [12]. He concluded that the data also indicated an anisotropy of the speed of light. The results of these two experiments have been critically discussed by many scientists, as they indicate deviations from the widely-accepted theory of special relativity and Lorentz invariance.

10.1.5 Slowing the Speed of Light in Free-Space

In the free-space of the cosmos, where the refractive index $n \equiv 1$, the speed of light is considered as invariant, traveling at the cosmic speed c. However, several investigators in the United Kingdom have measured a reduced group velocity of light in free-space using spatially structured photons [13]. A structured Bessel light beam was created in free-space using a diffractive axicon element with a plane wave input, as shown in Fig. 10.2. In the Bessel beam region conical wave fronts were specified by wavefront propagation vectors $k_0 = 2\pi/\lambda$ at an angle α with a vector k_z in the z-direction, such that $k_z^2 + k_x^2 + k_y^2 = k_0^2$. For $k_r << k_0$, a radial wave vector k_r can be

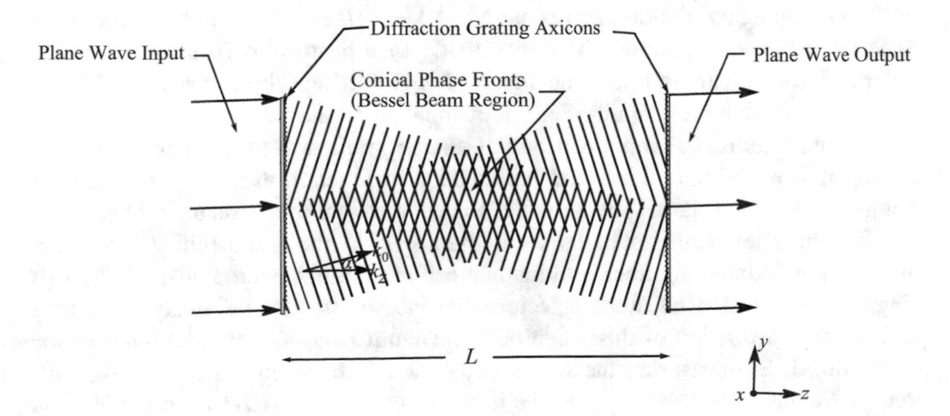

Fig. 10.2 Creation of a structured light Bessel beam in free-space using two diffraction grating axicons separated by a distance L. (Adapted with permission of AAAS from Ref. [14]; permission conveyed through Copyright Clearance Center, Inc.)

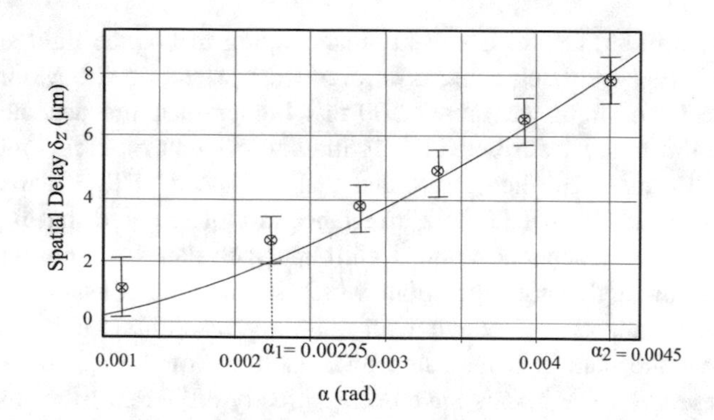

Fig. 10.3 Plot of calculated spatial delay $\delta_z = (L/2)\alpha^2$ versus α, where $L = 800$ cm. Calculated delays are $\delta_z = 2.0\,\mu\text{m}\,(\alpha_1 = 0.00225\,\text{rad})$ and $\delta_z = 8.1\,\mu\text{m}\,(\alpha_2 = 0.00450\,\text{rad})$. Several measured delay values are also shown. (Adapted with permission of AAAS from Ref. [14]; permission conveyed through Copyright Clearance Center, Inc.)

defined as $k_r = \sqrt{2k_0(k_0 - k_z)}$. Axicons can be designed to produce various α values, where $\alpha = k_r/k_0$. An output diffractive axicon can then convert the structured beam back to a plane wave. In the free-space Bessel beam region the group velocity v_g in the z-direction is given by $v_g = (1 - \alpha^2/2)c$, reducing the v_g value compared to free-space. The slight group velocity reduction from the cosmic speed c was expected to produce a time delay of about 30 fs over a travel distance $L = 1$ m. For example, if $\alpha = 0.0045$, then $v_g = \left[1 - (0.0045)^2/2\right]c = 0.9999899c$, producing a spatial delay $\delta_z = (L/2)\alpha^2$ of about 8 μm. Figure 10.3 shows the relationship between spatial delay and α values.

In the experiment a single parametric down-conversion BBO crystal light source produced entangled photon pairs centered at $\lambda = 710$ nm. The photons were separated and the idler photon was sent directly to a beamsplitter through a coiled polarization-maintaining fiber. The signal photon was sent through a variable path length polarization-maintaining fiber, and inputted to the free-space Bessel structured beam region. The input and output axicon gratings were pixilated reflective spatial light modulators that could be encoded to function as a simple diffraction grating, axicon, or lens. The travel distance through the structured beam was $L = 80$ cm. The exiting photons were then sent to the beamsplitter through a polarization-maintaining fiber, and the beamsplitter outputs were registered by two single-photon avalanche diode detectors. To measure the spatial delay associated with a speed reduction of this magnitude, quantum Hong-Ou-Mandel interference was utilized. By measuring the spatial separation δ_z between the H-O-M dips for various values of α, resulting changes in v_g were calculated relative to the H-O-M dip position for a collimated beam traveling at speed c. Although this experiment was performed in a thin spatial slice of the cosmos, it clearly demonstrated the occurence of a reduction of light speed in free-space for non-planar light waves.

10.1.6 Waves Exceeding the Speed of Light in Free-Space

Some of the earlier reports on superluminal behavior of electromagnetic waves in free-space involved radio waves and microwaves [14, 15]. In 1991 Gaikos and Ishii measured the leading edge of microwave pulses in free-space and a waveguide, and associated this edge with a superluminal phase velocity [16]. Their measurements indicated a resultant phase velocity $v_p = c/\cos\theta$, where θ is the observation angle of a propagating TEM wave. However, several other investigators have challenged the experimental accuracy of these measurements and their resultant contradiction of Maxwell's equations. Other researchers performed actual measurements of RF pulsed signals, showing they could not travel faster than light, and transported energy at the group velocity v_g [17]. Other investigators performed mathematical simulations of experiments that could lead to possible misinterpretation of the faster-than-light measurements [18]. The results of these differing reports and rebuttals are still being studied.

An experiment demonstrating superluminal propagation of light in free-space over distances on the order of one meter was performed by Mugnai and associates at the Italian National Research Council in Firenze in 2000 [19]. Microwaves of 3.5 cm wavelength were sent through a narrow, ring-shaped opening onto a 50 cm diameter focusing mirror. The mean diameter of the ring d was fixed at either 7 or 10 cm. Point light sources originating at positions on the ring were focused at half-angle θ and entered a horn receiver to a detector. See Fig. 10.4a. The distance L between the ring source and the horn receiver aperture was varied between 30 and 140 cm, giving θ values between 17° and 23°. The 8.6 GHz source was rapidly modulated with rectangular pulses detected in the horn receiver, with the waves moving along the z-axis with a superluminal velocity $v = c/\cos\theta$, where c is the cosmic light speed. A series of arrival times T at the detector delays was measured with an accuracy of ≈10 ps, and using adjacent (L, T) data pairs, a series of delay times ΔT was obtained. A series of velocity points was then calculated from $v = \Delta L/\Delta T$. To maximize the superluminal effect, the 10 cm diameter ring aperture was used to produce $\theta \approx 23°$. As shown in Fig. 10.4b, the light velocity was about 5–7% above c at $L \approx 40$ cm, but decreased as L increased. Beyond a propagation distance of about 1 m, the light speed approached the cosmic speed c and the observed superluminality was negligible. In 2001 a theoretical explanation for these free-space superluminal results was reported by researchers in Brazil [20].

In 2004 John Singleton at Los Alamos Laboratory described how the envelope of the polarization current density $\mathbf{J}_P = \partial \mathbf{P}/\partial t$ could act as an extended source to produce superluminal electromagnetic radiation [21]. The electric polarization vector **P** appears when the Ampère-Maxwell law (Eq. 2.8) is applied to a dielectric material. This type of extended source has no rest mass and avoids the Einstein limit on exceeding the cosmic speed c. To demonstrate this, a prototype machine

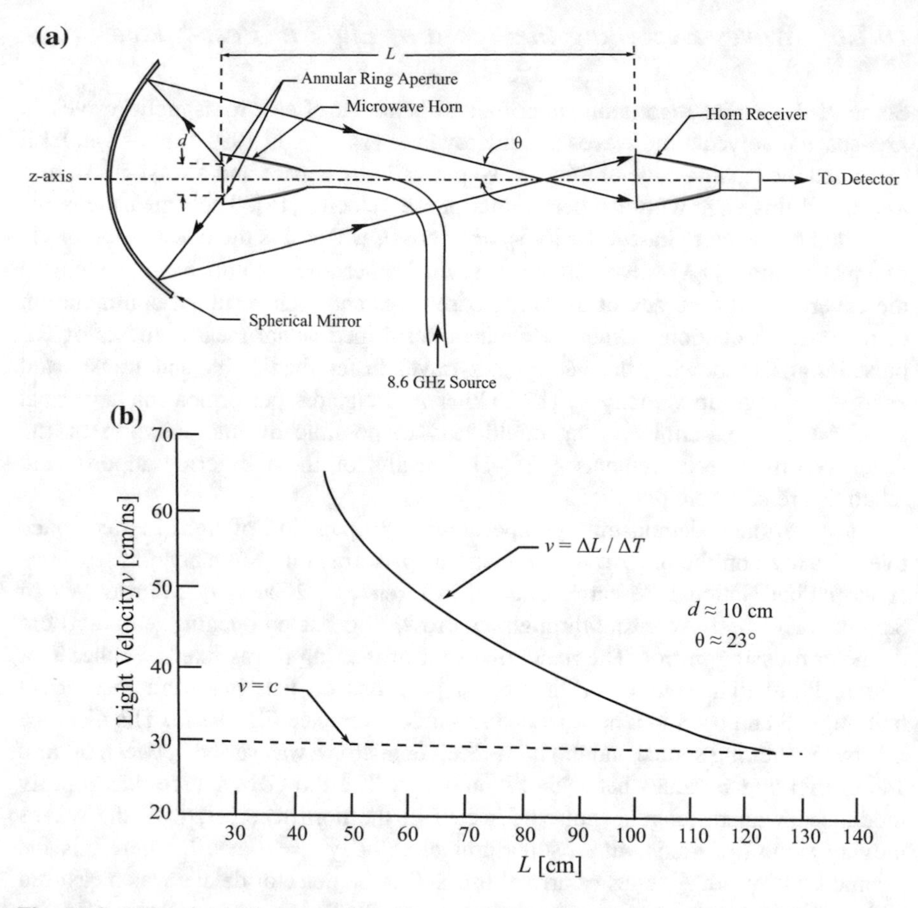

Fig. 10.4 **a** Experimental setup to demonstrate superluminal velocity of light in free-space using microwave pulses. (This figure adapted with permission from Ref. [20]. Copyrighted by the American Physical Society). **b** Plot of velocity v versus separation L, showing the superluminal region and the approach to the cosmic velocity c as L exceeds one meter

called a polarization synchrotron was constructed at Los Alamos [22]. A section of this device is illustrated in Fig. 10.5a, b. The polarization synchrotron source consisted of a curved dielectric sheet of alumina (Al_2O_3, dielectric constant $\approx$10), with a radius of curvature r = 10.025 m, mounted on a continuous ground electrode. An array of 41 separated electrodes, each driven by a varying voltage, covered a portion of the top of the dielectric. As the electrodes were sequentially activated, separation of the alumina ions along the moving electric field formed a polarization current density $\mathbf{J}_P$ which moved along the dielectric. The curvature of the dielectric produced centripetal acceleration of $\mathbf{J}_P$ and Cherenkov type radiation was emitted. Emission of superluminal light with speeds up to 8c has been reported.

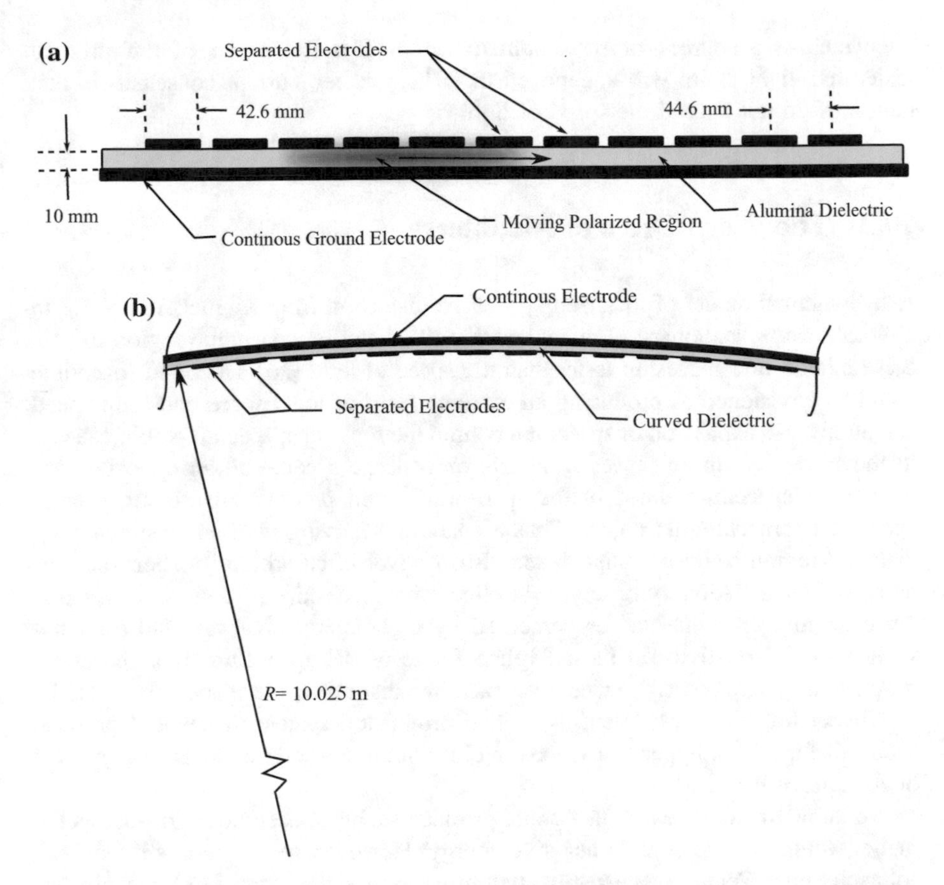

Fig. 10.5 **a** Sectional view of the polarization synchrotron superluminal light source with moving polarized region. **b** Sectional view of the curved polarization synchrotron producing centripetal acceleration of the polarized region

10.1.7 Particles Exceeding the Speed of Light in Free-Space

The possibility of relativistic particles that can exceed the speed of light c has been analyzed by Feinberg using quantum field theory [23]. These particles would be non-interacting and spinless, and satisfy the principle of relativity.

Some actual evidence from the OPERA Collaboration Team was presented in 2011, indicating that neutrinos may have exceeded the cosmic speed of light c [24]. Neutrinos from the CERN particle accelerator traversed a 730 km path under the Alps from Geneva, Switzerland to a detector in Gran Sasso, Italy. Using accurate measurements from over 15,000 neutrino events, the OPERA team in 2011 concluded that the average arrival time was about 60 ns faster than expected for travel at the cosmic speed of light c. However, other researchers questioned these results as a violation of special relativity, and further analysis by the OPERA team uncovered two sources of measurement error [25]. Retraction of the original faster-than-light claim was announced in 2012, and the current consensus is that neutrinos do not exceed the speed of light *c*.

10.2 The Warp Drive of Alcubierre

A hypothetical model of localized spacetime distortion from Miguel Alcubierre in 1994 allowed a stationary observer outside the distorted spacetime region to perceive motion of a spaceship faster than the speed of light [26]. Distorted spacetime could be envisioned as producing an extreme form of local space and light speed anisotropy. An expansion of spacetime behind the moving spacecraft would cause it to move away from an object at an arbitrarily large speed, and an opposite contraction of spacetime ahead of the spacecraft could produce an arbitrarily large speed in approaching an object. Thus a spaceship moving with an accompanying distorted region could accomplish round-trip travel in an arbitrarily short time, as perceived by a stationary observer. In effect, the spaceship traverses a longer distance for any time interval, as perceived by a stationary observer, and does not violate special relativity in the reference frame of the spacecraft. If a spacecraft propulsion system could produce this spacetime distortion, the spacecraft would be effectively traveling faster than light. The propulsion system that would produce such a distortion for hyper-fast space travel is commonly referred to as a *warp drive* or *Alcubierre drive*.

No known material exists that could produce such a spacetime distortion, as the matter would need to contain negative energy. However, researchers at the NASA Johnson Space Center are currently attempting to produce a proof-of-concept warp drive propulsion system. One approach is to fold spacetime around a photon so it travels over a greater distance at the speed of light. Another investigation uses "quantum-thruster" physics, which would generate the negative vacuum energy necessary for the required distortion of spacetime. Time will tell whether these approaches lead to a practical warp drive propulsion system for hyper-fast space travel.

10.3 Dark Matter and Gravitational Lensing

In 1933 astrophysicist Fritz Zwicky, using a Schmidt telescope, was studying the movement of galaxies in the COMO galaxy cluster about 320 million light-years from Earth. Measuring Doppler shifts, he found the velocities of the galaxies to greatly exceed expected values. Using a classical theorem describing the effect of gravitation on orbiting objects, the total mass of the COMA cluster was calculated. Measurement of the total light output from the cluster produced an estimation of the expected stellar mass of the cluster. Surprisingly, the gravitational mass exceeded the stellar mass by a factor of about ten! Zwicky proposed that a high proportion of the COMO cluster mass emitted no radiation, and called this invisible mass *dark matter*. In the late 1970s astronomer Vera Cooper Rubin measured the rotation properties of several spiral galaxies, confirming that their rotational properties were higher than anticipated—thus advancing the argument for the existence of dark matter [27].

As predicted by Einstein and measured by Eddington (Sect. 3.9.7), the gravitational mass of the Sun warps the space-time fabric in its vicinity, causing a curvature of light from stellar objects that pass near its surface. This effect also occurs on a cosmic scale for distant galaxies and galactic clusters. In 1937 Zwicky suggested that a *gravitational lensing* effect could be used as a measurement tool to detect dark matter in deep space. In practice, a foreground cluster galaxy acts as a lens for a galaxy cluster background source, where the background source could be several billion light-years behind the foreground lens. This lensing effect can be classified as strong or weak. The rare strong lensing effect occurs when the mass density of the gravitational lens is high, causing the background galaxy cluster to appear as arcs, multiple images, or even a complete ring. Using strong lensing viewed through the Hubble space telescope, J. Anthony Tyson and team members at the University of California Davis constructed a mass map of the CL0024+1654 galactic cluster [28]. When the mass density of the gravitational lens falls below a critical value, the more common weak gravitational lensing occurs, and produces stretched (sheared) and magnified (converged) images of the background source cluster. Weak lensing is very useful in detecting the distribution of dark matter over large areas of the cosmos. Using data from145,000 background galaxies, in 2008 David M. Wittman and co-investigators determined the dark matter distribution at several half-degree view angles, and the presence of *cosmic shear* [29].

Gravitational lenses function as natural achromatic cosmic telescopes. The additional magnification, used in conjunction with earth-based and orbiting space telescopes, can enhance the observation of small faint galaxies in the distant universe. An excellent review has been written on gravitational lensing by Richard S. Ellis in 2010, close to the 90th anniversary of the landmark Eddington solar light-bending measurements [30]. Current estimates of the composition of the universe are: 4% normal matter, 21% dark matter, and 75% of the even more mysterious *dark energy*.

10.4 The Cosmological Beginning of the Photon

Although the evolution and properties of the physical universe are based on sound scientific principles, the conclusions drawn are often speculative. The Big Bang theory of the origin and expansion of the universe is generally accepted, but specific details concerning the Big Bang and predictions for the ultimate fate of the physical cosmos, and in particular the photon, are less certain. The photon first made its appearance in the *Lepton Epoch* (about 1 s to 3 min after the Big Bang), where the collision of dominating electrons (leptons) and positrons (anti-leptons) created energetic photons which further reacted with these electrons and positrons. In the subsequent *Photon Epoch* (about 3 min to 240,000 years later) photon radiation dominated the universe, although the photons continued to interact with other particles. In the following period (about 240,000–380,000 years), photons became *decoupled* from other energetic particles, and traveled as non-interacting particles. Around this time, at a universe temperature of about 3000 K, the stable atoms hydrogen and helium were formed and the universe became transparent to radiation. The cosmic background radiation (CBR) became visible as a blackbody spectrum.

10.5 Temperature of the Universe and Wien's Law

Cosmic microwave background (CMB) radiation was accidentally discovered by Penzias and Wilson in 1964, and represents the oldest observable radiation in the universe [31]. Due to red shift of light radiation caused by an expanding universe, the wavelength of this radiation has increased from short gamma ray wavelengths near the time of the Big Bang to microwave wavelengths at the present time. The weak CMB radiation is present throughout space. For their discovery of the cosmic microwave background, Penzias and Wilson were awarded the 1978 Nobel Prize in Physics.

In 1989 the Cosmic Background Explorer (COBE) satellite was launched to perform CMB measurements. Using the far-infrared absolute spectrophotometer (FIRAS), the CMB photon spectrum was accurately measured to 0.005% accuracy [32]. The measured spectrum accurately followed a blackbody curve as predicted by Big Bang theory with a peak wavelength $\lambda_{max}^{\text{current}} \approx 1.063$ mm ($f_{max} = 160.2$ GHz). See Fig. 10.6. For this groundbreaking work on measuring the blackbody spectrum of the CMB, John Mather received the Nobel Prize in Physics in 2006 (shared with colleague George Smoot for his measurements on the anisotropy of the CMB). Assuming the CMB temperature represents the current temperature of the universe T_{univ}, application of Wien's law thus predicts $T_{univ} = \frac{2.897\times10^{-3}}{1.063\times10^{-3}} = 2.725$ K. A recent determination of the temperature of the CMB using data from the far-infrared spectrophotometer on the COBE satellite has obtained a value $T_{\text{CMB}} = 2.7260 \pm 0.0013$ K [33].

The temperature of the universe shortly after the Big Bang has been estimated to be about two trillion degrees. When the universe cooled to a temperature of about 3000 K, neutral hydrogen atoms were formed. Photons were no longer scattered by electrons, became *decoupled*, and streamed freely through space. The expanding universe then became transparent to radiation. At the time of this photon decoupling the peak wavelength of the blackbody curve was $\lambda_{max}^{\text{decoupling}} \approx \frac{2.897\times10^{-3}\text{m K}}{3\times10^{3}\text{ K}} \approx 9.66\times10^{-7}$ m. This is in the near-infrared region of the electromagnetic spectrum. The expansion factor of the universe from that time to the present can then be

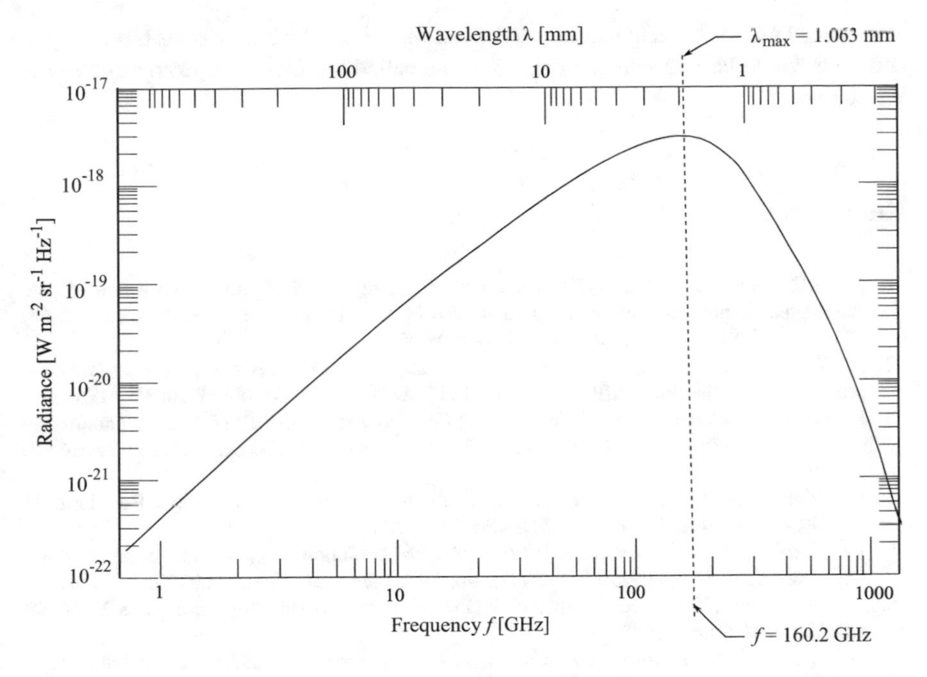

Fig. 10.6 Cosmic microwave background radiation spectrum as measured by the COBE satellite

approximated from the ratio $\frac{\lambda_{max}^{current}}{\lambda_{max}^{decoupling}} = \frac{1.063\text{mm}}{9.66\times10^{-4}\text{mm}} \approx 1100$. Since the universe is still expanding, we can expect a further drop in the current CMB temperature.

10.6 The Cosmological Fate of the Photon

In one scenario of the evolution of the universe, the physical cosmos continues expanding until the end, or forever. At the present time, about 13.6 billion years after the Big Bang, the estimated size of the universe is about 10^{24} km across. Jumping forward in time to about 10^{14} years, the greatly expanded universe consists mainly of black holes, which have swallowed most remaining celestial matter [34]. There is also a collection of elementary particles, including photons. However, over a period of about 2.7 billion years the black holes lose mass due to the emission of positive mass particles, and lose energy due to a type of blackbody radiation emission known as *Hawking radiation*. They eventually shrink and get so hot that they explode or evaporate. At a time of 10^{100} years after the Big Bang, the universe might consist of separated photons, neutrinos, electrons, and positrons. In this scenario the photons would have an enormous red shift wavelength, and the temperature of the universe would be close to absolute zero. Probably in this era of low-energy photons, a degenerate state would exist in which all matter has decayed and time would be fragmented. This could be called the *Dark Era* or the *Age of the Weak Photon*.

References

1. MAGIC Collaboration Team, Probing quantum gravity using photons from a flare of the active galactic nucleus Markarian 501 observed by the MAGIC telescope. Phys. Lett. B **68**, 253–257 (2008). doi:10.1016/j.physletb.2008.08.053
2. R. Wagner, Exploring quantum gravity with very-high-energy gamma-ray instruments—prospects and limitations. AIP Conf. Proc. **1112**, 187–197 (2009). doi:10.1063/1.3125781
3. Fermi LAT and Fermi GBM Collaborations, Fermi observations of high-energy gamma-ray emission from GRB 080916C. Science **323**(5922), 1688–1693 (2009). doi:10.1126/science.1169101
4. J.K. Webb et al., Search for time variation of the fine structure constant. Phys. Rev. Lett. **82** (5), 884–887 (1999). doi:10.1103/PhysRevLett.82.884
5. J.K. Webb et al, Further evidence for cosmological evolution of the fine structure constant. Phys. Rev. Lett. **87**(9), 09130-1-4 (2001). doi:10.1103/PhysRevLett.87.091301
6. A. Alfonso-Faus, The speed of light and the fine structure constant. Phys. Essays **13**(1), 46–49 (2001). doi:10.4006/1.3025
7. D. Hils, J.L. Hall, Improved Kennedy-Thorndike experiment to test special relativity. Phys. Rev. Lett. **64**(15), 1697–1700 (1990). doi:10.1103/PhysRevLett.64.1697
8. H. Müller et al., Modern Michelson-Morley experiment using cryogenic optical resonators. Phys. Rev. Lett. **91**, 020401–020404 (2003). doi:10.1103/PhysRevLett.91.020401
9. S. Herrmann, et al., Test of the isotropy of the speed of light using a continuously rotating optical resonator. Phys. Rev. Lett. **95**, 15040-1-6 (2005). doi:10.1103/PhysRevLett.95.150401
10. M. Nagel, et al., Direct terrestrial test of Lorentz symmetry in electrodynamics to 10^{-18}. *Nature Comm.* **6**, 8174-1-6 (2015). doi:10.1038/ncomms9174
11. D.Y. Gezari, Lunar laser ranging test of the invariance of *c*. *arXiv: 09 12.3934*, 1–16 (2009)
12. R.T. Cahill, Lunar laser-ranging detection of light-speed anisotropy and gravitational waves. Prog. Phys. **2**, 31–35 (2010)
13. D. Giovannini et al., Spatially structured photons that travel in free-space slower than the speed of light. Science **347**(6224), 857–860 (2015). doi:10.1126/science.aaa3035
14. H.W. Milnes, On electrical signals exceeding the velocity of light. Toth-Maatian Rev. **2**(4), 870–890 (1984)
15. T.K. Ishii, G.C. Giakos, Transmit radio messages faster than light. Microwaves and RF **30**, 114–119 (1991)
16. G.C. Giakos, T.K. Ishii, Rapid pulsed microwave propagation. IEEE Microwave Guided Wave Lett. **1**, 374–375 (1991)
17. W. Su, I.M. Bessieris, S.M. Riad, Velocity of an RF pulse signal propagating in a waveguide. IEEE Microwave Guided Wave Lett. **2**(6), 255–256 (1992)
18. M.E. Adamski, A. Abramowicz, M.T. Faber, How fast can radio messages be transmitted? IEEE Antennas Propag. Mag. **36**(4), 94–96 (1994). doi:10.1109/74.317809
19. D. Mugnai, A. Ranfagni, R. Ruggeri, Observation of superluminal behaviors in wave propagation. Phys. Rev. Lett. **84**(21), 4830–4833 (2000). doi:10.1103/PhysRevLett.84.8430
20. W.A. Rodriguez, D.S. Thober, A.L. Xavier, Causal Explanation for observed superluminal behavior of microwave propagation in free apace. Phys. Lett. A **284**(6), 217–224 (2001). doi:10.1016/S0375-9601(01)00316-4

21. A. Ardavan et al., Experimental observation of nonspherically-decaying radiation from a rotating superluminal source. J. Appl. Phys. **96**(8), 4614–4631 (2004). doi:10.1063/1.1787591
22. J. Singleton, et al., Experimental demonstration of emission from a superluminal polarization current—a new type of solid-state source for MHz–THz and beyond. in *Joint 29th Int. Conf. on Infrared and Millimeter Waves and 12th Int. Conf. on Terrahertz Electronics* (2004), pp. 591–592
23. G. Feinberg, Possibility of faster-than-light particles. Phys. Rev. **159**(5), 1089–1105 (1967). doi:10.1103/PhysRev.159.1089
24. G. Brumfiel, Particles break light-speed limit. Nature, (September 22, 2011). doi:10.1038/news.2011.554
25. E.S. Reich, Flaws found in faster-than-light neutrino measurements. Nature, (February 27, 2012). [doi:10.1038/nature.2012.10099]
26. M. Alcubierre, The warp drive: hyper-fast travel within general relativity. Class. Quant. Grav. **11**, L73–L77 (1994). doi:10.1088/0264-9381/11/5/001
27. V. Rubin, W.K. Ford Jr., N. Thonard, Rotational properties of 21 Sc galaxies with a large range of luminosities and radii from NGC 4605 (R = 4kpc) to NGC 2885 (R = 1224kpc). Astrophys. J. **238**, 471–487 (1980). doi:10.1086/158003
28. J.A. Tyson et al., Detailed mass map of CL0024 + 1654 from strong lensing. Astrophys. J. **498**, L107–L110 (1998). doi:10.1086/311314
29. D.M. Wittman et al., Detection of weak gravitational lensing distortions of distant galaxies by cosmic dark matter at large scales. Nature **405**, 143–148 (2000). doi:10.1038/35012001
30. R.S. Ellis, Gravitational lensing: a unique probe of dark matter and dark energy. Philos. Trans. A **368**, 967–987 (2010). doi:10.1098/rsta.2009.0209
31. A.A. Penzias, R.W. Wilson, A measurement of excess antenna temperature at 4080 Mc/s. Astrophys. J. **142**, 419–421 (1965). doi:10.1086/148307
32. J.C. Mather et al., A preliminary measurement of the cosmic background radiation spectrum by the Cosmic Background Explorer (COBE) satellite. Astrophys. J. **354**, L37–L40 (1990). doi:10.1086/185717
33. D.J. Fixsen, The temperature of the cosmic background radiation. Astrophys. J. **707**(2), 916–920 (2009). doi:10.1088/0004-637X/707/2/916
34. F.C. Adams, G. Laughlin, A dying universe: the long term fate and evolution of astrophysical objects. Rev. Mod. Phys. **69**, 337–372 (1997). doi:10.1103/RevModPhys.69.337

Appendices

Appendix 3A Lorentz Contraction of a Modern Spacecraft

In 2006 the *New Horizons* spacecraft left the Earth at a velocity of 16.26 km/s (36,373 mi/hr) to Pluto and the Kuiper Belt at the edge of the solar system.The gravity assist near Jupiter temporarily increased the velocity to 22.78 km/s (51,000 mi/hr) relative to the Sun. As of September 2011, the gravitational attraction of the Sun has slowed down the spacecraft to about 15.64 km/s (34,989 mi/hr) with a position just beyond Uranus.The fastest spacecraft, Voyager I, is travelling at about 17.164 km/s (38,400 mi/hr) relative to the Sun. For spacecraft, we must always specify the speed relative to some celestial body, such as the Sun, the Earth, or a nearby planet. We can calculate the Lorentz factor γ at an average spacecraft speed of 16,986 km/s (38,000 mi/hr) relative to the Earth. This speed represents the current technology limits of spacecraft velocity using rocket launching. Then

$$\gamma = \frac{1}{\sqrt{1 - (16986\,\mathrm{m/s})^2/(2.998 \times 10^8\mathrm{m/s})^2}} = 1.011869, \tag{3A.1}$$

and the spacecraft has undergone a length contraction of $\approx$1.2 %.

Appendix 3B A Derivation of $E = mc^2$

There are various approaches to deriving the equation $E = mc^2$. This derivation method uses the relationships between energy E, relativistic velocity v, momentum p, and a force F acting on a moving mass particle. The incremental energy dE required to move the particle an incremental distance x is given by $dE = Fdx$, where F is defined as $F \equiv \frac{dp}{dt}$. This leads to $dE = \left(\frac{dp}{dt}\right)dx = dp\left(\frac{dx}{dt}\right) = vdp$.

D.F. Vanderwerf, *The Story of Light Science*,
DOI 10.1007/978-3-319-64316-8

Since $p = mv$, $dp = mdv + vdm$, we obtain

$$dE = mvdv + v^2 dm. \tag{3B.1}$$

The relativistic mass $m = \gamma m_0 = m_0\left(1 - \frac{v^2}{c^2}\right)^{-1/2}$, where m_0 is the so-called *rest mass* where $v = 0$.

Then $\frac{dm}{dv} = \frac{m_0 v}{c^2}\left(1 - \frac{v^2}{c^2}\right)^{-3/2}$, which can be put in form $\frac{dm}{dv} = \frac{mv}{c^2}\left(\frac{c^2 - v^2}{c^2}\right)^{-1} = \frac{mv}{c^2 - v^2}$, or $mvdv = (c^2 - v^2)dm$. Substituting $mvdv$ in Eq. (3B.1) gives $dE = (c^2 - v^2)dm + v^2 dm$, or

$$dE = c^2 dm. \tag{3B.2}$$

The difference form of this equation is

$$\Delta E = c^2 \Delta m, \tag{3B.3}$$

and is a form of Einstein's famous equation that is valid for any $v < c$, and also for $v = 0$. It relates changes in mass to changes in energy, and vice-versa. If the Δ-values are summed over the mass limits, then the total contained energy E is

$$E = mc^2, \tag{3B.4}$$

which is the well-known form of the equation. The energy equation can then be expressed in terms of momentum and velocity as $E = \left(\frac{p}{v}\right)c^2$. However, for a particle at rest, where $p = v = 0$, this equation gives an indeterminate result.

Appendix 3C Time Dilation and Muon Lifetime Calculations

For the experiment of Rossi and Hall we note that the classical expected muon travel time from a 15 km height H to Earth ground is $t = \frac{H}{v} = \frac{1.5 \times 10^4 \text{ m}}{2.98 \times 10^8 \text{ m/s}} = 50.3\ \mu\text{s}$. The expected fraction of surviving muons at ground-level would then be

$$\frac{N(t)}{N_0} = \exp\left(-\frac{t}{\tau}\right), \tag{3C.1}$$

where $N(t)$ is the number of muons detected per second at ground level after travel time t, and N_0 is the number of muons created per second at the source in the atmosphere. The calculated fraction of surviving muons is 1.18×10^{-10}, indicating a very low survival rate. However, actual ground-level measurements indicated a much higher survival rate.

We now make a relativistic calculation by assuming that the muon moving reference frame travel time t' is dilated by the relationship $t' = t\sqrt{1 - v^2/c^2} = t/\gamma$, when viewed from the stationary Earth frame. For a muon velocity $v = 0.994c$, thc Lorentz factor $\gamma = 9.14$. Then $t' = 5.5\ \mu s$. Using Eq. (3C.1) with $t' = 5.5\ \mu s$, the observed muon survival fraction observed on Earth $\frac{N(t')}{N_0} = 0.082$, which is a much higher survival ratio. The calculation can also be considered from the viewpoint of the moving muon reference frame, where a height contraction in the Earth reference frame occurs at the muon velocity v. Then $H = \frac{H'}{\gamma} = \frac{1.5 \times 10^4}{9.14} = 1641\ \mathrm{m}$. The observed time of travel for the Earth reference frame is $t = \frac{1641\ \mathrm{m}}{2.98 \times 10^8\ \mathrm{m/s}} = 5.5\ \mu s$, the same as the previous time dilation.

The actual measurements of Rossi and Hall consisted of muons per unit time on a detector at the top of Mount Washington and at sea level. The measurements indicated ≈570 muons/h at the mountain top and ≈400 muons/h at sea level. Using a 2000 m height, the travel time t in the Earth reference frame is $t = \frac{2000\ \mathrm{m}}{2.98 \times 10^8\ \mathrm{m/s}} = 6.71\ \mu s$, and the travel time t' in the moving muon reference frame is $t' = t/\gamma = 0.734\ \mu s$. The half-life $\tau_{1/2}$ of a muon is ≈1.5 μs, where $\tau_{1/2} = \tau \ln 2$. Then the fraction of muons $N(t)$ surviving from a population N_0 over travel time t is then given by

$$\frac{N(t)}{N_0} = \exp\left(-\frac{t \ln 2}{\tau_{1/2}}\right) \tag{3C.2}$$

Using the relativistic time dilation t' of the muon reference frame gives

$$N(t') = (570\ \text{muons/hr}) \exp\left(-\frac{(0.734\ \mu s)(0.693)}{1.5\ \mu s}\right) = 406\ \text{muons/hr} \tag{3C.3}$$

The measured muon arrival rate at sea level (400 muons/hr) is thus close to the rate predicted using relativistic time dilation (406 muons/hr).

Appendix 4A Derivation of Wien's Displacement Law from Planck's Law

Wien's displacement law in Eq. (4.1) can be restated as

$$\lambda_{max} T = \text{constant} = 2.8978 \times 10^{-3}\ \mathrm{m\,K} \tag{4A.1}$$

The maximum wavelength λ_{max} at a given temperature T can be found by setting the derivative of $I(\lambda, T)$ in Eq. (4.4) to zero, and solving for λ, to obtain λ_{max}:

$$\frac{dI(\lambda, T)}{d\lambda} = 2hc^2 \frac{d}{d\lambda}\left(\frac{1}{\lambda^5 e^{hc/\lambda kT} - 1}\right) = 0. \tag{4A.2}$$

Using the general relations $\frac{d}{d\lambda}\left(\frac{f_1}{f_2}\right) = \frac{f_2\frac{df_1}{d\lambda} - f_1\frac{df_2}{d\lambda}}{f_2^2}$ and $\frac{d}{d\lambda}\left(e^{f_3}\right) = e^{f_3}\frac{df_3}{d\lambda}$, where f_1, f_2, and f_3 are differentiable functions containing λ, then

$$\frac{dI(\lambda,T)}{d\lambda} = 2hc^2\left[\frac{-5\lambda^{-6}\left(e^{hc/\lambda kT}-1\right) - \lambda^{-5}\left(e^{hc/\lambda kT}\right)\left(-\frac{hc}{kT}\lambda^{-2}\right)}{\left(e^{hc/\lambda kT}-1\right)^2}\right] = 0, \quad (4A.3)$$

giving

$$\left(\frac{hce^{hc/\lambda kT}}{kT\lambda^7\left(e^{hc/\lambda kT}-1\right)^2} - \frac{5}{\lambda^6\left(e^{hc/\lambda kT}-1\right)}\right) = 0. \quad (4A.4)$$

Multiplying this equation by $\lambda^6\left(e^{hc/\lambda kT}-1\right)$ then yields

$$\frac{hce^{hc/\lambda kT}}{kT\lambda\left(e^{hc/\lambda kT}-1\right)} - 5 = 0. \quad (4A.5)$$

Rearranging terms gives

$$\left(\frac{hc}{\lambda kT}\right)\frac{e^{hc/\lambda kT}}{\left(e^{hc/\lambda kT}-1\right)} - 5 = 0$$

This transcendental equation can be solved for $\left(\frac{hc}{\lambda kT}\right)$ by an iterative process, yielding $\left(\frac{hc}{\lambda kT}\right) = 4.9651142317\ldots$ Then

$$\lambda_{max}T = \frac{hc}{(4.965)k} \approx \frac{(6.626\times10^{-34}\ \text{J s})(2.998\times10^{8}\ \text{m/s})}{(4.965)(1.381\times10^{-23}\text{J/K})}, \quad (4A.6)$$

or $\lambda_{max}T \approx 2.897\times10^{-3}\text{m K} = 2.897\times10^{6}\text{nm K}$, which is Wien's displacement law.

Appendix 5A Combustion-Based Light Sources

Besides sunlight, the most available source of light produced on Earth is from *combustion*, commonly referred to as *fire*. Combustion requires a fuel and an oxidant producing an exothermic chemical reaction that produces heat and light. Naturally produced fires, such as brush or forest fires produced from lightning strikes, were certainly observed by early man and perhaps utilized in an opportunistic manner. However, the actual production and containment of fire for heat and illumination is considered to be one of the great advances of mankind. Evidence supports the habitual use of controlled fire by man to have originated about 300,000 to 400,000 years ago. Controlled production of fire at that time included the campfire to more portable lighting devices such as the torch.

The first lamps appeared in the period around 70,000 BC, and consisted of natural burning materials soaked in natural fuels such as fish, animal and vegetable oils, and stuffed in cavities of rock or shell. Wicked candles using whale fat were first used in China around 200 BC. The light output of a bare candle flame is only about 13 lumens. The luminous intensity of a candle defines the term *candela* in SI units, which is still used in photometry, although being determined in a more precise manner. In the mid 1700s, oil refined petroleum liquid fuels such as kerosene led to a series of portable incandescent lighting devices known as lanterns or oil lamps. These were wicked devices enclosed in a glass chimney with some control of the fuel supply. Innovations were the Argand hollow wick lamp, patented in 1780, and the *Welsbach Mantle*, patented by von Welsbach in 1885. This cylindrical shaped cloth Welsbach mantle is infused with thorium and cesium oxides, and is placed above the flame to produce a more intense and whiter light over a larger area. This mantle is still used today in camping lanterns.

In 1792 gas lamps were first used for illumination when William Murdoch used coal gas to light his house in Germany. In the early 1800s, gas lighting was used extensively for house and street lighting, both in the United States and internationally. Short-lived pyrotechnic chemical incendiary flares are still in use today for illumination and signaling. However, these oil, gas, and incendiary light sources represent the limit of achievable illumination without the use of electrical voltage or current. In fact, there was no means for these devices to be further developed for bright and sustained illumination.

Appendix 6A Multiple Laser White Light Illumination

White light illumination can be produced by simultaneous illumination by a set of four narrow bandwidth solid-state lasers, each having a blue, green, yellow, or red color. An experiment was performed to determine whether the four RGBY lasers could satisfactorily provide acceptable color rendering of an object compared to other types of white light sources. Commercially-available laser diodes and solid-state diode pumped lasers were used, having the following wavelengths and maximum powers: red (635 nm, 800 mW), yellow (589 nm, 500 mW), green (532 nm, 300 mW), and blue (457 nm, 300 mW). The four laser beams were integrated using four chromatic beam combiners, sent through multiple ground glass diffusers to reduce speckle, and the combined beam was expanded to illuminate the test object. To test whether the four colored lasers would produce illumination similar to an incandescent lamp, or a set of wider bandwidth light emitting diodes, either warm, cool, or neutral white types, a focus group experiment was conducted in 2011 at Sandia National Laboratories. Two side-by-side scenes of a bowl of fruit were separately and randomly illuminated by different source types, and forty-one observers of varying age groups were asked to compare the comfort level of each view. The observations were randomized, and the power level of the lasers was adjusted to match the light level of the comparison sources. No statistically significant difference in the viewing comfort level was found between the lasers and neutral white LED illumination. However, there was a statistically significant viewing preference for the multiple laser white light illumination over illumination using warm or cool white LEDs, or the incandescent lamp. This opened up the possibility of using more electrically efficient solid-state lasers for general illumination purposes.

Appendix 9A Circuit Model Calculations for Deutsch's Algorithm

Qubits $|0\rangle$ and $|1\rangle$ are prepared and enter two quantum Hadamard gates $H\,(|0\rangle)$ and $H\,(|1\rangle)$ producing the superposed quantum state

$$|\Psi\rangle = H(|0\rangle)H(|1\rangle) = \left(\frac{1}{\sqrt{2}}|0\rangle + \frac{1}{\sqrt{2}}|1\rangle\right)\left(\frac{1}{\sqrt{2}}|0\rangle - \frac{1}{\sqrt{2}}|1\rangle\right) \quad (9A.1)$$

or

$$|\Psi\rangle = (|00\rangle - |01\rangle + |10\rangle - |11\rangle)/2. \quad (9A.2)$$

This superposed state is then processed by a special f-CNOT gate U_f incorporating $f\,(x)$. This gate performs the following operations on each qubit: $|ab\rangle = |a\rangle[|b + f(a)\rangle]$. Applying this to each term in Eq. (9.21) gives

$$|\Psi\rangle_f = [|0\rangle\{|0 + f(0)\rangle - |1 + f(0)\rangle\} + |1\rangle\{|0 + f(1)\rangle - |1 + f(1)\rangle\}]/2 \quad (9A.3)$$

By using appropriate values of $f\,(0)$ and $f\,(1)$, a set of four entangled qubit pairs is obtained. Finally, Hadamard gates are applied to each qubit pair in the set $|00\rangle, |01\rangle, |10\rangle, |11\rangle$, where

$$H(|0\rangle H(|0\rangle = (|00\rangle) + |01\rangle + |10\rangle + |11\rangle)/2 \quad (9A.4)$$

$$H(|0\rangle H(|1\rangle = (|00\rangle) + |10\rangle - |10\rangle - |11\rangle)/2 \quad (9A.5)$$

$$H(|1\rangle H(|0\rangle = (|00\rangle) + |10\rangle - |10\rangle - |11\rangle)/2 \quad (9A.6)$$

$$H(|1\rangle H(|1\rangle = (|00\rangle) - |01\rangle - |10\rangle + |11\rangle)/2. \quad (9A.7)$$

The output will be a qubit pair, eg. either $|0\rangle|0\rangle$ or $|1\rangle|1\rangle$. Since the output qubits are entangled, only the first qubit in each pair is measured as $|0\rangle$ or $|1\rangle$. A measurement of $|0\rangle$ indicates that $f\,(0) = f\,(1)$, and $f\,(x)$ is constant. A measurement of $|1\rangle$ indicates that $f\,(0) \neq f\,(1)$, and $f\,(x)$ is balanced.

Timeline of Some Notable Achievements in Light Science

ca. 500 BC Pythagoras—concepts on the nature of light
1676 Rømer—astronomical speed of light measurement
1690 Huygens—wave theory of light publication
1704 Newton—corpuscular theory of light publication
1725 Bradley—astronomical speed of light measurement
1792 Murdoch—first gas lamp for illumination
1800 Davy—first arc-discharge lamp for illumination
1801 Young—double-slit interference experiment
1849 Fizeau—terrestrial speed of light measurement
1850 Swan—first working carbon filament incandescent lamp
1850 Fizeau and Foucault—terrestrial speed of light measurement
1851 Fizeau—speed of light in moving medium measurement
1855 Weber and Kohlrausch—relationship between electricity, magnetism, and light
1862 Foucault—speed of light measurement in water
1865 Maxwell—Maxwell's equations, electromagnetic wave equations
1879 Michelson—terrestrial speed of light measurement
1880 Lorentz—theory of frequency dispersion in dielectric media
1881 Rayleigh—definition of wave and group velocities for light waves
1886 Hertz—experimental generation of electromagnetic waves
1887 Michelson and Morley—speed of light and source motion experiment
1887 Hertz—first observation of photoelectric effect
1890 Lorentz—length contraction of moving object and velocity transformations
1895 Roentgen—discovery of X-rays
1900 Planck—quantum energy states of blackbody radiation, Planck's constant
1902 Lenard—experimental analysis of photoelectric effect
1904 Just and Hanaman—first working tungsten filament incandescent lamp
1905 Einstein—relativity of mass and time, invariance of light speed
1905 Einstein—derivation of $E = mc^2$
1905 Einstein—proposal that light is a quantum particle
1916 Einstein—theory of stimulated absorption and emission
1919 Eddington and Dyson—confirmation of bending of light by the Sun

D.F. Vanderwerf, *The Story of Light Science*,
DOI 10.1007/978-3-319-64316-8

1923 Compton—X-ray scattering is a quantum phenomenon
1924 Bose—quantum derivation of Planck's law using photon particles
1924 Einstein—prediction of the Bose-Einstein condensate state of matter
1927 de Broglie—concept of wave-particle duality of light
1933 Zwicky—proposal of invisible dark matter
1940 Ives and Stilwell—detection of transverse Doppler shift of light
1941 Rossi and Hall—experimental confirmation of time dilation
1948 Casimer—darkness experiment between conducting plates
1950 Feynman—formulation of quantum electrodynamics for light-matter interactions
1960 Maiman—first operational ruby crystal laser
1961 Snitzer, Hicks—demonstration of single core fiber laser
1962 Holonyak—first visible light emitting semiconductor diode
1964 Babcock and Bergman—terrestrial confirmation of light speed invariance
1964 Bell—derivation of Bell test for entangled quantum particles
1970 Basov—invention of excimer laser
1972 Hall and Barger—determination of currently accepted value for the speed of light
1976 Madey—demonstration of free-electron laser
1978 Wheeler—proposal of delayed-choice photon experiment using double slit
1979 Brecher—astronomical confirmation of light speed invariance
1982 Moulton—development of Ti:sapphire laser
1986 Granger, Roger, and Aspect—verification of single-photon interference
1994 Alcubierre—proposal of warp drive for faster-than-light travel
1994 Giesen—development of thin disk laser
1995 Cornell and Wieman—production of first observable Bose-Einstein condensate
1999 Hau—reduction of light speed to 17 m/s in a Bose-Einstein condensate
2000 Kim, et al.—delayed-choice quantum eraser experiment
2001 Paul, et al.—generation of first train of laser attosecond pulses
2004 Singleton, et al.—generation of superluminal radiation in vacuum.
2006 Summers, et al.—first tunable luminescent photonic crystal
2009 Noginov—first demonstration of a nanolaser
2012 Shalm—demonstration of three-photon quantum entanglement
2012 Bohnet, et al.—first prototype superradiant laser
2012 Ma, et al.—demonstration of quantum teleportation over 143 km distance in space
2015 LIGO Consortium—detection of gravitational waves using squeezed light detector

A Selection of Additional Readings

Arthur S. Eddington, *The Nature of the Physical World*, Macmillan Company, New York (1929).

Michael I. Sobel, *Light*, The University of Chicago Press, Chicago (1987).

Arthur Zajonc, *Catching the Light: The Entwined History of Light and Mind*, Oxford University Press, New York (1993).

Ralph Baierlein, *Newton to Einstein: the Trail of Light: An Excursion to the Wave-Particle Duality and the Special Theory of Relativity*, Cambridge University Press, Cambridge, United Kingdom (2002).

Richard P. Feynman, *QED: The Strange Theory of Light and Matter*, Princeton University Press, Princeton, New Jersey (2006).

Daniel Fleisch, *A Student's Guide to Maxwell's Equations*, 1st ed., Cambridge University Press, New York (2008).

Chandra Roychoudhuri, A.F. Kowalski, Katherine Creath, Ed., *The Nature of Light: What is a Photon?*, 1st ed., CRC Press, Boca Raton, Florida (2008).

Jeff Hecht, *Understanding Lasers: An Entry-Level Guide*, 3rd ed., IEEE Press, Piscataway, New Jersey (2008).

Rudiger Pashotta, *Field Guide to Lasers*, SPIE Press, Bellingham, Washington (2009).

Brian Cox and Jeff Forshaw, *Why Does* $E = mc^2$*? (And Why Should We Care?)*, Da Capo Press, Cambridge, Massachusetts (2009).

Anton Zeilinger, *Dance of the Photons: From Einstein to Quantum Teleportation*, 1st ed., Farrar, Straus and Giroux, New York (2010).

Ernest Freeberg, *The Age of Edison—Electric Light and the Invention of Modern America*, 1st ed., Penguin Books, New York (2013).

D.F. Vanderwerf, *The Story of Light Science*,
DOI 10.1007/978-3-319-64316-8

Index

A
Acubierre warp drive, 310
Ampère's law, 14
anti-laser, 128
  coherent perfect absorber, 128
  time-reversed lasing, 128
arc-discharge lamp, 61
  electrodeless gas discharge lamp, 63
  long-arc lamp, 62
  short-arc lamp, 61
  vortex-stabilized arc discharge lamp, 62
Aristotle, 1

B
Babcock and Bergman experiment, 45
Bartholinus, Erasmus, 2
  Polarized light observation, 2
Big Bang, 311
  Lepton Epoch, 311
  Photon Epoch, 312
biological laser, 128
Bose, Satyendra Nath, 137
  Bose-Einstein condensate, 139
  Bose-Einstein statistics, 138
  boson, 138
  Fermi-Dirac statistics, 138
  thermal deBroglie wavelength, 138
boson quantum sampling, 280
  boson sampling problem, 280
  bosonic coalescence, 283
  evanescent coupling, 284
  permanent Per, 281
  tritter, 283
  unitary transformation matrix, 283
  waveguide deformation, 285
  waveguide fabrication, 282
  waveguide rotation, 285
boson sampling
  fiber optic sampler architecture, 281
Bradley, James, 6
  speed of light measurement, 6
  stellar aberration, 7

C
Casimir effect, 192
  Casimir, Hendrick, 192
  Casimir force, 193
  dynamical Casimir effect, 194
  Josephson junction, 195
  motion induced radiation, 194
  superconducting quantum interference device, 195
  virtual photon, 192
Compton, Arthur, 55
  Compton effect, 56
  Compton shift equation, 56
cosmic light speed, 301
  anisotropy, 303
  fine structure constant, 302
  gamma-ray burst, 302
  Lorentz invariance, 301
  Lorentz symmetry, 304
  Planck length, 301
  polarization synchrotron, 308
  quantum foam, 301
  reduced group velocity, 305
  superluminal phase velocity, 307
cosmic microwave background, 312
  radiation, 312
  temperature, 312
  Wien's law, 312
cosmic photon fate, 313
  Dark Era, 314
  Hawking radiation, 313
Coulomb's law, 13
coupled lasers, 126

D.F. Vanderwerf, *The Story of Light Science*,
DOI 10.1007/978-3-319-64316-8

D

dark matter, 310
 gravitational lensing, 311
 strong lensing, 311
 weak lensing, 311
dark pulse laser, 125
 bright soliton, 126
 dark soliton, 125
de Broglie, Louis, 57
 de Broglie wavelength, 57
delayed-choice, 174
 delayed-choice quantum eraser, 182
 double-slit delayed-choice, 175
 idler photon, 182
 Mach-Zehnder interferometer
  delayed-choice, 176
 photon particle nature, 178
 photon wave nature, 178
 photon wave-particle transition, 187
 principle of complementarity, 176
 signal photon, 182

E

Einstein, Albert, 31
 Doppler frequency shift, 32
 general theory of relativity, 37
 gravitational redshift, 38
 mass/energy equivalence, 36
 photoelectric effect, 53
 photon, 53
 principle of relativity, 31
 special theory of relativity, 31
 stimulated emission, 55
 time dilation, 32
electromagnetic spectrum, 19
Empedocles, 1
Euclid, 1
Euler, Leonhard, 3

F

Faraday, Michael, 15
 Faraday effect, 15
 Faraday's law of induction, 15
fiber laser, 95
 doped fiber amplifier, 97
FitzGerald, George, 28
 FitzGerald contraction, 28
Fizeau, Armand Hippolyte, 3, 8
 moving medium experiment, 23
 speed of light measurement, 8
flying qubit, 199
Foucault, Léon, 3, 9
 Foucault and Fizeau apparatus, 9
 speed of light measurement, 9
free-electron laser, 118
 self-amplified stimulated radiation, 119
Fresnel, Augustin-Jean, 3

G

Galileo Galilei, 4
 speed of light measurement, 4
Gauss' law of magnetism, 14

H

Hertz, Heinrich, 18
 Hertz effect, 49
 photoelectric effect, 49
Hooke, Robert, 2
Huygens, Christiaan, 3
 Huygens-Fresnel principle, 3

I

Ibn al Haythen (Alhazen), 1
incandescent lamp, 63
 carbon filament lamp, 64
 tungsten filament lamp, 64
 tungsten-halogen lamp, 64
invisibility cloaking, 158
 carpet cloak, 162
 dielectric cylindrical ring cloak, 162
 electromagnetic cloaking, 158
 free-space invisibility cloak, 165
 mantel cloaking, 167
 metamaterial cloaking, 159
 optical cloaking, 160
 plasmonic cloaking, 164
 spacetime cloak, 167
 transformation optics, 158
Ives and Stilwell experiment
 transverse doppler shift, 39
Ives-Stilwell experiment, 39

K

Kennedy-Thorndike experiment, 38
Kepler, Johannes, 2

L

laser, 75
 argon ion laser, 86
 CO2 laser, 86
 dye laser, 89
 excimer laser, 87
 helium-neon laser, 84
 longitudinal nodes, 77
 Nd:YAG laser, 83
 optical pumping, 77
 population inversion, 76
 pulsed ruby laser, 81

Q-switching, 81
spontaneous absorption, 76
spontaneous emission, 76
stimulated emission, 76
temporal and spatial coherence, 80
transverse electromagnetic modes, 79
laser diode, 91
edge-emitting laser diode, 91
quantum well laser diode, 91
vertical cavity surface emitting laser, 93
vertical external cavity surface emitting laser, 94
Lenard, Phillip, 49
photoelectric measurement apparatus, 50
light-emitting diode, 65
organic light emitting diode, 67
photonic crystal LED emission, 70
quantum dot light-emitting diode, 68
semiconductor light-emitting diode, 66
Lord Rayleigh, 51
group velocity, 133
phase velocity, 133
Rayleigh-Jeans radiation law, 52
ultraviolet catastrophe, 51
Lorentz, Hendrik, 28
frequency dispersion theory, 133
Lorentz factor, 28
Lorentz-FitzGerald contraction, 28
Lorentz transformations, 29

M
matter creation, 229
Breit-Wheeler pair, 229
bremsstrahlung, 229
two photon collision, 229
Maxwell, James Clerk, 15
electromagnetic equations, 15
electromagnetic wave equations, 17
metamaterial, 151
negative refractive index, 152, 153
split-ring resonator, 152
Michelson, Albert, 10, 24
Michelson interferometer, 24
Michelson-Morley experiment of 1887, 25
speed of light measurement, 10
Minkowski, Hermann, 33
Einstein-Minkowski spacetime, 33
light cone, 33
spacetime, 33
Morley, Edward, 24

N
nanolaser, 108
nano-conical waveguide, 112
relativistic oscillating mirror, 113
SPASER, 108
surface plasmon polaritron, 108
Newton, Isaac, 2
corpuscular theory of light, 2
emission theory, 2
law of refraction, 2
luminiferous ether, 2

O
Oersted, Hans Christian, 14
Ohm's law, 13

P
Pardies, Ignace-Gaston, 3
particle entanglement, 179
Bell's inequalities, 179
Bell test, 179
CHSH inequality, 199
eight-photon entanglement, 211
Einstein-Poldolsky-Rosen paradox, 179
four-photon entanglement, 209
Greenberger-Horne-Zeilinger state, 206
Hardy scheme, 202
non-coexistent photon entanglement, 211
polarization entanglement, 179
Schrödinger cat state, 206
single-particle nonlocality, 199
single-photon entanglement, 199
six-photon entanglement, 211
spontaneous parametric down-conversion, 183
three-photon entanglement, 206
photon angular momentum, 249
circular polarization, 250
OAM fiber transmisson, 251
orbital angular momentum., 249
photon free-space transmission, 250
spin angular momentum, 250
twisted light, 249
photon double-slit experiments, 173
single-photon interference, 173
wave-particle paradox, 173
photonic crystal, 155
negative refraction, 155
negative refractive index, 156
photonic crystal superlattice, 156
zero refractive index, 156
photon observation, 223
quantum non-demolition, 223
photon quantum interference, 186
photon quantum Interference
Hong, Ou, and Mandel, 186
Hong-Ou-Mandel dip, 186

Hong-Ou-Mandel effect, 187
photon size, 230
gauge boson, 231
gauge invariance, 231
Heisenberg's uncertainty principle, 231
Planck length, 231
Standard Model, 231
photon trajectories, 227
Heisenberg's uncertainty principle, 227
strong position measurement, 228
weak momentum measurement, 228
Planck, Max, 51
Planck's constant, 52
Planck's radiation law, 51
quanta, 52
Plato, 1
polarization of light, 20
Ptolemy, 1
pulsed laser, 104
chirped mirror, 105
fast laser, 104
Gires-Tournois interferometer, 105
group velocity dispersion, 105
Kerr lens mode-locking, 105
mode-locking, 104
pulsed X-ray free electron laser, 119
semiconductor saturable absorber mirror, 104
soliton, 105
ultrafast laser, 104
Pythagoras of Samos, 1
Pythagorean theorem, 1

Q

quantum cascade laser, 100
quantum communication, 235
BB84 protocol, 238
Bell-state, 235
correlated photons, 240
decryption, 238
eavesdropping, 240
encryption, 238
free-space QKD communication, 248
quantum dense coding, 236
quantum key distribution (QKD), 238
quantum computation algorithm, 286
circuit model, 286
cluster state, 287
Deutsch-Josza algorithm, 288
Deutsch's algorithm, 286
Shor's algorithm, 288
quantum computation efficiency, 286
BQP problem class, 286
NP-complete problem class, 286
NP problem class, 286
P problem class, 286
PSPACE, 286
quantum computing, 277
controlled NOT gate (CNOT), 278
controlled-phase CSIGN gate, 278, 279
Hadamard gate, 277
Hadamard transform, 277
KLM theory, 278
linear optics quantum computation, 278
modular universal scalable ion trap quantum-computer, 288
nonlinear sign (NS) gate, 279
photon information carrier, 277
single-photon logic gate, 277
Toffoli controlled CNOT gate, 279
quantum correlation of photons, 197
single-photon particle anticorrelation, 198
quantum electrodynamics (QED), 189
Feynman diagram, 190
Feynman, Richard, 189
probability amplitude, 189
quantum internet, 270
encrypted communication, 271
photonic quantum network, 270
quantum processing node, 270
quantum storage node, 270
quantum mechanical beamsplitter, 180
ket notation, 180
spatial quantum state, 181
quantum memory, 252
atomic ensemble, 253
atomic frequency comb, 255
bulk diamond, 257
crystal-based quantum memory, 254
diamond crystal, 256
polarization-entangled photon pair, 253
room temperature, 256
rubidium vapor cell, 258
single-photon, 253
quantum random number generator (QRNG), 272
detector photon position coordinate, 274
50/50 quantum beamsplitter, 272
quantum vacuum random phases, 276
quantum teleportation, 258
Bell-state measurements, 261
detector photon arrival times, 274
energy-time entanglement, 263
entanglement swapping, 268
free-space, 265
hybrid qubits, 261
long-distance fiber, 264
path entanglement, 261

photonic quantum network, 269
polarization entangled photons, 260
quantum repeater protocol, 267
time-bin entanglement, 263
quantum tunneling, 213
evanescent coupling, 215
frustrated total internal reflection, 215
Goos-Hänchen shift, 216
Hartman effect, 213
single-photon tunneling, 214
superluminality, 216
quantum Zeno effect, 226

R
random illumination laser, 124
photon degeneracy, 124
refractive index, 133
anomalous dispersion, 135
group delay, 135
group delay dispersion, 135
group refractive index, 135
phase refractive index, 135
photonic crystals, 144
Roentgen, Wilhelm, 18
discovery of X-rays, 18
Rømer, Olaus, 6
speed of light measurement, 6

S
single-photon detector, 246
superconducting nanowire, 246
single-photon interaction, 217
Rydberg blockade, 218
Rydberg state, 218
single-photon light source, 218
single-photon light source, 240
antibunching, 241
coupled-resonator optical waveguide (CROW), 244
diamond nanocrystal, 241
organic molecule emission, 246
quantum dot source, 244
second-order autocorrelation function, 242
whispering gallery mode resonator, 242
slow light, 135
Bose-Einstein condensate, 140
dispersion compensated slow light, 145
electromagnetic induced transparency, 137
photonic crystals, 144
resonant light propagation, 143
stimulated Brillouin scattering, 135
Stokes photon, 135
Stokes resonance, 136
zero-dispersion slow light, 145
Snell, Willebrord, 2
squeezed light, 220
amplitude-squeezed light, 220
gravitational wave detector, 221
Laser Interferometer Gravitational Wave Observatory, 221
phase-squeezed light, 220
phasor diagram, 220
quantum noise, 221
squeezed vacuum, 220
stopped light, 146
Sun, 44
blackbody radiation, 60
gravitational bending of light, 44
photosphere, 59
solar constant, 60
solar irradiance, 60
source of energy, 59
superluminal light, 149
causality violation, 148
in dispersive medium, 149
tachyon, 148
using four-wave mixing, 150
superradiant laser, 128
synchrotron radiation, 71

T
thin disk laser, 98
time dilation, 41
GPS Satellite Clocks, 43
moving atomic clocks, 42
muon lifetime, 41
timelike curve, 290
causal loop paradox, 290
classical computation, 294
cloned quantum states, 292
closed timelike curve, 290
consistency paradox, 290
grandfather paradox, 290
open timelike curve, 290
time-reversed light, 147
time-reversed Light
using photonic crystals, 147
tunable laser, 102
Ti:sapphire laser, 103

U
ultra-high power laser, 114
chirped pulse amplification, 114

optical parametric chirped pulse amplification, 115

V
visible color laser, 121
colloidal quantum dot vertical cavity emitting laser, 121

W
wave-particle duality, 57
Weber, Wilhelm Eduard, 15
Weber and Kohlrausch experiment, 15
white light laser, 121
supercontinuum laser, 121
Wien, Wilhelm, 50
Wien's displacement law, 51
Wien's radiation law, 51

Y
Young, Thomas, 3
double-slit interference experiment, 4

Printed in the United States
By Bookmasters